Tending the Wild

The publisher gratefully acknowledges the generous contribution to this book provided by the David B. Gold Foundation and the General Endowment Fund of the University of California Press Associates.

M. KAT ANDERSON

Tending the Wild

Native American Knowledge
and the Management of
California's Natural Resources

University of California Press
BERKELEY   LOS ANGELES   LONDON

A portion of the royalties go to Native Americans
to support the rekindling of their land management
practices in certain natural areas of California.

University of California Press
Berkeley and Los Angeles, California

University of California Press, Ltd.
London, England

First paperback printing 2006

Parts of chapter 7 appeared in a different form in "The Fire, Pruning, and Coppice Management of Temperate Ecosystems for Basketry Material by California Indian Tribes," published in *Human Ecology* 27 (1999):79–113, reprinted with kind permission of Springer Science and Business Media. Parts of chapter 10 appeared in a different form in "At Home in the Wilderness," published in *News from Native California* 6 (1992):19–21, used with permission from Heyday Books and Heyday Institute. Parts of chapter 11 appeared in the article "California's Endangered Peoples and Endangered Ecosystems," published in *American Indian Culture and Research Journal* 21 (1997):7–31, reprinted through permission of the American Indian Studies Center, UCLA © Regents of the University of California.

Library of Congress Cataloging-in-Publication Data

Anderson, Kat, 1955–.
   Tending the wild: Native American knowledge and the management of California's natural resources / M. Kat Anderson.
     p.   cm.
   Includes bibliographical references and index.
   ISBN 978-0-520-24851-9 (pbk : alk. paper)
    1. Indians of North America—Agriculture—California.
   2. Indians of North America—California—History.   3. Human ecology—California.   4. Nature—Effect of human beings on—California—History. I. Title.
   E78.C15A676   2005
   333.7'089'970794—dc22   2004017995

Manufactured in the United States of America

13   12   11   10   09   08
10 9 8 7 6 5 4 3

This book is printed on Natures Book, which contains 50% post-consumer waste and meets the minimum requirements of ANSI/ NISO Z 39.48–1992 (R 1997) (*Permanence of Paper*).

To my parents, Jan and Andy, and to the original teachers—the native plants, animals, and ancestors of California

Plants are thought to be alive, the juice is their blood, and they grow. The same is true of trees. All things die, therefore all things have life. Because all things have life, gifts have to be given to all things.

WILLIAM RALGANAL BENSON, *Pomo,*
in E. M. Loeb, Pomo Folkways

Contents

PART III. REKINDLING THE OLD WAYS

Illustrations

Maps

Figures

Tables

Preface

The idea that would become the foundation of this book—that indigenous people's stewardship of the land carries important lessons for us in the modern world—germinated in my mind as I stood in a Mexican farmer's fields in summer 1983. Growing alongside the many kinds of crop plants were a variety of native herbs and trees. Insects buzzed and clicked, and birds chattered. The land smelled good and radiated beauty. The farmer was using the land quite intensively, yet much of the natural plant and animal diversity remained. He explained to me the importance of diversifying crops, using locally available resources, retaining overstory trees, and planting vegetation that harbored beneficial insects that would feed on the "bad" insects. All these farming techniques were starting to catch on among organic farmers in the United States and to be taught in the new field of agroecology at American universities—and they all had roots in the farming methods of the indigenous people of this and other regions.

I began to wonder about the ways in which the native peoples of my home state, California, had managed wildland resources and altered the composition and structure of plant communities to meet their needs for food, clothing, and shelter. Had we underestimated their knowledge and ingenuity, their ability to transform their environment, just as we had underestimated the value of the indigenous-based traditional farming methods used around the world? I became determined to find out and decided to focus my graduate study in Wildland Resource Science at the University of California, Berkeley, on what I began to call the "indigenous resource management" practiced by the first peoples of California. I spent many months over a seven-year period, 1986–92, in the Sierra Nevada living in towns such as El Portal, Arnold, and North Fork and interviewing individuals from the Southern, Central, and Northern Sierra Miwok; the Mono from Dunlap, North Fork, and Auberry; and the Choynumni, Chukchansi, Wukchumni, and Yawlumne Yokuts tribes. In later years I spent time among the Pit River and Washoe. I also interviewed longtime non-Indian residents, many of them from established ranching families.

The more elders I talked with, the more I appreciated the importance of what I was learning. The fieldwork involved capturing elders' memories, their stories of how the land used to look and feel and how it differed from what one sees today. I was excited to hear their descriptions of the Old Ways of relating to nature—especially the management techniques (e.g., burning, pruning, tilling, weeding, and selective harvesting) that they had learned from their parents, grandparents, and great-grandparents and that they were still practicing today.

Several important insights were revealed to me as I talked with elders and accompanied them on plant gathering walks. The first of these was that one gains respect for nature by *using* it judiciously. By using a plant or an animal, interacting with it where it lives, and tying your well-being to its existence, you can be intimate with it and understand it. The elders challenged the notion I had grown up with—that one should respect nature by leaving it alone—by showing me that we learn respect through the demands put on us by the great responsibility of using a plant or an animal.

Many elders I interviewed said that plants do better when they gather them. At first this was a jarring idea—I had been taught that native plants were here long before humans and did best on their own without human interference—but it soon became clear to me that my native teachers were giving me another crucial gift of insight. California Indians had established a middle ground between the extremes of overexploiting nature and leaving it alone, seeing themselves as having the complementary roles of user, protector, and steward of the natural world. I had been reading about how various animals' interactions with plant populations actually benefited those plants—how grizzly bears scattered the bulblets of *Erythronium* lilies in the process of rooting up and eating the mature bulbs, how California scrub jays helped oaks reproduce by losing track of some of the acorns they buried—and it seemed plausible that the many generations of humans in California's past had played a similar role. If it was true that native plants did better with our help, it meant that there was a place for us in nature.

About halfway into the years of fieldwork, I began to ask native elders, "Why are many plants and animals disappearing?" Their answers, which always pinned the blame on the absence of human interaction with a plant or an animal, began to add up to a third major insight: not only do plants benefit from human use, but some may actually *depend* on humans using them. Human tending of certain California native plants had been so repetitive and long-term that the plants might very well have become adapted to moderate human disturbance. This idea had a very practical corollary: the conservation of endangered species and the restoration of historic ecosystems might require the reintroduction of careful human stewardship rather than simple hands-off preservation. In other words, reestablishing the ecological associations between people and nature might be appropriate in certain areas.

As I rediscovered the Old Ways in California, I began to look more broadly at indigenous practices around the world. I found that some of the judicious harvesting and horticultural practices of California indigenous cultures were remarkably similar to those of native peoples in South America, Australia, and Africa. I was intrigued that parallel resource use and management systems had developed independently on different continents. For example, Australian Aborigines and California Indians both used cattail (*Typha* spp.) for cordage and other purposes, burning off tule marshes to recycle dead material and spur new growth. Ancient peoples in Egypt cultivated flax *(Linum usitatissimum)* to encourage it to produce long, straight stems with longer bast fibers good for cordage and textiles, much like the first Californians tended patches of dogbane *(Apocynum cannabinum)* so that its stems would produce fibers suitable for weaving belts, tumplines, feather capes, skirts, and many other items. There were also important parallels in food production systems. The Mojave in southern California and the ancient Incas of South America sowed and cultivated the edible seeds of a pigweed in the genus *Chenopodium*, and many California Indian tribes echoed the practice of the indigenous peoples of Chile by eating the oily seeds of species in the genus *Madia* and caring for populations of the plants. The Paiute tribe of Owens Valley broadcast edible seeds of chia *(Salvia columbariae)* to thicken stands and introduce the plant to irrigated areas, and we have evidence that chia was sown and regularly cultivated along with corn in ancient Mexico.

These parallels indicated to me that our human forebears everywhere did not just passively gather food and basketry materials but actively tended the plant and animal populations on which they relied. There was no clear-cut distinction between hunter-gatherers—the category into which most California Indians had been tossed—and the more "advanced" agricultural peoples of the ancient world. Moreover, California Indians had likely completed the initial steps in the long process of domesticating wild species, something that ancient peoples in other lands had always been given sole credit for.

Sometime in 1993, while I was finishing my dissertation, I came to realize that what the California Indian elders had been teaching me was so important that it needed to be shared with others. Eventually I decided that writing a book about the enormous influence of Indian stewardship on the California landscape would serve the interests of indigenous people and enable me to give something back.

Most of the existing written information on how California Indians had cared for the land through burning, tilling, sowing, and pruning was not easily accessible: it was widely scattered in handwritten diaries, notecards in shoeboxes, and correspondence files in many libraries and museums. Even more important, the unwritten knowledge possessed by the Indian elders I was interviewing was endangered: this last generation of people taught directly by grandparents and great-grandparents born before the Gold Rush was made up of elders in their sixties, seventies, eighties, and nineties, and they were beginning to pass away, taking with them a wealth of cultural knowledge about how to relate to the earth. Some of these elders had grown up on traditional foods and still spoke their respective languages. I wanted to put some of the valuable information about Indian stewardship of the land in one place and preserve it for posterity.

Now that the book is being published, it is my fervent hope that certain benefits will be realized. First, I hope that greater understanding of the stewardship legacy left us by California Indians will foster a paradigm shift in our thinking about the state's past—particularly with regard to wildland fire—and the necessity of prescribed burning in the management of the state's natural resources today. Second, increased appreciation for the diverse indigenous cultures of California could lead to collaborative projects that reestablish access to the land and maintain culturally significant plant resources for the perpetuation of native traditions. Third, experiments and cross-disciplinary studies to determine the thresholds of harvest for native plant species and to assess the degree to which particular ecosystems and plant species are dependent on indigenous disturbance regimes could be launched. In this way human use can become integrated into conservation and restoration work. Fourth, I would like to encourage people to pursue studies in natural history and ethnobiology,[1] both of which emphasize tactile contact with and direct learning from nature and indigenous peoples. And fifth, we desperately need to foster a new vision of human–nature relationships and the place of humans in the natural world. Those who peopled California before the arrival of Westerners are some of the best teachers.

Acknowledgments

Through the years of fieldwork, library and museum research, and writing, I have been helped, supported, and encouraged by many people and institutions. First and foremost, I give my warm thanks to the indigenous people of California who shared their intimate knowledge and wisdom about plant gathering and management with me. It took courage, trust, and optimism to share information with one more non-Indian in the hope that this research would benefit their cultures. Barbara Bill, Clara Charlie, Pauline Conner, Ruby Cordero, Bill Franklin, Ron Goode, Kathy Heffner, Lucille Hibpshman, Hazel Hutchins, Marshall Jack, Virginia Jeff, Nellie Lavell, Sylvena Mayer, Avis Punkin, Willard Rhoades, James Rust, Clara Jones Sargosa, and Francys Sherman taught me the importance of rediscovering and restoring the Old Ways of relating to nature, and they have renewed my faith in humanity's ability to live compatibly with nature.

Among the California Indians I interviewed, special thanks go to Margaret and Thane Baty, Edith, Lydia, Martha, and Melba Beecher (especially Melba), Norma Turner Behill, Rosalie and Robert Bethel, Virgil and Linda Bishop, Florence and Nick Brocchini, Stanley Castro, Lorraine and Paul Cramer, Justin Farmer, Lalo (Hector) and Darlene Franco, Jay and Mary Johnson, Rowena and Dale Lilly, Gladys and Jake McKinney, William Pink, and Grace Tex for their openness, generosity, and kindness. I always felt welcome in their homes, and their trust and eager support of this field-work affirmed its rootedness in Indian communities. Gladys McKinney took me under her wing and spent much time introducing me to her relatives and friends, for which I am deeply grateful. Norma Turner Behill spent many hours with me discussing the experiences of her childhood, youth, and adult life and the numbers of plant parts required to make baskets; she contributed greatly to this book. Commendably, Rosalie Bethel boldly stood up for my field research and volunteered to be interviewed, despite a climate of skepticism at a board of directors meeting. William Pink led me on many memorable ethnobotanical journeys that involved the judicious harvest, preparation, and eating of parts of some of our native plants. I wish to thank longtime residents of the Sierra Nevada from whom I learned much: Leroy Brown, Betty Jamison, Jim McDougald, Rose Mitchell, and Buster Riedel.

This book has been greatly enriched by conversations and outings with many people who were not quoted directly. I am most grateful to the following California Indians and non-Indian longtime residents: Bill Airola, Jeri, Lottie, and Wilshire Alec (Big Sandy Rancheria Band of Western Mono Indians), Vivian (Katie) Appling (Miwok/Paiute), Ross Baker, Phillip Bennett (Washoe), Dave Bowman (Mono), Ed Bowman (Mono), Floyd Buckskin (Pit River), John Burrows, Iliene Cape (Western Mono), Forest Clingan, Helen Clingan, Lois Conner (North Fork Mono/Chukchansi), Paul Cramer,

George Dick (Wobonuch Mono), Ramona Dick (Washoe), Julie Dick Tex (Mono), Rosemary Faulkner, Dewey Fischer, Jack Fischer, Jennie Franco (Wukchumni Yokuts/Paiute), Ulysse Goode (Mono), Frances Grigsby (Mono), Margaret Hammond (Chukchansi Yokuts), John Hawksworth, Marge Hawksworth, Madelina Henry (Washoe), Mina Jackson (North Fork Mono), Les James (Southern Sierra Miwok/Yokuts), Ray Jeff, Rema John (Washoe/Miwok), Jean Kirpatrick, Galen Lee (North Fork Mono), Annie Lewis (North Fork Mono), Francis Lewis (Choynumni/Chukaymina Yokuts), Oswald Lombardi, Mandy Marine (Mono/Maidu), Dora Mata (Central Sierra Miwok), Pete McDonald (North Fork Mono), Jim McNamara, Dan McSwain (North Fork Mono), Ella McSwain (North Fork Mono), Ray Neilsen, Francis Nieto (Yawlumne Yokuts/Paiute), Merk Oliver (Yurok), Julia Parker (Coast Miwok/Kashaya Pomo), Dinah Pete (Washoe), Lyda Peyron (Wukchumni Yokuts), Rubie Pomona (North Fork Mono), Amy Rhoan (Paiute), Virginia Riley (Western Mono/Chukchansi), Rusty Rolleri, Karen Sargosa (Southern Sierra Miwok/Chukchansi Yokuts), Eddie Sartuche (Wukchumni), Mike Skenfield, John Snyder, Eva Soracco, Mary Spears (Southern Sierra Miwok/Yokuts), Hayden Stephens, Cornelia Stevenot, Charlie Stone, Ethel Temple (North Fork Mono), Nacomas Turner (Mono), Lucy Tyron, Agnes Vera (Wukchumni/Paiute/Bankalachi/Yawlumne), Annie Voitich, Jack Voitich, Marada Wade (Washoe), George Wessel (Northern Sierra Miwok), Nellie Williams (North Fork Mono), and Betty Wooster.

Recognition and appreciation are given to Yosemite National Park, the Yosemite Association, and Yosemite Research Center for funding much of the fieldwork on which this book is based. I am grateful to Scott Carpenter, former Park Archaeologist, and Jan Van Wagtendonk, Research Forester with the U.S. Geological Survey, for their recognition of the importance of cross-disciplinary research and their instrumental backing of my studies. I am indebted in particular to Scott Carpenter, who saw the promise in this work from the very beginning. Other generous grants that supported the research for this book came from the California Native Plant Society, Hardman Foundation, Institute of American Cultures, University of California, Los Angeles, and Soroptimist International.

I am deeply indebted to my dissertation committee members—Robert Ornduff, Arnold Schultz, and William Waters at the University of California, Berkeley—for their support and guidance during my field and experimental research.

My appreciation goes to Donald Worster, Director of the Hall Center for the Humanities at the University of Kansas, and to the Rockefeller Foundation for a fellowship in Nature, Culture, and Technology that enabled me to explore the wildland management practices of other tribes in the United States for comparison with California. I gained many insights into Western and Native American ways of thinking about nature and the human place in the natural world through lively debates and informal discussions with Worster and his students, as well as students at Haskell Nations University, a nearby tribal college. Half of the book was written during a postdoctoral fellowship at the University of California, Los Angeles, in 1995–96, at the American Indian Studies Center. Duane Champagne, former director, and Lynn Gamble, former research coordinator, encouraged and supported me wholeheartedly during the fellowship.

This book has benefited immensely from discussions with and comments from my colleagues Michael Barbour, Craig Bates, Robert Bettinger, Frederica Bowcutt, Judith Carney, Dave Egan, Victor Golla, Leanne Hinton, Michael Horn, William Jordan III, Judith Lowry, James Rawls, Nancy Turner, and Eric Wohlgemuth. Craig Bates sent citations and appropriate journal entries to me, participated in frequent conversations about wild plant management, and reviewed the manuscript at length, offering major insights and advice that improved it considerably. Other colleagues contributed their expert knowledge by reading parts of the manuscript: Susan Bicknell, James Cornett, Richard Dodd, Philip Rundell, Leslie Sauer, Mark Stromberg, Tedmund Swiecki, and James West. I have also been inspired by the work and writings of Janis Alcorn, Wendell Berry, Bob Fry, Aldo Leopold, Henry Lewis, Dennis Martinez, John Muir, Gary Nabhan, David Peri, Darrell Posey, Gary Snyder, Omer Stewart, Jan Timbrook, and Nancy Turner. Special thanks to Gary Nabhan, who has been a major visionary and creative force behind the field of ethnobiology.

Many of our Western teachers—entomologists, zoologists, pathologists, botanists, and mycologists—have accumulated a lifetime of knowledge about the natural world. They are "the elders" in Western culture. Their expertise was invaluable in the writing of this book, and I owe them a great deal. Entomologists Cheryl Barr, Steve Haydon, and Jerry Powell assisted me in identifying insects. Joanna Clines, Travis Columbus, Lincoln Constance, Ellen Dean, Steve Edwards, Jim Effenberger, Glenn Keator, Barry Prigge, Roger Raiche, Warren Roberts, and Wayne Roderick answered many questions regarding native plants, gave me valuable sources, and identified plant specimens. Richard Dodd answered my many questions on plant architecture and how plants activate adventitious shoots or epicormic branches after a disturbance such as flooding, burning, or pruning. Joan Humphrey and David Yee answered my questions about birds and updated the bird nomenclature. Bob Raabe taught me about various kinds of diseases that attack oak acorns, tule stems, and other plants. Kathy Ann Miller helped me update the seaweed nomenclature. Tedmund Swiecki answered my questions about insects that attack oak acorns. James Patton answered my questions about wildlife and checked the accuracy of animal scientific names. David Arora, Michelle Seidl, and Isabel Tavares answered my many questions on mushrooms, gave me mushroom citations, and identified mushroom specimens. Margaret Mathewson generously shared her knowledge on the plant architecture necessary for the manufacture of basketry, provided plant harvesting information, and shared her slide library with me.

Working with the editors at the University of California Press was a wonderful collaboration. Thanks go to Doris Kretschmer for her mentorship and belief in the book. Her visions and key decisions were vital to the book. Dore Brown, with the greatest care and understanding, shepherded the book through its many stages of production, and suggested a structural revamping that proved to work extremely well. Eric Engles, developmental editor extraordinaire, became intimately familiar with the book, making it much more accessible to the reader through superb editing, tightening, and reorganization. Sheila Berg refined the writing even more by adding her own expert language editing.

A special thanks is owed to Dave Rowney, who always eagerly assisted me in any computer-related problems, and to the highly talented Claudia Graham at Mediaworks, University of California, Davis, for creating all of the line drawings.

I owe immeasurable thanks to my parents, Jan and Andy Anderson, for providing me with a home and office for many years when my finances were wearing thin, when I continued to do fieldwork in the Sierra Nevada, set up ecological field experiments, and write. Without their logistical, financial, and moral support, I would have stopped pursuing this work. Mom and Dad also insightfully gave me my first tape recorder when I was eight. I have the deepest appreciation for my relationship with my sister, Jandy, and give thanks to her, to her husband, Larry, and to their children, Whitney and Savannah, for brightening my life. To my brother Scott, my best friend, I say thank you for buying me the computer and printer that I used for writing the book.

Heartfelt gratitude goes to cherished friends who nurtured me while I wrote: Jan Allen, David Birt, Claudia Delman, Hermine Holtland, Greg House, Henry House, Cindy Jarrett, Vicki Kramer, Bahia Mar, Dr. Maoshing Ni, Bruce Stevens, Nancy Switzer, Helene Wagner, and Margo Wallace. My deep appreciation goes to Scott Peterson for his personal support and understanding. Thanks also to the other three members of the Society of Four Ethnobotany Club—Jennifer House, Judith Lowry, and Lillian Vallee—for our long conversations, listening to and critiquing my ideas for chapters, and our experiments with edible and other useful plants.

Patricia Schaefer provided wonderful home-cooked meals, companionship, and a charming rustic cabin along the rushing Merced River for the summer and fall months of my fieldwork with the Southern Sierra Miwok in 1987. The friendly staff of Calaveras Big Trees State Park graciously gave me housing in 1990 and again in 1992 while I conducted regional research. I wish to acknowledge Craig Engel, former superintendent, and Wayne Harrison, fire ecologist, for providing logistical support and technical expertise for my fieldwork. Thanks are due also to the Minarets Work Station staff, especially FMO Ralph Taylor, U.S. Forest Service Mariposa Ranger District, who gave me housing in 1991 in North Fork to facilitate my ethnographic and ethnohistoric research.

Fred Cuneo was a superb networker in Calaveras County and took much time to introduce me to the ranchers and longtime residents. Judy Marvin introduced me to Indians in the Sierra Nevada, contributed knowledge of archival resources, and shared her wonderful home. Dotty Theodoratus spurred me on to pursue a career in ethnobotany, gave me the names of Native Americans to contact, and graciously granted me access to her extensive library and archives. Ray Garamendi, Carmen and Jim Hickling, Verna Johnston, Kelly Kindscher, Margaret Molarsky, Scott Murphy, and Wayne Roderick also generously shared their extensive libraries with me or donated ethnobotanical books.

I owe a considerable debt to the staff of the various museums, herbaria, historical societies, libraries, and photo archives from which I gathered information, including the Bancroft Library, Calaveras County Museum and Archives, Calaveras Historical Society, California State Indian Museum, California State Library, Grace Hudson Museum, Holt-Atherton Center for Western Studies at the University of the Pacific, Humboldt State University Library, Huntington Library, Jepson Herbarium, Mariposa Museum and History Center, North Fork Library, Oakland Museum, Old Timers' Museum in Murphys, Phoebe Hearst Museum, Plumas County Museum, Santa Barbara Museum of Natural History, Smithsonian Institution, Southwest Museum and Braun Library, Tuolumne Historical Society, University of California, Berkeley, Forestry Library, Yosemite Museum, and Yosemite National Park Research Library. I wish to single out Suzanne Abel-Vidor, Dot Brovarney, and Sherrie Smith-Ferri (Grace Hudson Museum), Mary Elizabeth Ruwell and staff (Smithsonian Institution), Lorrayne Kennedy (Calaveras

County Museum and Archives), Linda Eade, Jim Snyder, and Mary Vocelka (Yosemite National Park Research Library), Norma Craig (Yosemite National Park Slide and Photo Library), Maria Dawson and Lisa Deitz (University of California, Davis, Anthropology Museum), and Linda Agren and Jan Timbrook (Santa Barbara Museum of Natural History), all of whom devoted considerable time to tracking down obscure references and photographs relating to ethnobotany and Native American wild plant management. Richard Buchen, former Reference Librarian at the Southwest Museum, Los Angeles, was extremely helpful in finding photographs, obscure unpublished materials, and applicable published references.

The scientific names for plants are taken from the *Jepson Manual* (Hicks 1993) and NRCS PLANTS database at http://plants.usda.gov.

Many other people took a special interest in the various phases of this book (including research, creation of graphics, reference searches, accounting, interviews, setting up experiments in the field, and writing), and I thank them for their assistance: Karen Adams, Steve Anderson, Chuck Bacchus, Barbara Balen, Eloise Barter, James Barry, Letty Barry, Michelle Berditschevsky, Harold Biswell, Thomas Blackburn, Chris Boyer, Linda Brennan, Edward Castillo, Helen and Forest Clingan, Don Conner, Bob Cooke, Ben Cunningham-Summerfield, Kimberly Cunningham-Summerfield, Susan D'alcamo, Lee Davis, Shelly Davis-King, Julie Ann Delgado, Eric Edlund, Al Elsasser, Corinne Elwart, Joe Engbeck, Nancy Evans, Glenn Farris, Rosemary Faulkner, Catherine Fowler, Rich "Oso" Garcia, Frank Gault, Dave Glober, George Gray, Louise Greenlaw, John and Marge Hawksworth, William Hildebrandt, Dan Holmes, Bill Horst, Pauline and John Howell, Patty Johnson, Elaine Joyal, Jeannine Koshear, Kathleen Kraft, Louise Lacy, Robert Laidlaw, Frank Lake, Bruce Lawler, Judy Lee, Richard Lerner, Susan Lindstrom, Dorothy and Ed Loomis, Gary Maniery, Malcolm Margolin, Robert Martin, Helen McCarthy, Deborah McConnell, Neil McDougald, Steve Medley, Jim Meyer, Edra Moore, Claudia Mueller, Joe Mundy, Larry Myers, Bud and Carol Olson, Penny Otwell, Peter Palmquist, Breck Parkman, Jean Perry, Scott Pinkerton, Lorrie Planas, Don Potter, Leroy Radanovich, David Raymond, Paul Rich, Victoria Shoemaker, Bruce Smith, Neil Sugihara, Sonia Tamez, Tom Tanner, Lou Thoman, Tanis Toland, Fred Velasquez, Muriel Waters, Kenneth Whistler, Mary Willson, and Fred York.

My hope is that this book accurately portrays the past and present indigenous traditional ecological knowledge and land management practices in California. While the completeness of the work is largely due to the contributors named above, any shortcomings rest on my shoulders alone.

Half of the royalties from this book will be donated to two groups: the California Indian Basketweavers Association, P.O. Box 2397, Nevada City, CA 95959; and the journal *News from Native California*, P.O. Box 9145, Berkeley, CA 94709, which will use the money to establish a fund for Native American writers. I encourage readers to make donations to these two organizations as well.

Note on Languages, Territories, and Names of California Indian Tribes

The native peoples of California can be grouped in several different ways. The map that appears in this book on page 36 is linguistic, identifying groups of people that spoke the same language or dialect. Anthropologists have most commonly grouped the native peoples of California in this way, because classification by language is relatively straightforward, and because peoples speaking the same language tend to have other cultural traits in common. However, the bonds between people that share a language are often less important than those created by other sorts of relationships. This was especially true in most of native California, where the nature of trade, warfare, cooperation, and almost every other kind of interaction depended on whether the individuals involved belonged to the same sociopolitical group, not whether they spoke a common language. There is, thus, a disconnect between the linguistic map used here and the groupings that determined day-to-day behavior in native California, since the individual language groups often comprised many independent sociopolitical groups. In most cases, everyone in a sociopolitical unit spoke the same language, but speaking the same language did not ensure membership in the same sociopolitical unit. The picture in native California is further complicated because sociopolitical groups differed in kind and composition across the state, and because these groups were displaced and disrupted by European and American contact and occupation.

Prior to contact, many language groups, particularly in northern California, were subdivided into politically and economically independent groups consisting of several unrelated families that collectively owned and defended a specific tract of land. Alfred Kroeber called these groups "village communities" or "tribelets," to distinguish these political entities from the more familiar Plains tribes, which were much larger and more unified. Tribelets spoke a particular dialect, but they were independent from adjoining tribelets that spoke the same dialect. In other places, particularly in southern California and the San Joaquin Valley, otherwise independent sociopolitical units were somewhat unified by moieties that divided the whole of society into two ritually interdependent halves, forcing them to cooperate in giving first fruit rites and other ritual affairs. In still other places, notably northwestern California, there were no tribelets at all, nor indeed were there any sociopolitical units larger than the family. This diverse sociopolitical landscape was drastically transformed by European and American occupation. Many groups disappeared altogether; many independent groups joined to form new social units. Contributing to this joining and mixing was the policy of the federal government to grant services (through the BIA) only to groups designated as Federally Recognized Tribes (such as the Big Valley, Hopland, and Robinson Rancherias among the Pomo, for example).

All this explains why the linguistic map in this book does not show all the groups mentioned in the text. Some of the tribal names in use today reflect traditional geographic designations, referring to a characteristic feature or landmark of the place inhabited by the group. The Pit River tribe, for example, takes its name from that major river, yet on the linguistic map the tribe is also called by the name of its language, Achumawi. In this case there is a one-to-one correspondence between language and political unit. Because the language groups shown on the linguistic map frequently comprised multiple independent units, and because so few of these were documented by ethnographers, it is impossible to present a comprehensive sociopolitical map, which might show five to six hundred groups. See Robert Heizer's *Languages, Territories and Names of California Indian Tribes* for a more detailed discussion and presentation of data and sources.

Because some groups had no specific name for themselves (after all, they knew who they were), they were given the names used by their neighbors, with predictable results when neighboring tribes were not on friendly terms! More common are the names imposed by anthropologists. The Karuk, for example, have created their name by varying the spelling of *Karok*, the term used by anthropologists for the tribe's language. *Karok*, meaning "upriver," was not in use as a tribal name aboriginally. The neighboring Yurok, in turn, take their name from the anthropological name for their language, which was derived from the Karok term *yurok*, meaning "downriver."

To further complicate matters, there are many variations on the spelling of almost every language group and tribal name. The Choynumne are also referred to as Choinumni and Choynimni. Frequently this variation is intentional, as tribal groups change the spelling of terms used by anthropologists. Thus, the Washo of the anthropologist has become the Washoe. Karok/Karuk, Yawelmani/Yawlumne, and Kato/Cahto are other examples of this change. For various tribal names and spellings, see the synonymies in the *Handbook of North American Indians* (volumes 8, 10, 11, and 12 cover California, Southwest, Great Basin, and Plateau tribes).

In this book, when two tribal memberships are listed for an informant (Mono/Dumna, for example), the person claims both language affiliations because one or both parents have that ethnic background.

This note is based upon discussions with and comments from anthropologist Robert Bettinger and linguists Victor Golla and Leanne Hinton.

Introduction

The New World is in fact a very old world. The mountain forests, broad inland valleys, oak-studded hills, and deserts of the region now called California were thoroughly known, celebrated in story and song, named in great detail, and inhabited long before European explorers sailed along the west coast of North America for the first time. Every day of every year for millennia, the indigenous people of California interacted with the native plants and animals that surrounded them. They transformed roots, berries, shoots, bones, shells, and feathers into medicines, meals, bows, and baskets and achieved an intimacy with nature unmatched by the modern-day wilderness guide, trained field botanist, or applied ecologist.

The first European explorers, American trappers, and Spanish missionaries entering California painted an image of the state as a wild Eden providing plentiful nourishment to its native inhabitants without sweat or toil. But in actuality, the productive and diverse landscapes of California were in part the outcome of sophisticated and complex harvesting and management practices.

California Indians protected and tended favored plant species and habitats, harvested plant and animal products at carefully worked out frequencies and intensities, and practiced an array of horticultural techniques. Through coppicing, pruning, harrowing, sowing, weeding, burning, digging, thinning, and selective harvesting, they encouraged desired characteristics of individual plants, increased populations of useful plants, and altered the structures and compositions of plant communities. Regular burning of many types of vegetation across the state created better habitat for game, eliminated brush, minimized the potential for catastrophic fires, and encouraged a diversity of food crops. These harvest and management practices, on the whole, allowed for sustainable harvest of plants over centuries and possibly

1

thousands of years. In other words, California Indians were able to harvest the foods and basketry and construction materials they needed each year while conserving—and sometimes increasing—the plant populations from which they came.

During the course of their long history in California, Indians so exhaustively explored the plant kingdom for its uses and so thoroughly tested nature's responses to human harvesting and tending that they discovered how to use nature in a way that provided them with a relatively secure existence while allowing for the maximum diversity of other species. In the context of the entire continuum of possible human interactions with nature, ranging from exploitation and human-designed environments to hands-off preservation, this relationship between the indigenous people of California and the natural world represented a middle way, a calculated, *tempered use* of nature. *Tending the Wild* explores how California Indians managed economies that occupied this middle portion of the continuum. It recasts them as active agents of environmental change and stewardship, shattering the hunter-gatherer stereotype long perpetuated in the anthropological and historical literature of California.

The terms "hunter-gatherer" and "forager," inaccurate anthropological labels assigned to most California Indian groups, connote a hand-to-mouth existence. They imply that California Indians dug tubers, plucked berries, and foraged for greens in a random fashion, never staying in any one place long enough to leave lasting human imprints. But as *Tending the Wild* demonstrates, the indigenous people of California had a profound influence on many diverse landscapes—in particular, the coastal prairies, valley grasslands, and oak savannas, three of the most biologically rich plant communities in California. Without an Indian presence, the early European explorers would have encountered a land with less spectacular wildflower displays, fewer large trees, and fewer parklike forests, and the grassland habitats that today are disappearing in such places as Mount Tamalpais and Salt Point State Park might not have existed in the first place.

A Tended Wilderness

Through twelve thousand or more years of existence in what is now California, humans knit themselves to nature through their vast knowledge base and practical experience. In the process, they maintained, enhanced, and in part created a fertility that was eventually to be exploited by European and Asian farmers, ranchers, and entrepreneurs, who imagined themselves to have built civilization out of an unpeopled wilderness. The concept of California as unspoiled, raw, uninhabited nature—as wilderness—erased the indigenous cultures and their histories from the land and dispossessed them of their enduring legacy of tremendous biological wealth. As the environmental historian William Cronon notes, "The removal of Indians to create an 'uninhabited wilderness'—uninhabited as never before in the human history of the place—reminds us just how invented, just how constructed, the American wilderness really is."[1]

John Muir, celebrated environmentalist and founder of the Sierra Club, was an early proponent of the view that the California landscape was a pristine wilderness before the arrival of Europeans. Staring in awe at the lengthy vistas of his beloved Yosemite Valley, or the extensive beds of golden and purple flowers in the Central Valley, Muir was eyeing what were really the fertile seed, bulb, and greens gathering grounds of the Miwok and Yokuts Indians, kept open and productive by centuries of carefully planned indigenous burning, harvesting, and seed scattering.

Of course, there were some places that had little or no intervention from native peoples, and these would qualify as true wilderness under the modern definition. The subalpine forests, the drier desert regions of southern California, the lower salt marsh areas, the beach and dune communities, and the alkali flats and serpentine balds with widely spaced plants do not burn readily; nor do they support large numbers of economically useful plants. In addition, there were areas that were off limits to burning because their favored plants were not fire-tolerant or the terrain was too rugged, or for other reasons.[2] In general, however, most of the plant communities in California were influenced in varying degree by Indian management.[3]

California Indians did not distinguish between managed land and wild land as we do today. The word for wilderness is absent from many tribal vocabularies, as is the word for civilization.[4] "Viewed retrospectively," writes Max Oelschlaeger in *The Idea of Wilderness*, "the idea of wilderness represents a heightened awareness by the agrarian or Neolithic mind, as farming and herding supplanted hunting and gathering, of distinctions between humankind and nature."[5]

Interestingly, contemporary Indians often use the word *wilderness* as a negative label for land that has not been taken care of by humans for a long time, for example, where dense understory shrubbery or thickets of young trees block visibility and movement. A common sentiment among California Indians is that a hands-off approach to nature has promoted feral landscapes that are inhospitable to life. "The white man sure ruined this country," said James Rust, a Southern Sierra Miwok elder. "It's turned back to wilderness" (pers. comm. 1989). California Indians believe that when humans are gone from an area long enough, they lose the practical knowledge about correct interaction, and the plants and animals retreat spiritually from the earth or hide from humans.[6] When intimate interaction ceases, the continuity of knowledge, passed down through generations, is broken, and the land becomes "wilderness."

Indigenous Resource Management

Resource management is not a modern invention. Indigenous people in California and elsewhere have practiced the roots of this applied discipline for millennia. Our California landscapes, a reflection of historical processes, both natural and cultural, bear the indelible imprint of a medley of management techniques. The major aim of this book is to shed new light on the diverse ways in which native peoples of California very purposefully harvested, tended, and managed the wild—pruning tobacco patches, burning willow to discourage insect pests, allowing for rest periods between sedge rhizome harvests, and maintaining plants with edible seed in the understories of open lower montane forests.

The foundation of native peoples' management of plants and animals was a collective storehouse of knowledge about the natural world, acquired over hundreds of years through direct experience and contact with the environment. The rich knowledge of how nature works and how to judiciously harvest and steward its plants and animals without destroying them was hard-earned; it was the product of keen observation, patience, experimentation, and long-term relationships with plants and animals. It was a knowledge built on a history, gained through many generations of learning passed down by elders about practical as well as spiritual practices. This knowledge today is commonly called "traditional ecological knowledge."[7]

The traditional ecological knowledge of California Indians and the techniques they used to manage nature are still retrievable. The historical literature contains many descriptions of Indian practices and former landscapes, before they were completely transformed by Euro-American settlement. Archaeological findings provide information on diet, tools, and demographics. Phytolith studies and fire scar data can tell us about patterns of indigenous burning and the former composition of plant communities. The growth pattern, form, and age of plant material used for the weapons and baskets in museum collections can tell us how the plants were cultivated in nature. Ecological field studies of the responses of plants to burning, pruning, or digging can also tell us much about indigenous management techniques and their effects. Finally, native people themselves still retain a

great deal of the knowledge of their ancestors. Even today, Bodega Miwok/
Dry Creek Pomo women gather edible peppernuts *(Umbellularia californica)* along stream banks; Yokuts men dig yerba mansa *(Anemopsis californica)* tubers for medicine in wind-riffled valley grasslands; Cahuilla women
pluck long golden flowering stalks from deergrass *(Muhlenbergia rigens)*
tufts along desert washes for their baskets.[8] Interviews of these people—
especially the elders, whose grandparents lived before the Gold Rush—yield
valuable and rich information about how and when areas were burned,
which plants were eaten and used for basketry, and how those plants were
managed.

Tending the Wild uses all these diverse sources of information to make
the case that indigenous land management practices were largely successful in promoting habitat heterogeneity, increasing biodiversity, and maintaining certain vegetation types that would otherwise have undergone successional change. In many cases, native harvesting and management strategies
were likely attuned to the reproductive biology of specific native plants and
grounded in sound ecological principles.

This is not to say that all actions of California's indigenous people proved
positive. The earliest humans in California may have been responsible, at
least in part, for the Pleistocene extinction of the region's megafauna. The
biologist Daniel Guthrie speculates that the earliest human settlers on San
Miguel Island in California's Channel Islands may have been involved in
the extinction of at least two of its wildlife inhabitants: the flightless goose
(Chendytes lawi) and the giant island mouse *(Peromyscus nesodytes)*. Other
research indicates that in later prehistory, California Indians may have overharvested certain animals. The research of the archaeologists Mark Raab and
Katherine Bradford suggests that indigenous people overharvested coastal
shellfish, especially black abalone *(Haliotis cracherodii)*, on San Clemente
and Santa Catalina Islands in prehistoric and historic times. And the archaeologists William Hildebrandt and Terry Jones have presented unmistakable
evidence that prehistoric hunting along the California coast led to the overexploitation of marine mammals.[9]

Certainly California Indians were effective predators and influenced the
distribution, abundance, and diversity of large mammals. "At times the intense intervention in non-human process by Indians resulted in depletions
of important resources, especially the larger animals," claims the geographer William Preston. "By late Pre-Columbian times, many of the larger
species of animals were constrained demographically and spatially by the
subsistence requirements of the native dwellers." He postulates that the large
numbers of deer, elk, antelope, beaver, and otter reported around the time of

Spanish missionization were a result of the relatively sudden diminishment of native hunting: "Their populations simply irrupted as their chief predator, the California Indians, were reduced by protohistoric plague."[10]

Very little is known about the impact of native harvesting on the flora. It is reasonable to assume, however, that the peoples migrating into what is now California more than ten thousand years ago undoubtedly experienced a learning curve, apprising the limits to resource use and then adjusting their harvesting and management from the lessons learned. At times, the result was landscape degradation and species reductions or extinctions, but over the long term, valuable lessons were learned about how to steward nature for future generations.

In general, accounts of the impact of native people on the land have been skewed in two almost contradictory ways. In some cases, these impacts are simply *assumed* to be negative. The possibility of beneficial influences, such as enhancing the numbers and diversity of other species, is seldom considered.[11] Then there is the old view that the population levels of Indians in California were so low, and their technologies so unadvanced that they had little or no impact on wild nature. Another version of this stance is the idea of the "conservation-minded Indian" put forth by some environmentalists. This view fosters a one-sided image of the California Indian as an ecological eunuch whose minimalist interventions on the environment served to guard nature's virgin treasures without despoiling or changing them. J. Donald Hughes expresses such a view in *American Indian Ecology:* "An Indian took pride not in making a mark on the land, but in leaving as few marks as possible: in walking through the forest without breaking branches, in building a fire that made as little smoke as possible, in killing one deer without disturbing the others."[12] The shallow image of the conservation-minded Indian who hardly uses, let alone influences nature and feels guilty about breaking a branch is perhaps based on a romantic notion stemming from Euro-American longings to have those same tendencies rather than on serious research into indigenous lifeways. California Indians have never advocated leaving nature alone.

Restoration

Learning about the ways in which the indigenous people of California appropriated plants and animals for cultural uses while allowing them to flourish can help us to change the ways in which we interact with nature today. Following the indigenous example, we can move beyond knowing and celebrating nature only through the view of a camera lensfinder, the end of a

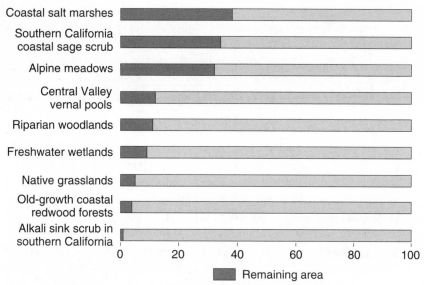

Figure 1. Remaining area of selected California ecosystems, as a percentage of total original acreage. Sources: Robert Holland pers. comm. 2005; Mark Stromberg pers. comm. 2005; Noss et al. 1995; R. F. Noss, ed., *The Redwood Forest* (Island Press, 2000).

tape measure, or the stroke of a paintbrush on canvas. We can begin to see the possibility of becoming part of localized food webs once again, being full participants in nature, and restoring and reinhabiting damaged lands.

Many of the state's native ecosystems—which contain plants of cultural significance to native people, give California its uniqueness, and act as reservoirs of precious biodiversity—are vanishing. (See Figure 1.)[13] Some temperate habitats in California are being eliminated more rapidly than most tropical rain forests and stand to lose as great a proportion of their species. With exceedingly diverse soil, topography, and climate, California harbors 25 percent of the biological diversity in the continental United States. Since the 1850s, at least twenty animal species and thirty-four plant species native to the state have gone extinct. For example, the San Joaquin Valley tiger beetle *(Cicindela tranquebarica)*, the Santa Barbara song sparrow *(Melospiza melodia graminea)*, and the Los Angeles sunflower *(Helianthus nuttallii* subsp. *parishii)* have vanished forever. Although extinction is a natural process, modern humans have driven the rate of extinctions today to about one hundred times the natural rate.[14] Dwindling biodiversity is linked to contemporary land uses, which cause degradation, fragmentation, and outright loss of habitat.[15]

A primary way that we have responded to the loss of biodiversity, the degradation of ecosystems, and the endangerment of particular species is by setting aside land and protecting it from virtually all human influences. The assumptions behind this strategy are apparent in the way that we define wilderness. According to the Wilderness Act of 1964 (Public Law 88–577), wilderness is "an area where the earth and [its] community of life are untrammeled by man, where man himself is a visitor who does not remain. An area of wilderness is further defined to mean in this Act an area of undeveloped Federal land retaining its primeval character and influence, without permanent improvements or human habitation, which is protected and managed so as to preserve its natural conditions and which . . . generally appears to have been affected primarily by the forces of nature, with the imprint of man's work substantially unnoticeable."

Much of what we consider wilderness today was in fact shaped by Indian burning, harvesting, tilling, pruning, sowing, and tending. This fact suggests an alternative way of conserving the lands that have so far largely evaded (or have somewhat recovered from) the impact of modern society: manage them by applying the traditional ecological knowledge and traditional resource management practices of California's indigenous peoples.

Although setting aside areas as wilderness is still absolutely necessary given our population numbers, there are compelling reasons to protect, restore, and manage some "wild" lands by following a model other than the hands-off wilderness model. Under what could be called the "indigenous management model," we can re-create specific human–ecosystem associations within designated areas and use them to restore and maintain these areas as they were when first visited by non-Indians. David Egan, editor of *Ecological Restoration*, defines this kind of ecological restoration as "[t]he practice of re-establishing the historic plant and animal communities of a given area or region and renewing the ecosystem and *cultural* functions necessary to maintain these communities now and into the future" (pers. comm. 1996). The indigenous management model can also be extended to the more settled and human-influenced rangelands, forests, and residential open space mosaics surrounding the state's wild lands, with indigenous management principles helping landowners, gardeners, farmers, and resource managers to better manage, restore, and use their lands.[16]

Wildland areas can also be co-managed with California Indian tribes, with the goal of restoring, maintaining, and enhancing the natural resources important to their cultures. Such arrangements could be beneficial to all stakeholders, particularly the Indian tribes, whose cultures are endangered in part because of the obstacles they face in continuing their cultural traditions: lack of access to gathering sites and the degradation of plant quality, soil fertility, and biological diversity.[17]

Renewal

The cultures of the indigenous people of California are rooted in a belief that nature has an inherent ability to renew itself, to cause the return of the geese, the regrowth of the plants with edible bulbs, the germination of next year's crop of wildflowers. This belief is reflected in the names of the lunar cycles and the annual ceremonies welcoming the return of particular animals and dances honoring the ripening of acorns and other crops. But native peoples also believe that renewal cannot happen in the absence of appropriate human behavior toward nature.

The idea that nature has a capacity for renewal as long as humans allow it to occur is not foreign to either Western culture or modern science. The English language is laden with words whose forgotten meanings point to nature's capacity for renewal and instruct us on how to live with nature. For example, the word *resource*, which now connotes ownership and production for profit, comes from the old French feminine past participle *resourdre*, which meant "to rise again."[18] The word *horticulture*, which comes from the roots *hortus* ("to garden") and *culture* ("to take care of, worship, cultivate, respect"), essentially means "to garden with respect."[19] The visionary forester Aldo Leopold wrote of "the renewal capacity of the earth" and the need for human relationships with nature that *preserve* this capacity. Ecologists point out that large human disturbances that do not mimic perturbations in nature have the effect of simplifying ecosystems and drastically reducing the land's capacity for self-renewal. California has many examples: agricultural fields with excessive salinity from irrigation in the Central Valley; overgrazed areas in pinyon–juniper woodlands; vast clear-cut areas in the Sierra Nevada and along the North Coast. In these and other places, ecosystem processes and structures have been so damaged that the land can no longer be used for farming, grazing, or timber harvesting without expensive technological inputs.[20]

Finding ways to use and live in the natural world without destroying its renewal capacity is one of the major challenges facing modern-day Californians, just as it was for the people who migrated here more than ten thousand years ago. The detailed descriptions of the land use and management practices of California Indians contained in *Tending the Wild*—the results of thousands of years of experimentation, adaptation, and ingenuity—can help us to meet this challenge. With a better understanding of the California that untold generations of Indians created, and the ways in which they brought it about and maintained it, we can reinhabit California as more circumspect stewards.

Part I

CALIFORNIA AT CONTACT

CHAPTER 1

Wildlife, Plants, and People

No country in the world was as well supplied by Nature, with food for man, as California, when first discovered by the Spaniards. Every one of its early visitors have left records to this effect—they all found its hills, valleys and plains filled with elk, deer, hares, rabbits, quail, and other animals fit for food; its rivers and lakes swarming with salmon, trout, and other fish, their beds and banks covered with mussels, clams, and other edible mollusca; the rocks on its sea shores crowded with seal and otter; and its forests full of trees and plants, bearing acorns, nuts, seeds and berries.

TITUS FEY CRONISE, *The Natural Wealth of California* (1868)

California is a land of superlatives. It has the highest mountain peaks, the largest, oldest, and tallest trees, the rivers of the greatest variety, and the most diverse Indian tribes found in the coterminous United States. California harbors the smallest bird on the continent north of Mexico, the calliope hummingbird *(Stellula calliope)*, and the largest flying bird, the California condor *(Gymnogyps californianus)*. From the summit of Telescope Peak in the Panamint Range, one can face east to see the lowest point on the American continent, in Death Valley, and then turn around to gaze at the highest point of land in the United States outside of Alaska, the summit of Mount Whitney, 14,501 feet above sea level.[1]

Not unlike that of an isolated island, the plants, animals, landscapes, and native peoples of California have a distinctness and unusual diversity that casts them apart from the rest of the mainland. This was apparent to every European visitor during the period of early exploration. One-third of the state's 6,300 native plant species are endemics and grow nowhere else on earth. It has nineteen of the ninety oak *(Quercus)* species that grow in the United States. And it contains nearly all of the world's approximately sixty species of manzanitas *(Arctostàphylos* spp.) and forty-three of the forty-five species of California-lilacs *(Ceanothus* spp.).[2]

In 1542 one hundred languages resonated across California's myriad landscapes—one quarter of the 418 native languages that existed within the borders of the present-day United States. Alfred Kroeber, the father of California anthropology, split the state into six major Native American culture areas, which reflect the state's tremendous variety of lifestyles. The archaeologist Michael Moratto states, "Such cultural, linguistic, and biological variations bespeak a long and rich prehistory in this part of the Far West."[3]

Early European explorers and settlers were universally impressed not just by California's diversity but also by the sheer abundance of its wildlife. Jean-François de Galaup, Comte de La Pérouse, a French seafarer, described California in 1786 as a land of "inexpressible fertility." He and others were taken with the prodigious congregations of wildlife: rookeries of seals, shoals of fish, pods of whales, flocks of birds, and herds of pronghorn antelope. The immense numbers of tule elk in the Central Valley, for example, rivaled ungulate numbers in Africa's Serengeti.[4]

Thomas Jefferson Mayfield, a white man who came to live with the Choynumni Yokuts in the San Joaquin Valley in the 1850s when just a boy, vividly described this overflowing abundance of wildlife: "Thousands of bandtail pigeons" came "in flights that would sometimes shut out the sun like a cloud. They piled into the nearest trees until there was not a single place for another pigeon to sit."[5]

Padre Pedro Muñoz, a member of Gabriel Moraga's Spanish military expedition through the San Joaquin Valley in 1806, observed thousands of California tortoise-shell butterflies (Nymphalis californica) on September 27, possibly at what is now Mariposa Creek, and jotted in his diary: "This place is called [the place] of the mariposas [butterflies] because of their great multitude, especially at night and morning. . . . One of the corporals of the expedition got one in his ear, causing him considerable annoyance and no little discomfort in its extraction."[6]

James Carson, a sergeant in the U.S. Army, wrote of the biological wealth of the Tulare Plains in the Great Central Valley between 1846 and 1852: "[S]wan, geese, brant, and over twenty different descriptions of ducks . . . cover the plains and waters in countless myriads from the first of October until the first of April, besides millions of grocus [sandhill cranes], plover, snipe and quail. The rivers are filled with fish of the largest and most delicious varieties, and the sportsman and epicurean can find on the Tulares everything their hearts can desire."[7] By the mid- to late 1800s, dozens of travel guides were written to attract new settlers to California, including Felix Paul Wierzbicki's 1849 book, California: A Guide to the Gold Region, and Charles Nordhoff's California: A Book for Travellers and Settlers, published in 1873. The alluring descriptions were designed to captivate the newcomer eager for a fresh start in life. But in many cases the advertisements were not exaggerated, because the truth worked just as well. The American-born traveler and writer Bayard Taylor wrote his fiancée from California in 1849:

> I cannot express to thee how I have been charmed with this country. Its pure, cloudless sky; its spring-like airs, always filled with the odor of balmy shrubs and grasses; its vast plains that stretch away like seas with forest islands and shores; its mountain ranges, which the wild oat cover with cloth of gold and which loom through the violet haze; its deep-cloven ravines, its shores and sparkling seas impress me like some new-created world.[8]

Early California was a massive flower garden. John Muir dubbed the state "the Pacific land of flowers," and he compared the frothy white flowers in the Sierra foothills to "patches of unmelted snow" and the wavy hills of yellow composites to "abundant, divine, gushing, living plant gold, form-ing the most glowing landscape the eye of man can behold."[9] These densely growing native wildflowers and grasses of hundreds of varieties such as bro-diaeas, yampah, mule ears, farewell-to-springs, lilies, balsam root, tarweeds, evening primroses, wild ryes, deergrass, and California bromes at one time covered large areas of ground not just in open grasslands but also in the open understory of California's coniferous forests, oak woodlands, chapar-ral, and pinyon-juniper forests, forming the bulk of the plant diversity in these communities. (See Figure 2.)

The goldfields (*Lasthenia californica*) were among the first wildflowers to appear in large, dense carpets in grasslands and open woodlands. Red maids *(Calandrinia ciliata)* were common throughout the state in the lower elevations, forming battalions of magenta pink. Charles Greene spotted red maids along with other wildflowers in the San Joaquin Valley and published this species account in 1892: "Eschscholtzias [sic] flamed in places, and nemo-philae repeated the blue of the skies. Mallows and calendrinias [sic] made more beautiful red on the sod than we had been looking upon; and besides these there were a multitude more of flowers, red and white, and yellow and blue." Central Valley farmers still find red maids in their agricultural fields—but their status has changed; they are listed in *Weeds of the West.*[10]

Masses of fragrant blue lupines *(Lupinus nanus, L. bicolor,* and *L. mi-cranthus)* and California poppies *(Eschscholzia californica)* were particularly notable, as they blanketed hills and valleys through large portions of the state. Blue and gold, the colors of the University of California, are said to have been chosen originally because of the great abundance of blue lupines and golden poppies in the vicinity of the first campus in Berkeley, founded in 1868.[11]

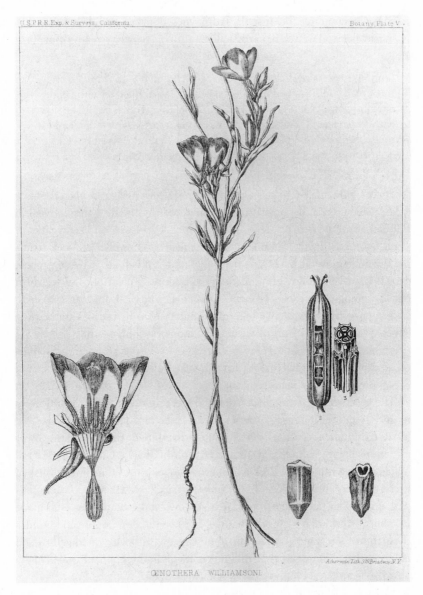

OENOTHERA WILLIAMSONI.

Ackerman Lith. 379 Broadway, N.Y.

Figure 2. William's clarkia *(Clarkia williamsonii,* formerly *Oenothera williamsoni)*
was one of many abundant wildflowers noted during the Pacific Railroad Survey
of 1853 under the charge of Lieutenant R. S. Williamson, for whom the wildflower
was named. It was new to science and first collected as a botanical specimen at Fort
Miller (now Millerton Lake) in July during the survey by Dr. A. L. Heermann.
Drawing by Charles Koppel, assistant civil engineer.

California poppies set a tilted mesa north of Pasadena aglow with their blooms in spring, serving as a beacon to ships more than twenty-five miles away. Seeing hillsides covered with these flowers, early Spanish explorers named the coast "the Land of Fire," appropriate in a literal as well as a figurative sense because of the hot, arid summers and the frequency of fires ignited by lightning strikes.[12]

The Role of Natural Disturbance

According to the plant ecologist Michael Barbour, "Late summer and early fall fires were an expected natural event in many California vegetation types below six thousand feet elevation. The same acre of ground could be expected to burn every ten to fifty years. Fire was uncommon only in deserts and at high elevations. California plants evolved with fire as a natural environmental factor over millions of years. As a result, not only do many California species survive fires, but some require fire in order to complete their life cycle or to remain vigorous" (pers. comm. 2004). The ecologist Richard Vogl has postulated that fire helped to shape three-fourths of California's vegetation.[13]

Carl Purdy, an astute horticulturist, made the connection between wildflower abundance and fire in his description of a stagecoach ride from Petaluma to Ukiah in 1870: "The trip was through lovely country, at its loveliest in mid-May. Brush fires had kept the hillsides open, cultivation did not cover much of the land, and we passed through a long succession of wild flower gardens. There were masses of a single flower covering acres, or even at times, hundreds of acres. This wonderful flower show was surpassed only by that vast one that John Muir described as adorning all of the uplands of the great interior valleys of the San Joaquin and Sacramento. That was a circuit of a thousand miles of bloom. Now, one has to go clear to the southern end of the San Joaquin Valley to see any spring flower show that is at all comparable."[14]

Fire was not the only natural occurrence shaping the landscape. Spring floods scoured watercourses and deposited silt, small and large mammals dug in the soil, storms felled trees, and torrential rains caused landslides. Each of these common events is known by plant ecologists as a natural disturbance, defined as "any relatively discrete event in time that disrupts ecosystem, community, or population structure and changes resources, substrate availability, or the physical environment." Having evolved with these erratic or episodic perturbations, many plant species not only tolerate them, but require them to complete their life cycles or to maintain dominance.[15]

Ecological studies from all over the world, in both aquatic and terrestrial ecosystems, bear out the important ecological role of disturbance in the development and maintenance of forest, shrubland, grassland, and wetland habitats. In many instances, moderate to medium disturbance promotes habitat heterogeneity and allows for greater diversity of plants and animals. For example, small mammal activity increases the abundance and diversity of geophytes (perennials with bulbs, corms, or tubers); wave action contributes to biodiversity in the rocky intertidal zone; fires maintain biologically rich grasslands; and the alternate filling and draining of lakes, marshes, and estuaries supports vast populations of aquatic life and waterfowl. Some scientists suggest that pyrodiversity (the diversity in frequency, scale, season, and type of fire) leads to great biodiversity of plant species and vegetation types.[16]

Disturbance is a recurrent feature of virtually every vegetation type in California. In fact, it is now accepted that perturbations are *required* for the rejuvenation of many plant populations and ecosystems. According to a hypothesis put forth by the ecologist Joseph Connell, disturbances that occur at intermediate intensities and frequencies promote the greatest biological diversity.[17]

Many perennial species have underground or otherwise protected organs that enable them to regenerate and spread after the aboveground parts are burned, grazed by herbivores, disrupted by landslides, or moved by digging rodents. Through the process of vegetative reproduction, they can multiply their parts or create replicas of themselves. Many shrubs and trees, for example, can resprout from suppressed epicormic or adventitious buds along their trunks or roots. Sometimes these new shoots arise on shrubs from large, gnarly underground burls, particularly after a fire. This adaptation is referred to as crown-sprouting, and it is characteristic of many chaparral shrubs. Some plants—ferns, sedges, cattails, tules, rushes, certain grasses, milkweeds, and dogbanes—die back to woody roots each year and arise anew the next year. They send out horizontal stems under the ground, or rhizomes, that enable the plants to survive and reestablish themselves and even expand the portion of the site they occupy. Still other perennials have bulbs or corms that produce offsets—tiny bulblets or cormlets containing the beginnings of a new plant genetically identical to the parent—that are spread more readily when disturbed.[18]

Wildlife

The California encountered by the Portuguese explorer Juan Rodríguez Cabrillo in 1542, the English explorer Sir Francis Drake in 1579, and the Franciscan missionary Junípero Serra in 1769 was an altogether different place from their tamed European homelands. Much of Europe had already been ecologically degraded centuries before, its wildlands deforested, mined, and overgrazed by goats, sheep, and cattle.[19] California, in contrast, was not a human-dominated landscape; its sights, sounds, and smells dwarfed Europeans and put them in awe of nature's grand show. The animals often took center stage: the large mayfly swarms hovering over streams in springtime, the orange-black clouds of hundreds of thousands of monarch butterflies in autumn, or the harsh "wah" calls of a hundred thousand white-fronted geese in winter could hardly go unnoticed.

LARGE MAMMALS

Grizzly bears *(Ursus arctos)*, the largest of California's terrestrial mammals, historically ranged from Siskiyou and Humboldt Counties in northern California to San Diego County in the south, at least two-thirds of the state. John Bidwell spotted a group of sixteen grizzlies in the Sacramento Valley in 1841 and said "grizzly bears were almost an hourly sight, in the vicinity of the streams, and it was not uncommon to see thirty to forty a day." The grizzly bear population has been estimated at ten thousand, or one every fifteen square miles. Many California place-names reflect the former abundance of bears: Bear Valley on Highway 4 above Arnold, and Bear Creek Gulch in San Mateo County, for example. Many bear trails crisscrossed chaparral thickets, and numerous tracks could be seen at springs. Large paths worn half a foot below the surface of the earth appeared in the alluvium of flat valleys. Grizzly bears would come down to the coast at night and feast on beached whales.[20]

As many as half a million tule elk *(Cervus elaphus nannodes)* fed on the lush grasses and forbs of the valley grassland. Herds as large as one thousand to three thousand were reported. Historically, tule elk in central California ranged over the entire San Joaquin and Sacramento Valleys and adjacent foothills and through the Livermore and Sunol Valleys across to the Santa Clara Valley. Don Sebastián Vizcaíno spotted abundant tule elk when he landed at Monterey on December 10, 1602. In 1848 the traveler James Lynch witnessed the San Joaquin plains covered with tule elk "as far as the eye could reach." The animals' heads rose in surprise at the approach of his regiment, and the multitude of horns reminded him of a "young forest." Edward Bosqui spotted tule elk between Merced and Stockton and said, "At times we saw bands of elk, deer and antelope in such numbers that they actually darkened the plains for miles, and looked in the distance like great herds of cattle." During the Gold Rush, well after the first European settlement, elk might be seen in bands of forty or fifty, grazing on the edge of the marshes, near Stockton. Their whistles could be heard nearly a mile away.[21]

Pronghorn antelope *(Antilocapra americana)* were common in the Central Valley from at least the Sutter Buttes southward to the desert, and there were many thousands of antelope in what are today Los Angeles and Orange Counties. The Spanish soldier Pedro Fages, during his inland exploration of the country between Monterey and the head of the Bay of San Francisco in November 1770, jotted in his diary: "We crossed it [probably Santa Clara Valley] at a cost of three leagues' [march], seeing on the way many herds of antelopes, some of them exceeding fifty." In 1874 John Muir wrote, "The antelope is quite abundant in the plains and open timber to the north of Shasta. One of the fleetest and most graceful of all wild animals, he ranges not only the open valleys but the pine woods, and feeds upon grasses. In flocks of a hundred or more they are still seen almost any day by the vaqueros of the region."[22]

The now extinct black jaguar *(Panthera onca)* roamed the South Coast Ranges between San Francisco and Monterey. Mountain lions *(Panthera concolor)* roamed from sea level to 11,000 feet elevation and screamed at night near Indian village campfires. Their lairs might be in a rocky cavern on a mountainside, under an uprooted tree, or in heavy underbrush. Gray wolves *(Canis lupus)* occurred along the eastern edge of the state and in the Central Valley. The graceful bighorn sheep *(Ovis canadensis californiana and O. c. cremobates)*, their huge horns "rising like the upturned roots of dead pine-trees," occupied large areas of mountainous California. They could be seen in bands of fifty or more. They reared their young in oval-shaped hollows of inaccessible crags, above the nesting rocks of eagles.[23]

BIRDS

As recently as two centuries ago, perhaps tens of millions of birds lived in California or passed through on an annual basis as migrants, nesters, or over-winterers. Fish-eating bald eagles *(Haliaeetus leucocephalus)* were at one time common as nesters along California's rivers and lakes. Alcatraz Island in San Francisco Bay was named after a Spanish word for "pelican," and with good reason. A French visitor, Auguste Duhaut-Cilly, described their numbers vividly in 1827: "A rifle shot we fired across these feathered legions made them rise in a dense cloud with a noise like that of a hurricane." Hundreds of pairs of cormorants occupied rookeries in Tule Lake, Eagle Lake, and Clear Lake and along the flooded banks of the San Joaquin and Sacramento Rivers. Condors, whose wingspans approach ten feet, were not an uncommon sight in south-central California in the earliest years of Euro-American settlement. Hunters would spot these birds in the vicinity of recently killed game. At least one European settler spotted four of them jointly dragging a dead young grizzly bear weighing perhaps one hundred pounds.[24]

As an area along the Pacific Flyway, California was a major destination and stopover point for dozens of species of migratory birds. Fox sparrows *(Passerella iliaca)* and golden-crowned sparrows *(Zonotrichia atricapilla),* for example, traveled thousands of miles to winter in California's coastal woodlands. Arctic terns *(Sterna paradisaea)* stopped in California on their way between the Arctic, where they nested, and Antarctica, where they spent the winter (their round-trip annual journey is 33,000 miles, requiring them to be "on the wing" most of the time). Hundreds of thousands of sooty shearwaters *(Puffinus griseus)* stopped in California before turning back to cross the Pacific Ocean, linking California to New Zealand.[25]

Some migratory species made use of California's elevational extremes, moving from lowlands to highlands and back down again. The ornithologist Scott Weidensaul says, "This kind of altitudinal migration is common in North America among many Western mountain species—gray-crowned rosy finches *(Leucosticte tephrocotis)* and dark-eyed juncos *(Junco hyemalis)* among them—moving to lower elevations in the winter, then back up to alpine meadows and forests in summer."[26] Common mergansers nested by the crystal clear waters of the high Sierra and then spent their winters at the ocean or the larger lakes of the Central Valley. Sierra Nevada rosy finches *(Leucosticte tephrocotis dawsoni)* bred at 14,000 feet on Mount Whitney. Mountain quail *(Oreortyx pictus)* scurried on foot to lower ground to avoid the heavy Sierran snows, while blue grouse *(Dendragapus obscurus)* moved in the opposite direction, climbing higher in the winter to feed on an exclusive diet of fir needles.

Many kinds of birds, such as quail, occurred in extremely large concentrations. According to the ornithologist William Dawson, in the mid-1870s, flocks of from one thousand to five thousand California quail *(Callipepla californica)* were considered commonplace.[27] Walter R. Welch published this account of quail: "In 1867 we moved to a ranch located between 'Spanishtown,' now called Half Moon Bay, and San Gregorio, on the coast side of San Mateo County. There I saw quail by the thousands everywhere; every canyon, gulch and ravine contained quail . . . and the whole country seemed to be alive with them."[28] The naturalist José Longinos Martínez mentions the "plague" of quail at the missions. La Pérouse, who reached Monterey in 1786, says this of the California quail: "The coppices and plains are covered with small grey crested partridges, which live in society like those of

PERDRIX, MALE ET FEMELE, DE LA CALIFORNIE.

Figure 3. Jean-Louis-Robert Prévost, drawing of a male and a female California quail, from Jean-François de Galaup, Comte de La Pérouse, *Voyage de la Pérouse autour du monde*, 1793. Flocks of valley quail could number from one thousand to five thousand. Gift of E. E. and D. A. Eyre in memory of Florence Atherton Eyre, North Baker Research Library. Courtesy of the California Historical Society, FN-30508.

Europe but in coveys of three or four hundred. They are fat and of excellent taste." (See Figure 3.) Nordhoff said: "[I]n favorable seasons the whole plain [the Central Valley] is alive with the pretty top-knotted quail." In the lowlands of California, large bands of several thousand California quail fed on the tender leaves of clovers in spring and built their nests in a deep tangle of grass and briars and often beneath a pile of brush or dead cuttings.[29]

The Great Central Valley served as an important overwintering site for ducks, geese, and other waterfowl. Every winter, waterfowl flocked by the millions into the vast maze of waterways formed by the Sacramento–San Joaquin Delta; this elaborate labyrinth of splitting and merging distributaries made up "the richest ecosystem in the state."[30] Of all the displays of wildlife, the tremendous flocks of geese—snow geese, Ross' geese, white-fronted geese, and Canada geese—were perhaps the most striking in appearance. They could appear in thick cloudlike congregations, so large that the roar of their wingbeats as they took flight was deafening. Wild geese were so common that an American officer who shot and presented them to his hostess in the early 1850s was considered stingy.[31]

Dawson described the abundance of the snow and Ross's geese *(Chen hyperboreus,* now *caerulescens,* and *C. rossii):* "It is scarcely possible to exaggerate the number which frequented this region before the advent of the white man. It must have run into the millions, and may easily have reached the tens of millions. Practically the entire population of the North, breeding and bred on the Arctic shores of British America, in Banks Land and, presumably, upon the still undiscovered Hyperborean land mass, poured across the defiles of the Sierras in late September and early October, and covered the central California landscape as with a quivering white blanket." Adolphus Heermann wrote of them in the 1850s, "These birds often cover so densely with their masses the plains in the vicinity of the marshes as to give the ground the appearance of being clothed with snow."[32]

In his 1873 travel guide to California, Nordhoff entices the would-be traveler: "I have seen two square miles of geese feeding on a sheep pasture near a lagoon, and so tame I almost rode them down before they rose." Jeff Mayfield said that when he first came to the San Joaquin Valley in 1850, he saw so many white geese in flight that he was positive one band of them would cover four square miles when they landed.[33]

LIFE IN THE RIVERS

Salmon *(Oncorhynchus kisutch, O. tshawytscha, O. keta)* ran up every major river and creek from what is now the Smith River on the north to the Carmel River on the south, swimming against the current to return near where they were born. Thirty-one coastal and Central Valley rivers and hundreds of lesser creeks carried the lifeblood of millions upon millions of salmon and provided six thousand miles of spawning habitat. Joaquin Miller described the head of the Sacramento River as a "silver sheet" because salmon were so abundant. He had seen the stream "so filled with salmon that [it] was impossible to force a horse across the current."[34] A white man stated in the early 1900s, "[Y]ou could load wagons with salmon that got stalled on Mad River. . . . At the little sloughs near Arcata you could get salmon with pitch-forks and fork them on to the bank."[35] Bayard Taylor described salmon caught from the Sacramento River in 1850: "Salmon trout exceeded in fatness any fresh water fish I ever saw; they were between two and three feet in length, with a layer of pure fat, quarter of an inch in thickness, over the ribs. When made into chowder or stewed in claret, they would have thrown into ecstasies the most inveterate Parisian gourmand."[36]

Various species of trout *(O. clarki, O. mykiss, O. chrysogaster)* were plentiful in mountain brooks, rivers, and lakes. Theodore Van Dyke commented in 1886: "But a few years ago these fish were so plenty [*sic*] in these mountains and so tame that a failure in fishing was an impossibility for any one. They dashed in full confidence at the clumsiest bait upon the coarsest line. A fresh fish was often ready to take in a moment the same bait that had already caught two or three. And often six or eight trout were taken in succession from the same little pool or rapids."[37]

The rivers were full of mammals as well: river otters *(Lutra canadensis)*, mink *(Mustela vison)*, and beavers *(Castor canadensis)*. River otters' dens were located under steep riverbanks, and their loud, shrill voices or low coughs or grunts could be heard from every major central and northern California river. Rolling places, grassy areas where otters roll to dry off, and otter slides, paths worn smooth by otters sliding down streamside banks, were numerous along streams and sloughs. River otters and beavers also abounded in Kern and Buena Vista Lakes in the southern San Joaquin Valley and could be shot easily from a boat. River mussels, crayfish, frogs, newts, salamanders, turtles, and snakes were prolific, all indicators of healthy rivers and creeks.[38]

LIFE IN THE PACIFIC OCEAN

To someone standing on the shore at the dawn of Euro-American contact, the ocean was a virtual menagerie of life. Basking harbor seals *(Phoca vitulina)* covered offshore rock outcrops. Rafts of sea otters *(Enhydra lutris)* played and ate in the kelp beds. The flukes of whales, the dorsal fins of bottlenosed dolphins, and the heads of seals constantly broke what the Alaskan Yupik called the "skin of the world"—that line between water and air.

By the thousands and thousands, northern fur seals *(Callorhinus ursinus)* and Guadalupe fur seals *(Arctocephalus townsendi)* returned to the same rookeries on the Channel Islands year after year—usually the very islands where they were born. Alpha male northern elephant seals *(Mirounga angustirostris)* molted and defended their harems with shrieking cries. Gray whales *(Eschrichtius robustus)* born in the shallow lagoons and bays of Baja California migrated by the thousands to the Chukchi Sea in the Arctic and back again, a ten-thousand-mile annual journey. Many kinds of fish migrated along the coast or from deeper waters into shallower waters and back again.

Among the major coastal fish were bass, herring, halibut, smelt, sardine, flounder, rock cod, dogfish and other sharks, stingray, and sturgeon. In 1857 the *Crescent City Herald* reported that a massive school of smelt and sardines piled up a foot deep on the shore at Crescent City and extended three-fourths of a mile seaward. The fish were so numerous three men found it impossible to row a skiff through them.[39]

William Brewer described the abundant life in the tidepools at Pescadero near Monterey in 1861: "Shellfish of innumerable forms, from the great and brilliant abalone to the smallest limpet—every rock matted with them, stuck into crevices, clinging to stones—millions of them. Crustaceans (crabs, etc.) of strange forms and brilliant colors, scampered into every nook at our approach. Zoophytes of brilliant hue, whole rocks, covered closely with sea anemones so closely that the rock could not be seen—each with its hundred arms extended to catch the passing prey. . . . Every pool of water left in the rugged rocks by the receding tide was the most populous aquarium to be imagined."[40]

The plankton of San Francisco Bay was rich and abounded in shrimp. Tidepools along the Central Coast harbored abalone that could be easily taken at low tide. Dense beds of intertidal abalone were found in southern California. In his notes during a military reconnaissance in 1846, Lieutenant William Emory said that kelp *(Fucus giganteus)* was so thick in San Diego Bay that he mistook it for a low island.[41]

Enormous flocks of southern-bred pink-footed shearwaters *(Puffinus creatopus)* would annually flood California's coast and Channel Islands. As late as the 1920s, it was possible to see a million sooty shearwaters *(Puffinus griseus)* off the California coast feeding on millions of herring.

Plant Life

California's plant life was, and still is, exceptionally diverse in two senses. First, a very large number of plant species, many of them endemic, grow within the state's borders. The *Jepson Manual*, the authority on the state's flora, lists more than 5,800 species, and a California Native Plant Society inventory lists 6,300.[42] Second, these species grow together in particular patterns to form very many types of vegetation. This diversity can be seen in the way the vegetation changes as one moves from valley to mountain, from north to south, and from moist coast to arid, rain-shadowed desert.

Vegetation can be divided into types, each having a characteristic physical structure, or physiogonomy. Some of the major broad types are marsh, grassland, shrubland, woodland, and forest. Vegetation types are in turn divided into plant communities. A plant community is a distinctive grouping of plant species, defined as "an aggregation of living organisms having mutual relationships among themselves." Each plant community is distinguished by the presence of certain dominant or characteristic plant species.[43] For its size, California has one of the largest numbers of plant communities in the world, in part because of its varied climates, soils, and topographies.

Elna Bakker calls California "the great mosaic." In traveling the length of the state, one can be in the cool temperate rain forests of the northwest coast, cross open grassland in the Central Valley, trudge through chaparral in the foothills, walk across wet meadows in the high mountains, and finally weave among scattered shrubs under scorching 120 degree heat in the Mojave Desert.[44]

To cover all the vegetation types and plant communities in California would require an entire volume. Many books already do this very well. Instead, I sample the state's great vegetation diversity as it likely existed in pre-Columbian times—looking at some of the most significant plant communities in terms of acreage covered and the biodiversity they supported.

MARSHES

Coastal Salt Marshes

Coastal salt marshes were at one time much more extensive than today, forming important habitat rich in plant and animal life. Salt marshes occur in the upper intertidal zone of sheltered shallow bays, river mouths, and coastal lagoons. They are bordered, below the upper intertidal, by beds of sea grasses, both eel-grass (Zostera marina) and surf-grass (Phyllospadix spp.). In the marshes themselves grow stands of California cord grass (Spartina foliosa), which at the lower edges of the marsh are submerged for nine hours at a time. At the midmarsh zone, pickleweed (Salicornia bigelovii) abounds, sometimes in association with cord grass. Other common plants are jaumea (Jaumea carnosa) and arrow-grass (Triglochin concinna). The upper zone, where the vegetation is submerged for less than five hours at a time, is home to alkali heath (Frankenia salina), sea lavender (Limonium californicum), saltgrass (Distichlis spicata), and California sea-blite (Suaeda californica).[45]

These plants provided food and nesting and ground cover for myriad birds and mammals. Black-crowned night herons, in colonies of a thousand individuals, would bed down together in the rushes in San Francisco Bay area marshes. Waterfowl were so numerous in the Salinas marshes that boys throwing cords weighted at both ends could easily bring them down as they ascended.[46]

Brackish and Freshwater Marshes

Michael Barbour and colleagues describe brackish and freshwater marshes as "among the most productive ecosystems on earth."[47] According to the botanist Herbert Mason, the marshes that exist today represent only a fraction of what once originally prevailed in California, especially in the Central Valley.[48] At one time they occurred in many of the interior valleys, along river courses, creeks, and sloughs and bordering lakes, as well as in brackish areas landward of salty marsh areas along the coast. Common plant species included various tules (*Schoenoplectus americanus, S. validus, S. acutus, S. californicus,* formerly in the genus *Scirpus*), three species of cattails (*Typha latifolia, T. angustifolia,* and *T. domingensis*), sedges (many species of *Carex*), common reed *(Phragmites australis),* water plantain *(Alisma plantago-aquatica*), arrowhead (*Sagittaria* spp.), and yerba mansa *(Anemopsis californica).*

The Kern, Kaweah, and Kings Rivers, which drain the southern Sierra Nevada range, had no outlets and spilled into large natural lakes—Tulare, Kern, and Buena Vista—that extended for miles in all directions in the southern San Joaquin Valley. Tulare Lake, seven hundred square miles in extent, was the second largest lake west of the Mississippi River. On an 1830 map of upper California, José Maria Narváez labeled a large oval with irregular borders covering much of the Central Valley "Ciénegas o Tulares" (Marshes of Tules). Freshwater marsh areas along the lake borders and sloughs were immense. During wet years, a Kitanemuk man could paddle a tule balsa from Kern Lake up through connecting sloughs to Buena Vista Lake and on to Tulare Lake, eventually reaching the San Joaquin River. James Carson described the slough that conveyed the water from Tulare Lake into the San Joaquin River: "Its length, from its entrance into the San Joaquin to the edge of the tule beds of the lake, is about thirty-five miles. The tules at the lower end of the lake are some fifteen miles in width. Large flocks of white-fronted geese used Tulare Lake as a loafing ground."[49]

The lakes and surrounding marshes supported large populations of mammals, birds, and other animals. River otters and beavers abounded in Kern and Buena Vista Lakes, and tule elk dug into the tules and exposed themselves to moist mud to cool off. C. B. Linton reported two colonies of American white pelicans *(Pelecanus erythrorhynchos)* nesting on Buena Vista Lake in 1908: one of about 250 nests, on a small sandy island in the river mouth; the other of perhaps 500 nests, on the lake shore. Apparently, a large colony of thousands of white pelicans also nested on an island in Tulare Lake for many years.[50] Large numbers of Western pond turtles, freshwater mussels and clams, and myriad fish species could be found in the lakes. Of all the wildlife in the marshes, waterfowl were the most abundant. The canvasback, reputed to be the best-tasting duck by contemporary hunters, was found by the tens of thousands in the lakes and lagoons of southern California.[51]

GRASSLANDS

Pure grasslands, including coastal prairies, valley grasslands, vernal pools, and montane meadows, covered one-fifth of the state before 1850. After Euro-American settlement, they became the most productive rangelands and farmlands in North America.[52]

Coastal Prairie

Coastal prairie in aboriginal California was characterized by a mixture of broad-leaved herbs and native annual and perennial grasses. This grassland was discontinuous, occurring in large and small patches from the coast to 62 miles inland and up to elevations of 3,280 feet. In the south, it extended to what is now Point Lobos State Park, and in the north it reached beyond what is now the Oregon border. Among the important grass species in California's coastal prairies were Idaho and red fescues *(Festuca idahoensis* and *F. rubra)*, California oatgrass *(Danthonia californica)*, and bent grass *(Agrostis exarata)*. Characteristic broad-leaved species were Douglas iris *(Iris douglasiana)*, California buttercup *(Ranunculus californicus)*, and blue-eyed-grass *(Sisyrinchium bellum)*. Other species were yampah *(Perideridia kelloggii)*, goldfields *(Lasthenia* spp.), and tidy-tips *(Layia platyglossa)*. Recently, the ecologist Mark Stromberg and colleagues discovered that the species richness of the coastal prairies—averaging 22.6 species per square meter—is greater than that of any other grassland type in North America.[53]

Valley Grassland

Valley grassland is thought to have originally covered much of the Central Valley, as well as the lower elevations of the central and southern Coast Ranges. Like the pristine coastal prairie, it contained a mixture of annual and perennial plants. Most important were various bunchgrasses such as purple needlegrass *(Nassella pulchra)*, nodding needlegrass *(N. cernus)*, one-sided bluegrass *(Poa secunda* subsp. *secunda)*, and poverty three-awn *(Aristida divaricata)*. Deergrass *(Muhlenbergia rigens)* was another major associate. Prominent wildflowers possibly included miniature lupine *(Lupinus bicolor)*, blue dicks *(Dichelostemma capitatum)*, stinkbells *(Fritillaria agrestis)*, owl's-clover *(Castilleja* spp.), adobe-lily *(Fritillaria pluriflora)*, white broadiaea *(Triteleia hyacinthina)*, clovers *(Trifolium* spp.), goldfields *(Lasthenia fremontii, L. californica,* and *L. glaberrima)*, fiddleneck *(Amsinckia menziesii* var. *intermedia)*, red maids, and yellow carpet *(Blennosperma nanum)*. Today, this grassland exists only in small pockets and has largely been altered by the introduction of alien species, especially annual grasses.[54]

Within valley grassland are shallow basins with poorly drained soils that collect winter rainfall. These ephemeral wetlands are referred to as vernal pools. They support a rich variety of plants, which grow in concentric rings according to their tolerance for inundation. An average of thirty-five species is found per pool, with an estimated fifty community types among different pools throughout the state. Several species of meadowfoam (*Limnanthes* spp.) and downingia (*Downingia* spp.), as well as button-celery *(Eryngium vaseyi)*, willow herb *(Epilobium pygmaeum)*, and water-starwort *(Callitriche marginata)* are found here. At wet times of the year vernal pools are laden with animal life: fairy shrimp, delta green ground beetles, California tiger salamanders, and Pacific chorus frogs. John Muir was serenaded by "an immense crop" of frogs after late rains filled the vernal pools near Snelling. There were more than a million acres of these vernal pools in early aboriginal California in the Central Valley, on the western slopes of the Cascade Range and the Sierra Nevada, in the valleys of the Coast Ranges, and on the coastal terraces of southern California.[55]

The grasses and wildflowers of the valley grassland provided a vast food reservoir for insects, serving as both nectar and larval food plants to scores of butterflies, native bees, beetles, and flies. The white, lacy flowers of various species of yampah were important nectar sources to the anise swallowtail *(Papilio zelicaon)*. Owl's-clover, which occurred in great swaths, was an important nectar plant to the northern checkerspot *(Charidryas palla)* and the Leanira checkerspot *(Thessalia leanira)* butterfly.

One North Fork Mono/Chukchansi elder, Pauline Conner, reminisced about the time long ago when wildflowers covered vast areas of grassland and "butterflies were so thick they would come in clouds, and you could reach out and touch them. Sometimes they would land on you." These were the buckeyes *(Junonia coenia)*, anise swallowtails, and orange sulfurs *(Colias eurytheme)*. Acres and acres were clothed in owl's-clover (*Orthocarpus* and *Castilleja* spp.) and clovers *(Trifolium willdenovii* and *T. wormskioldii)*, important food plants to these butterflies.

SHRUBLANDS

Many types of shrublands cover the hills, plains, and coastal terraces of California, including coastal scrub along the coast and creosote scrub in the desert, but chaparral is the most extensive, covering about 8.6 million acres. Chaparral, composed of woody shrubs and small evergreen trees, is common on most hills and lower mountain slopes of the Coast Ranges, the western slopes of the Sierra Nevada, and the more southern mountains. Of all the state's plant communities, it is probably the most adapted to fire and passes "endlessly through cycles of burning and regrowth."[56]

Chaparral means "short woody vegetation" in Spanish. This plant community is rich in vascular plant species. There are at least nine major types of chaparral, each dominated by specific species. The most extensive type is chamise chaparral, dominated by chamise *(Adenostoma fascicula- tum)*. Another widespread chaparral type is ceanothus chaparral, dominated by buck brush *(Ceanothus cuneatus)*. Associated species include nude buck- wheat *(Eriogonum fasciculatum)*, scrub oak *(Quercus dumosa)*, and moun- tain mahogany *(Cercocarpus betuloides)*, all extremely important browse species for deer. A third type is mixed chapparal with toyon *(Heteromeles arbutifolia)*, scrub oak, chamise, buck brush, California coffeeberry *(Rham- nus californica)*, and silk tassel bush *(Garrya spp.)*. In addition to the 240 woody plant species found in chaparral, more than 600 annual and peren- nial herbs are present, many appearing in great abundance after a fire. These include red maids, farewell-to-spring *(Clarkia spp.)*, melic grass *(Melica spp.)*, gilia *(Gilia spp.)*, clover *(Trifolium tridentatum)*, and chia *(Salvia columbariae)*. Ecologists hypothesize that for many of these postfire species, biotic and abiotic conditions in mature chaparral have selected for seed dormancy mechanisms that delay germination until the first spring after fire.[57]

WOODLANDS

Riparian Woodland

Before dams and man-made levees, river boundaries surged and retreated with the seasons. Almost every spring the rivers and creeks in the lower el- evations spilled over their borders into overflow channels, dropping nitrogen- rich silt and sand—the nutrient influx that helped to make California's ri- parian corridors so productive. Flooding scoured the vegetation, removing dead and dying growth of trees, shrubs, and herbs and spurring new vege- tative growth. The riverbanks and overflow channels harbored a unique riparian flora that included a large variety of sedges *(Carex spp.)*, reeds *(Jun- cus spp.)*, grasses, deergrass, and deciduous trees and shrubs such as large cottonwoods *(Populus fremontii)*, redbuds *(Cercis orbiculata,* formerly *C. oc- cidentalis)*, willows *(Salix spp.)*, alders *(Alnus rhombifolia)*, maples *(Acer macrophyllum)*, and sycamores *(Platanus racemosa)*. Riparian woodlands, which before 1850 covered 900,000 acres in the Central Valley, teemed with animal life. Today only about 100,000 acres remain.[58]

Native walnut trees *(Juglans hindsii)*, measuring up to six feet in diameter and clear of branches for forty feet, lined the Sacramento River. These trees were the favored nesting sites of the now threatened Swainson's hawk *(Buteo swainsoni)*. Grand old California sycamores, with trunks five feet across, dotted the alluvial benches and river bottoms in the South Coast Ranges, the San Joaquin and Sacramento Valleys, the Sierra foothills, and parts of southern California. These splendid trees were overwintering sites for monarch butterflies in autumn and roosts for several kinds of bats. Birds such as the American kestrel *(Falco sparverius)* built their nests in deep cavities in hollow sycamores.[59] In the 1920s, Dawson found forty-one blue heron nests and twenty-eight night heron nests in one massive, seven-foot-diameter sycamore—what he called a virtual "heron village."[60] Sycamores often grew in the overflow channels of rivers and creeks and could withstand and even benefit from periodic inundations. After a flood, seedlings would sprout from the scoured earth in great numbers.

These trees were massive enough to create a home for a menagerie, from tiny wood-boring beetles to reptiles to large raptors and wading birds and small mammals. William Brewer commented as he hiked through Hospital Canyon near Mount Oso, "A few cottonwoods grow along the creek, and in them hundreds of cranes [great blue herons] have built their nests—great awkward birds, with their maltese-colored plumage, long slim necks, and longer slimmer legs. . . . Great numbers of other species of birds also congregate in these canyons."[61]

Lining many of the rivers and creeks in the Central Valley, above the overflow channels in the sandy flats, were huge valley oaks *(Quercus lobata)* that reached diameters of 8 to 12 feet and heights of 150 feet. Because they grow in the deep moist loam of alluvial or delta valleys, these oaks were dubbed the "old-time monarchs of the soil" by Willis Linn Jepson.[62]

Present-day Visalia, for example, was in the midst of a magnificent forest of these oaks. George Vancouver, a British naval captain who visited the San Francisco Bay area in 1792, noted that one valley oak near Santa Clara Mission measured fifteen feet in girth. However, he pointed out that the missionary fathers did not consider this tree of extraordinary size; other oaks were of even greater magnitude. He also mentioned that the timber from these oaks was "equal in quality to any produced in Europe."[63]

Valley oaks also created a microclimate for many kinds of understory plants and fungi that only survived under the trees. Thus, these giants of the plant world were like arks supporting a great diversity of life. William L. Finley reported in 1915 that near Crows Landing in Stanislaus County he found fifty to seventy-five nests of American egrets in the tops of large oak trees lining a small creek tributary to the San Joaquin River. He said that a much larger number had nested in the same vicinity two years before.[64]

Foothill Woodland

Foothill woodland covers more than three million acres in California (one-half of all oak-covered lands) and is so abundant that some ecologists have elected it the "state vegetation type." In the interior, drier foothills, deciduous blue oak *(Quercus douglasii)*, unique to California, is the dominant tree. This plant community is found below 3,500 feet on slopes bordering interior valleys from Los Angeles County to the head of the Sacramento Valley. Blue oaks may occur in nearly pure stands as dense woodland or open savanna; as a dominant in mixed stands that include gray or foothill pine *(Pinus sabiniana)*, interior live oak *(Q. wislizenii)*, valley oak *(Q. lobata)*, and/or coast live oak *(Q. agrifolia)*; or as a minor component in mixed stands of oaks and other hardwoods. Associated shrub species scattered in the understory include California-lilacs *(Ceanothus spp.)*, redbud *(Cercis orbiculata)*, yerba santa *(Eriodictyon californicum)*, manzanita *(Arctostaphylos spp.)*, coffeeberry *(Rhamnus californica)*, and poison oak *(Toxicodendron diversilobum)*. Closer to the Pacific Ocean, dominant oaks are the California black oak *(Q. kelloggii)* and Oregon oak *(Q. garryana)* in the north and the coast live oak everywhere else. Coast live oaks often attained a height of seventy feet, and some early accounts report that the canopy of certain trees sprawled outward from the trunk for a distance of sixty feet in every direction. These trees often grew with an absence of underbrush, and their woodlands were likened to "highly cultivated parks."[65]

FORESTS

Lower Montane Forest

The lower montane forests of all California mountain ranges form a distinct belt spanning the elevation between 3,200 feet and 6,900 feet. Ponderosa pine *(Pinus ponderosa)* and Jeffrey pine *(P. jeffreyi)* dominate the more xeric sites; white fir *(Abies concolor)* is found on more mesic sites. From historical archival records and early photographs, it is clear that at the beginning of Euro-American contact the central and southern Sierra Nevada in many areas was very open, featuring large-diameter trees, 40 to 60 feet apart, and minimal underbrush.[66]

Other tree species in the lower montane forest are giant sequoia *(Sequoiadendron giganteum)*, sugar pine *(Pinus lambertiana)*, incense cedar

(Calocedrus decurrens), Douglas-fir *(Pseudotsuga menziesii* var. *menziesii)* or bigcone Douglas-fir *(P. macrocarpa)*, and a major hardwood, California black oak. Sugar pine, ponderosa pine, and California black oak are well adapted to light and regular ground fires and are intolerant of deep shade. Black oak needs bare mineral soil to germinate. James Hutchings, one of the first non-Indian inhabitants of Yosemite Valley, described the openness of this forest community: "Large sugar-pine trees, *Pinus Lambertiana;* from five to ten feet in diameter, and over two hundred feet in height, devoid of branches for sixty or a hundred feet, and straight as an arrow, everywhere abound. . . . These forests are not covered up with a dense undergrowth, as [in] the East, but give long and ever-changing vistas for the eye to penetrate."[67]

Coastal Redwood Forest

Coastal redwoods *(Sequoia sempervirens)*, the tallest trees on earth (up to 369 feet), occur in a narrow belt five to thirty-five miles wide along the coast from the Oregon border to the southern boundary of Monterey County. Taking advantage of the moisture from the fog, which condenses on branches and drips to the ground, these trees thrive in California's mild humid marine climate. Plant companions include Douglas-fir, madrone *(Arbutus menziesii)*, colorful rhododendrons, several kinds of oaks, nutmeg *(Torreya californica)*, tan oak *(Lithocarpus densiflorus)*, several kinds of huckleberry *(Vaccinium ovatum, V. parvifolium)*, salal *(Gaultheria shallon)*, and California bay *(Umbellularia californica)*.[68]

Natural lightning fires are sparse along this belt, yet fire ecologists have found numerous fire scars on the long-lived stumps—suggesting that fires occurred in short-return intervals in many of these stands. Redwoods have durable, decay-resistant wood, and they are resilient in the face of flood or fire. In the latter case, they are protected by an insulating layer of bark and are able to sprout back. If the entire crown of the tree is fatally injured by fire, clones sprout from the base, forming a "family circle."[69]

Mixed Evergreen Forest

Mixed evergreen forest occurs along the upper edge of the redwood forest or the foothill woodland, in the North Coast Ranges, at elevations of 1,000 to 3,000 feet. Major species are Douglas-fir, maple, maul oak *(Quercus chrysolepis)*, California bay, tan oak, and California black oak. Understory species include hazelnut *(Corylus cornuta* var. *californica)*, mountain dogwood *(Cornus nuttallii)*, several wild lilacs *(Ceanothus* spp.), and yerba buena *(Satureja douglasii)*. Among the herbaceous plants are bear-grass *(Xerophyllum tenax)* on the drier sites and redwood sorrel *(Oxalis oregana)* and evergreen violet *(Viola sempervirens)* on the wetter sites. Clumps of ferns and epiphytic mosses grow commonly on tree trunks throughout these forests. In the southern and central Coast Ranges and the mountains of southern California, mixed evergreen hardwood species still form extensive forests on moist slopes, but Douglas-fir is much less prominent.[70]

People

Excluding desert and high-elevation areas, it was almost impossible for early Euro-American explorers to go more than a few miles without encountering indigenous people. Grass-thatched domical houses could be seen from southern coastal waters; redwood plank houses stretched along Pacific Northwest rivers; conical houses of incense cedar bordered meadows in the mountains of the western Sierra Nevada; clusters of saltbush huts dotted sagebrush clearings on the east side of the Sierra Nevada; tule mat–covered, wedge-shaped houses lined San Joaquin River delta waterways; and palm frond houses were nestled around springs in the Mojave Desert.

California's cultural diversity matched its biodiversity: it contained the most diverse native cultural groups of any other state or country of comparable geographic size from the Arctic to the tip of South America. (See Figure 4.) The state was thickly populated, and most anthropologists agree that north of Mexico City, California held the highest densities of people of any area of equal size in North America. Population density varied, from fewer than 0.08 people per square mile in desert regions to more than 1.49 per square mile in places near the Santa Barbara Channel.[71]

Estimates of California's total population vary from 133,000 to 705,000; about 310,000 is the most widely accepted number. The lowest estimates are thought by some demographers and archaeologists to be highly conservative, because Old World diseases had swiftly reduced populations before the arrival of settlers in many areas (William Hildebrandt pers. comm. 2004). In fact, when Cabrillo made the first recorded European visit to the west coast, in 1542, he noted in his journal that the native people (Kumeyaay and Tongva) spoke about and pantomimed bearded men on horses who lived somewhere to the north, suggesting that European contact was not new.[72]

Anthropologists have lumped diverse sociopolitical groups into sixty broad linguistic designations, such as Pomo, Achumawi, Yokuts, and Hupa, but this belies the true heterogenity of the state's indigenous peoples. (See Map 1.) What is assigned the term "Pomo" on the map, for example, is a language group that in reality is made up of seven mutually unintelligible languages. If these broad designations for each tribe were separated out, California

Figure 4. Alice Frank Spott, of Del Norte County in northwestern California, was a Yurok, one of the many diverse tribes of the state. Photograph taken in 1907. Courtesy of the Phoebe Apperson Hearst Museum of Anthropology and the Regents of the University of California, #15–3834.

would have perhaps as many as one hundred mutually unintelligible languages and many dialects. Alfred Kroeber listed between five and six hundred tribes as the number of sociopolitical groups that were autonomous and self-governing and encompassed a cluster of two or more separate villages led by a chief.[73]

It was not uncommon to encounter speakers of several different languages or dialects within a half-day's travel of one's village. "The Indians have never formed a national body," wrote Duhaut-Cilly, a Frenchman visiting California in 1827–28. "[E]ven their language undergoes great variations in very short distances; often those at one mission do not understand those of the nearest mission."[74] Father Geronimo Boscana wrote, "Almost every fifteen

Map 1. The territories associated
with California Indian language
groups. Names in bold represent a
language family of two or more
languages and multiple dialects.

or twenty leagues you find a distinct dialect; so different that in no way does one resemble the other." Thomas Henley, in an Indian Affairs report of 1854, commented that from "the San Joaquin northward to the Klamath there are some hundreds of small tribes."[75]

Ponder one hundred different ways to say blue elderberry *(Sambucus mexicana)*—such as 'angtáyu (Central Sierra Miwok), hunqwat (Cahuilla), ṭʰeqʰále (Kashaya Pomo), and kunuguvɨ (Kawaiisu). This great linguistic diversity reflects the tremendous length of time people have been here. Language, in part, shaped thinking and culture and brought a great variety of ways of using and tending the land. Tribal languages simulated the sounds of nature, reinforcing ties to places. The Luiseño word *putquivi*, the name for the olive-sided flycatcher *(Contopus borealis)*, imitates the bird's spirited song. The Choynumni Yokuts spoke sharply when they said *skée-til*, their word for squirrel, in imitation of the animal's bark.

California had been peopled for at least 12,000 to 13,500 years when European settlement began. The Tubatulabal say that they have always lived in the region where they are found today (the drainage area of the Kern and South Fork Kern Rivers, from their sources near Mount Whitney to approximately forty-one miles below the junction of the two rivers); many other tribes make a similar claim about their territories. In their own words, native people today say, "We've *always* been here." Some creation stories place the origins of plants, animals, people, and the earth in the very heart of the respective tribal territory. And often the center of the whole world was somewhere within the homeland. The Modoc believed (and still believe) that the center of the world is a hill on the eastern shore of Tule Lake. Migration stories are absent from the lore of many tribes.[76]

At the dawn of Euro-American settlement, there were dozens of villages lining every major river and creek, and great shell-mound deposits as well as explorers' logs provide evidence of large bayshore villages clustered around the margins of coastal estuaries.[77] In 1830 the fur trapper Alexander McLeod said of Bonaventura Valley, present-day Sacramento Valley, "The Indian population . . . is very great. It is impossible for me to give even an idea of their number. Several Villages that our route led us to pass close by each contained at least 1500 Men and every Creek or Lake where water could be found Indians were stationed at in great numbers as well in the low country as in the high land."[78]

Juan José Warner, who traveled through the San Joaquin Valley during the winter of 1832–33, noted, "The banks of the Sacramento and San Joaquin, and the numerous tributaries of these rivers, and the Tule Lake [probably Tulare Lake], were at this time studded with Indian villages of from one to twelve hundred inhabitants each. The population of this extensive valley was so great that it caused surprise, and required a close investigation into the nature of a country that without cultivation, could afford the means of subsistence to so great a community."[79]

The repertoire of subsistence activities included gathering, hunting, fishing, making fires, and quarrying stones for tools. Tribal territories often spanned several elevational zones encompassing a variety of plant communities. Gatherers and hunters visited every type of plant community—from beach and dune to coastal scrub, evergreen forest, sagebrush steppe, valley grassland, and pinyon-juniper woodland. The people obtained a diversity of plant and animal resources by following an annual cycle of population movements that coincided with seasonal availability of specific resources.

California's promontories, declivities, and unusual rock formations were infused with meaning. Every place was named. For example, Lake Tahoe, around which the Washoe lived, was called *Da ow a ga* ("giver of life") in their language.[80] Even modest knolls and insignificant peaks carried Indian names. Often places were named for their specific characteristics. Lindsay Creek and the adjacent prairie were called *Topōdērōs* ("Indian potato") in the Wiyot language, as many corms of *Brodiaea coronaria* were gathered there. False Klamath Rock was called *Wɛtc' atagasni* ("digging something") by the Tolowa, because edible roots once flourished at that spot. Abundant bulbs were dug by the Wintu at a spring at McCardle Place, and it was called *Ke-ten-ton* ("wild-potato-place"). In the Nomlaki language, some places carried the name of an abundant local animal, such as *Anunsawal* ("turtle spring"). A famous place in which to hunt deer at night was called *So-kut Men-yil* ("deer moon") by the Cahuilla.[81]

Areas now labeled simply "wilderness" or "national park" on topographic maps once encompassed ancient gathering and hunting sites, burial grounds, work stations, sacred areas, trails, and village sites, all making up what was home to hundreds of generations of California Indians. Being at home in a place meant that generation after generation of people were born, lived, and died in the same, familiar surroundings. C. Hart Merriam, one of the earliest biological scientists to describe California's Indian cultures, understood well the deep veneration that native people held for their birthplace: "In most parts of California, the greatest calamity that can befall an Indian is to be removed from the place where his father and mother lived and were buried." California Indians "are fond of home," wrote Robert Brown, "and if away for a short time from the locality where they have been born or brought up, soon weary to return."[82]

The anthropologist Julian Steward recorded and published in 1934 the life stories of two Owens Valley Paiutes, Jack Stewart and Sam Newland. Stewart said in his autobiography, "Few Indians leave their own country who do not return. I know of only one who stayed away." And the anthropologist George Foster wrote of Eben Tillotson (Yuki) describing his delight at walking over the ground where he had once lived: " 'I once killed a deer on that hillside. My, I remember how thick the berries used to grow on that slope.' He then explained how good it seemed to be on home ground again: it felt right and natural. The rocks and the trees knew him, and were glad to have him back; they were friendly toward him. One finds harmony in one's home that cannot exist in an alien place. It is best to die and be buried in the ground that knows a person, the ground that is waiting to receive home its children."[83]

Stephen Powers, a pioneering anthropologist, wrote about how intimately the Mattole of northwestern California and other tribes knew their homes—their places in the world:

> The boundaries of all tribes . . . are marked with the greatest precision, being defined by certain creeks, canyons, boulders, conspicuous trees, springs, etc., each of which has its own individual name. Accordingly, the squaws teach these things to their children in a kind of sing song. . . . Over and over, time and again, they rehearse all these boulders, etc., describing each minutely and by name, with its surroundings. Then, when the children are old enough, they take them around . . . and so faithful has been their instruction, that [the children] generally recognize the objects from the descriptions given them previously by their mothers.[84]

Many stories, myths, and sacred events were tied to odd-shaped rocks, certain mountain peaks, bends in rivers, or specific springs. The anthropologist William Simmons writes, "Through such legends and myths, Native Californians recognized rock formations, mountains, springs, rivers, soil colors, and other natural features as the signs of their predecessors' activities, inscribed in and imparting ancestral meanings to their physical landscapes."[85] "In the old ways, the flora and fauna and landforms are *part of the culture*," says Gary Snyder in *Practices of the Wild*. There is no compartmentalization of nature *from* humans.

Elaborate trade networks were established between coastal, plains, desert, and foothill peoples. The Sierra Miwok, for example, obtained red abalone (*Haliotis rufescens*) and purple olive *(Olivella biplicata)* shells through trade with the Ohlone (also known as the Costanoan). Shells of *Dentalium pretiosum* from as far away as Vancouver Island lie buried in Chumash archaeological sites along the south-central coast. Southwestern pottery shards can be found in Pacific Coast shell mounds.[86]

Many human trails were worn several feet deep from ancient use. In the desert and heavily wooded landscapes, trails often followed the courses of large rivers.[87] Footpaths almost always climbed over the tops of mountains in a straight line so as to cover the shortest distance possible, demonstrating the physical stamina of the indigenous people. So many trails crisscrossed the lower valleys that they caused non-Indian travelers to lose their way. The anthropologist Roland Dixon reported of the Maidu in the northern Sierra: "The whole Maidu area seems originally to have been crossed by a great number of well-beaten trails, connecting the different villages and hunting and fishing grounds." Well-traveled Indian trails such as the route over Pacheco Pass—linking the Salinas and San Joaquin Valleys—became horse trails, wagon roads, and, finally, highways.[88]

CHAPTER 2

Gathering, Hunting, and Fishing

I've always wondered why people call plants "wild." We don't think of them that way. They just come up wherever they are, and like us, they are at home in that place.

CLARA JONES SARGOSA, Chukchansi (1990)

California Indians depended deeply and directly on the breadth of the land for their livelihood. Superb natural historians, their knowledge of the natural world was grounded in ancient tradition and encompassed what today we call ornithology, entomology, botany, zoology, ichthyology, ecology, and geology. Through sharp observation, diligent experimentation, and insights from the spirit world, they unlocked the hidden attributes of California's flora and fauna. Evidence of indigenous people's intimate knowledge of the natural world comes from many quarters: the extraordinary array of human uses they assigned to native plants, animals, fungi, algae, and lichens; the active management techniques they used to maintain the health of these myriad species; their awareness of animal and plant behavior in response to harvesting and management; the preparation, cooking, and storage techniques they developed for utilizing native species; the manufacturing processes they devised to make products from plants and animals; the way they manipulated environmental conditions to influence plant properties.

The natural world served as grocery store, pharmacy, and hardware shop. It was the center of each cultural group's existence. Few plants, minerals, fungi, or animals remained unknown or unused, demonstrating how thoroughly the Indians of California had earned their nativeness to the land.

Utilizing California's Vast Flora

In aboriginal California, women were the ethnobotanists, testing, selecting, and tending much of the plant world, and men were the ethnozoologists, applying their intimate knowledge of animal behavior to skillful hunting,

41

fishing, and fowling. In most regions, vascular plants made up more than half of the diet; fish, fowl, mammals, mollusks, reptiles, amphibians, and insects were an important but smaller component. Because of this dependence on plant foods, women, the main plant harvesters and processors, were instrumental in ensuring the economic survival of their cultures.[1]

Of the approximately 6,300 flowering plants, gymnosperms, ferns, and fern allies native to California, hundreds to thousands occurred in each tribal territory, many of which were incorporated into the tribe's ethnobotany.[2] The vastness of the plant knowledge alone was stressed as early as 1891 by G. H. Harris when he wrote, "A complete history of the foods of the aborigines of North America would fill volumes."[3] That California Indians knew the flora intimately is suggested by their recognition and use of the subtlest distinctions in nature. For example, basket weavers among the Luiseño, Cupeño, Chumash, Tongva (Gabrielino), and Kumeyaay (Diegueño) knew that at the bottom of the stems of streamside rushes *(Juncus textilis)*, completely hidden under the leaf thatch shed by the overhanging sycamores and willows, was a deep brick red band of color, prized for basketry designs.

Many plant species of widespread distribution were used by different tribes, sometimes for similar purposes and sometimes for different ones. The California poppy *(Eschscholzia californica)* alleviated Yuki toothaches, fed Sierra Miwok stomachs, healed Wintu newborn babies' navels, and induced sleep among the Ohlone. The corms of blue dicks *(Dichelostemma capitatum)* were eaten wherever the plant occurred: Atsugewi women dug them on the volcanic slopes of Mount Lassen; Paiute women irrigated patches of them in the sandy soils of the Great Basin; Ohlone women uprooted them in the clay soils of the San Francisco Bay area.[4]

Native plants accompanied native people from birth until death. Strings of clamshell disk beads strung with Indian hemp *(Apocynum cannabinum)* hung from the cradleboards of Pomo babies and were laid on the funeral pyre upon death. Native plants were integrated into every kind of object used in daily living, from adornments to musical instruments and weapons. (See Figure 5, top.) A great variety of native plants were also utilized as foods and medicines. The Sierra Miwok, for example, relied on nearly 160 plant species for food and more than 110 plant species for medicines.[5]

Plants were gathered from below sea level to above the timberline. Surfgrass *(Phyllospadix torreyi)*, which grows below low-tide level in the waters of the Pacific Ocean, was harvested for Chumash skirts, and the sweet fruit of black elderberry *(Sambucus melanocarpa)* was plucked by hungry Washoe families at meadow edges approaching 11,000 feet. Every plant community in California—from beach dune to evergreen forest and pinyon-juniper

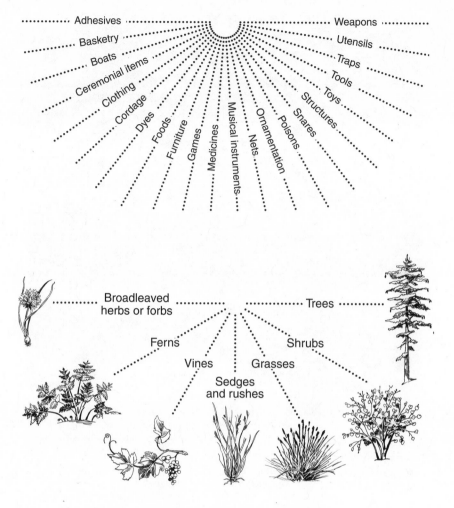

Figure 5. The rich variety of cultural uses for California's native plants that constituted the "material culture" of Indian tribes.

woodland—was visited by gatherers, and within those communities every type of plant life form was gathered. (See Figure 5, bottom.)

Every kind of plant part—underground bulbs, rhizomes, and roots; the oozing resin of trees; the sweet nectar of certain flowers; stems, bark, branches, shoots, leaves, thorns, flowers, seed pods, seeds, seed plumes, and cones—found a use. Baskets alone contained a marvelous diversity of plant parts: the red roots of the Joshua tree *(Yucca brevifolia)* made striking light-

Figure 6. Almost every species of plant had a use—even the hidden roots of six kinds of conifers (coast redwood, sugar pine, gray pine, ponderosa pine, Jeffrey pine, and Sitka spruce). Shown here is Annie Roberts, a Yurok woman, splitting the roots of Sitka spruce *(Picea sitchensis)* to make twined baskets. Courtesy of the Phoebe Apperson Hearst Museum of Anthropology and the Regents of the University of California. Photograph by B. F. White, 1930, #15–9016.

ning bolt designs in Kitanemuk baskets; the tough lemon-colored scapes of *Juncus cooperi* were split for Koso Panamint sewing strands; horsetail *(Equisetum arvense)* rhizomes were split by Ohlone weavers for black lacing; and the seed pod splints of devil's claw *(Proboscidea parviflora* subsp. *parviflora)* decorated Tubatulabal coiled gift baskets. (See Figure 6.)

Californian Indians even found uses for nonvascular plants (mosses, liverworts, and hornworts), fungi, algae, and lichens. For example, a moss *(Dendroalsia abietina)* provided the bedding material for Little Lake Pomo babies. Indian paint fungus *(Echinodontium tinctorium)*, which grows on red fir and mountain hemlock, was used as a brilliant orange-red dye by many tribes. Lace lichen *(Ramalina menziesii)*, a combination of algae and fungi, was used for baby diapers by the Kashaya Pomo. Submerged in water, it was said by the Kawaiisu to bring rain. Brown algae such as bull kelp *(Nereocystis luetkeana)*, red algae such as laver *(Porphyra* sp.), and green algae such as sea lettuce *(Ulva lactuca)* were eaten by coastal tribes.[6]

Food and Materials from California's Rich Fauna

Many of the 66 species of freshwater fishes, 46 amphibians, 96 reptiles, 563 birds, and 190 mammals that inhabited California were incorporated into the ethnozoologies of the tribes.[7] Among the animal parts used were bones, fur, organs, skin, tails, tendons, antlers, feathers, hooves, claws, teeth, meat, marrow, sinew, and horns. Weapons used to capture animals included the slingshot for killing small birds, the throwing stick for hunting rabbits, and bows and arrows, clubs, and lances for hunting terrestrial mammals. A skilled Wappo slinger could hit a fence post at a distance of seventy-five yards.[8] Harpoons, spears, bone gorges, fishnets, and hooks, lines, and sinkers were used for capturing fish.

Indians knew the animals intimately by studying their behaviors, walking the same paths, sharing a drink from the same watering holes, and sleeping in the same habitats. Animal identification came directly from signs— tracks, claw marks on trees, and crushed vegetation where animals had bedded down. Hunters wore clever disguises to stalk animals and whistled to imitate the distress calls of the young. These imitations were often so accurate that they brought forth a host of curious animals and hungry predators. Ingenious traps, snares, and weirs were used to catch birds, fish, and small mammals; for example, the Kumeyaay set up flat stone traps to catch kangaroo rats. For large mammals, hunters dug pits and covered them with boughs. The Pit River derived its name from the deep trapping pits dug by Indians along its banks. Choynumni Indians in the southern San Joaquin Valley would shoot both tule elk and antelope from hunting blinds when the animals came to Tulare Lake to water.[9]

Animal parts were incorporated into an array of cultural objects: adornments, baskets, ceremonial costumes, clothing, cosmetics, currency, dyes, foods, games, household utensils, medicines, musical instruments, poisons, tools, toys, weapons. Porcupine quills decorated basketry hats for the Karuk tribe, and antelope horns were fashioned into spoons by the Tubatulabal. The skins of mudhens were worn as gloves by the Pit River and Atsugewi. By boiling the horns of the bighorn sheep, the Panamint made a glue. In ancient times the Maidu and Pomo made cloaks for sacred dances from the wing and tail feathers of the California condor. Wooden combs of baleen, the fibrous material forming the sievelike mouth of plankton-eating whales, were combed through Tongva girls' hair. Abalone shell pieces were hung from the ears of the Salinans. The femurs of grizzly bears were fashioned by the Chumash into daggers and decorated with *Olivella* shells.[10]

A tremendous diversity of mammals provided food, including deer, elk, pronghorn antelope, bears, gray squirrels, and various kinds of rabbits. Among these, deer were the most important part of the diet for many tribes. The black-tailed jackrabbit *(Lepus californicus)* was so abundant on the San Joaquin plains near Tulare Lake in the 1850s that when the Choynumni conducted a rabbit drive with milkweed nets and throwing sticks, they could kill as many as two hundred in a forenoon. The rabbit meat was eaten, and the fur and skin were made into six-foot-square blankets, capes, and skirts. Many other tribes hunted rabbits as well. Birds killed for food included quail, bandtail pigeons, and various kinds of ducks and geese. Eggs were gathered from quail, geese, ducks, and turtles, then roasted in ashes and eaten. Le Conte noted that gull *(Larus* spp.) eggs were an important food item for the Mono Lake Paiute.[11]

Tribes living near the coast exploited the rich food resources of the Pacific Ocean. In 1775 Fray Benito de La Sierra wrote an account of the diet of the Indians of Little River and Trinidad Head, noting, "Mussels [probably *Mytilus californianus* and *M. edulis*] are very abundant and form their regular food. . . . What are to be found in great number are seals, to which the Indians are very partial, keeping in their houses a supply of the flesh and bladders of the oil. They preserve the flesh as we do ham and cook it by roasting slices over a fire, the grease being allowed to drip on to certain beaten herbs, the whole being eaten with great satisfaction."[12]

Large schools of fish frequented the coastal waters, feeding bird and human alike. The English naturalist Archibald Menzies wrote of the Chumash fishing off the coast of Santa Barbara in 1793: "[T]hey were always seen out by the dawn of day either examining their fish pots in the Bay or fishing in the middle of the Channel where they never failed to catch a plentiful supply of fish of different kinds particularly Boneto [Pacific bonito, *Sarda chiliensis*] & a kind of Herring [yellowtail, *Seriola lalandi*]." In northern California coastal waters, Indian men and women wielding cornucopia-shaped traps in the waves took surf smelt *(Hypomesus pretiosus)* by the thousands. Alfred Kroeber and Samuel Barrett reported, "The smelt were in former times so numerous that a man often got his net so full that he had to have help to carry it ashore—sometimes even so full that it was necessary to pour some of the fish back into the water for fear of tearing the net because of the weight of the fish." Probably the most important of all fish resources in central and

Figure 7. Chinook salmon *(Oncorhynchus tashawytscha)*, the largest and most numerous of California salmon, was an important food for both coastal and inland Indians. Courtesy of Peter E. Palmquist.

northern California, were salmon. At least five million pounds of salmon were consumed by Native Americans yearly.[13] (See Figure 7.)

A great variety of insects—grasshoppers, cicadas, ants, flies, crickets— were used for food and other items. Yellowjacket nests made of masticated cellulose were raided by members of many tribes for their edible larvae and pupae. The honeydew deposited by aphids on the stalks of cattails, tules, and common cane (probably *Phragmites australis*) was collected by the Owens Valley Paiute, Panamint Shoshone, Chemehuevi, Tubatulabal, and Salinan and made into sugary balls or cakes. Perhaps Honey Lake in northeastern California received its name from the honeydew the Indians collected along its shores. The Cahuilla and Panamint made a glue from the amber-colored gum deposited by a small insect on the bark of the creosote bush *(Larrea tridentata)*. When mixed with pulverized rock and heated, this sticky resinous gum fastened stone arrowheads to their shafts and mended pottery. Army worms collected from Oregon ash trees *(Fraxinus latifolia)* were eaten by the Pomo.[14]

Edible butterflies and moths, which were harvested in the larval and pupal stages, included the whitelined sphinx moth *(Hyles lineata),* eaten by the Ca-huilla, and the pandora moth *(Coloradia pandora),* eaten by tribes in the Sierra Nevada. According to Western Mono elders, the pandora moth, the larva of which feeds on ponderosa, sugar, and Jeffrey pine, used to be common on the west side of the Sierra Nevada, but now it is rare. The Mono elder Mel-ba Beecher can specify the exact mountainous areas where it occurred and where she and her family gathered the larvae, called *piagi.* Formerly, trenches were dug around the trees to trap the caterpillars as they descended. The pupae of the California tortoise-shell butterfly *(Nymphalis californica),* called *hooya,* were gathered by the Mono. The flavor was excellent, and one elder said they smelled "just like a pot roast" while cooking. The pupae were dried and stored for future use.[15] Many Indians recalled that the tortoise-shell butterfly was common in chaparral *(Ceanothus cuneatus)* areas at mid-elevations in the Sierra Nevada and would come every year. It is now seen once every decade or so, and in some areas it has not been sighted for forty years.

The Complexity of Gathering Knowledge

Because no one today gathers wild plants or hunts animals for all of his or her personal needs, it is difficult for us—even those of us who are secure in the wilderness—to realize the depth of knowledge required to comfortably stay alive in aboriginal California. Indigenous plant and animal gathering was carried out under an intricate system that took into account life-forms, species, plant properties, environmental site criteria, cultural uses, and strategies of cultivation. When combined, the levels of ethnobiological knowledge became exceedingly complex. Only by following the daily rou-tine of a Hupa fisherman, a Yokuts saltgrass collector, a Yahi hunter, or a Kumeyaay weaver would we begin to fully appreciate the keen memory, knowledge of reproductive biology, attention to environmental cues, and mimicking of animal behavior that was practiced by harvesters to live di-rectly from the land.

For instance, native plants important in Indian economies often grew on several sites within a variety of plant associations, each species and ecotype representing slightly different specific ecological requirements of the asso-ciation and the site. Thus, ripening times varied according to elevation, weather patterns, and microclimatic differences. Timing of plant harvesting to reap the maximum benefits from plant growth with a minimum of ani-mal or pathogen competition was critical. Memory, therefore, had to be long and accurate, and observation keen. If a village gave insufficient attention to animal ecology, weather, plant reproductive biology, or microclimatic variations, its winter food supply could be at risk, possibly meaning starvation.

CLASSIFYING ORGANISMS

Complex taxonomic classification systems were a foundation of California Indians' knowledge of the natural world. The ability to repeatedly distinguish a species by shape, color, size, smell, and other factors—and assign various uses to it—was crucial to a village's survival.[16]

Humans seem to have an innate ability to categorize organisms into groups based on varying degrees of similarity in their overall morphological structure, says the ethnobiologist Brent Berlin. He argues that human beings have an "inescapable and largely unconscious appreciation of the inherent structure of biological reality."[17] Indigenous groups all over the world classify plants and animals in a hierarchy—layers of groupings based on progressively finer features of distinction. There are broad groupings, such as "tree" or "bird"; intermediate categories that comprise certain plants or animals showing similar distinctive features, such as "pine tree" or "hummingbird"; and fine-textured groupings, equivalent to our species category, such as "Monterey pine" or "Anna's hummingbird."[18]

Unfortunately, few detailed studies of folk classification of plants and animals in California have been done.[19] Eugene Hunn's detailed study of the taxonomic knowledge of the Mid-Columbia Indians in Washington has led him to conclude that their "perceptual and analytical capacities" were "on a par with those of a modern-day professional botanist." From what is known of meticulously studied indigenous groups in other places, it is probable that California Indian classifications were every bit as sophisticated as Western schemes.[20]

With the binomial taxonomic system developed by the Swedish naturalist Carolus Linnaeus in the 1750s as their standard, Euro-Americans typically misjudged California Indians' classification systems. When Stephen Powers, a journalist, first entered Nisenan territory in 1872, he noted that the "savages have no systematic classification of botanical knowledge; there are no genera, no species."[21] Not being ethnobiologists by training, early observers such as Powers could not comprehend the possibility that Indian naming systems, so different from their own, could be every bit as accurate, finely grained by category, and precise as the Western system.

Merriam, a trained biologist and naturalist, drew the opposite conclusion from Powers. While wandering in the Sierra foothills to map the range of two manzanita species in 1910, he stumbled upon a little hut near a stream and asked an Indian woman where the lowland and upland manzanita grew. She bluntly told him that "all three [manzanitas] grow right here." Her daughter fetched specimens of the three species, one of which was unknown to Merriam and to his fellow botanists. As an acknowledgment to her and her tribe—the Miwok—Merriam named the new manzanita species *Arctostaphylos mewukka*.[22]

The genus *Arctostaphylos* has a reputation among taxonomists as being difficult to identify at the species level. Not only do manzanita species hybridize readily in areas where distinct species overlap, but there are more than fifty recognized species in the genus in California. Yet this did not prevent the Miwok from encoding the differences in separate names for each type that occurred in their territory. The Karuk in extreme northwestern California identified at least four types of manzanitas in their territory: *pahaav*, or green manzanita *(Arctostaphylos patula); fáthip*, or Parry manzanita *(A. manzanita); apúnfaath*, or pine-mat manzanita *(A. nevadensis)*; and *ohusukamfas (A. canescens)*.[23]

In assigning names to plants, native people paid attention not only to morphological characters but also to the places the plants grew and to what uses they were assigned. For example, the Pomo called the sedge *(Carex barbarae) Ka-hum*. The plant's rhizomes, also referred to as "white root" by weavers, were the chief source of weft (or sewing strands) for light-colored backgrounds in basket weaving. The length, color, and other characteristics of the rhizomes were dependent on moisture, soil, and elevational differences, and based on these differences, at least four types of *Ka-hums* were distinguished in the language: *Ka-pak ka-hum* was *Carex barbarae* from boggy ground; *Ko-di ka-hum* was found in rich, wet lowlands (its short, milky white rhizomes were valued for fine work); *Mi-tcat ka-hum* grew beside streams in sandy banks (and was best suited to large, coarse basket work); and *Bi-da ka-hum* grew in gravel next to creeks (its long, coarse, and strong yellow rhizomes were used for weft in utensils and storage baskets).[24]

KNOWING PLANT PROPERTIES

California Indians understood very well the adhesive, unctuous, nutrient, and chemical properties of plants. For instance, the Luiseño harvested the charred seeds of *enwish*, the wild cucumber (*Marah* sp.), for an oily ingredient that gave permanence to the paints used in pictographs. Some contemporary products have their origins in native discoveries, attesting to the efficacy of their assigned uses. Indigenous knowledge has made significant contributions to genetic conservation and modern pharmacology. Cascara sagrada *(Rhamnus purshiana)* bark, an important cathartic to the Modoc, Hupa, and virtually all indigenous peoples of the Pacific Northwest, is found in health food stores and pharmacies in commercial laxative preparations. Many tribes made fever remedies from the bark of willows; the Ohlone, for example, used *Salix laevigata* in this way. The bark contains the chemical constituent salicin—a white, crystalline glucoside—the original ingredient in aspirin. Chia *(Salvia columbariae)*, an important food to the tribes of central and southern California and parts of the Southwest, is recognized as a highly nutritious food and is currently sold in health food stores. The seed floss from milkweed *(Asclepias* spp.), used by California Indians for lining their children's cradles, is utilized today in jackets, comforters, and pillows.[25]

PAYING ATTENTION TO CYCLICAL CHANGES

Of necessity, Indians were closely attuned to cyclical changes in plants and animals—the succession of changes through which a plant passes in its development (vegetative growth, flowering, fruiting)—and the reproductive cycles and migration patterns of animals. Timing the harvesting of plants and animals to reap the maximum benefits added another dimension to the complexity of native people's knowledge. This richness of seasonal patterning was missed by most anthropologists in their ethnographic descriptions of native peoples.

Some greens, roots, and bulbs, such as the large, succulent roots of broom-rape *(Orobanche* sp.), dug by the Cahuillas, had to be harvested before the plants flowered.[26] Material from shrubs and trees for basketry was often gathered before vegetative processes began in spring or after they subsided in fall. For example, the wine red bark of redbud *(Cercis orbiculata)*, used by many tribes for basketry, had to be gathered in fall or winter, because in spring the bark cells have too much moisture, rendering the material unusable for creating red designs in baskets. If a perennial plant was used for more than one purpose, then harvesting of the same plants occurred at different times during the year, without destroying the plants the first time around. For instance, milkweed *(Asclepias* spp.) plants were gathered in the spring for their greens, in the late summer for their seed pods, and in the fall or early winter for their stems to make cordage. The seeds of many types of plants had to be picked within a week of ripening or they would blow away and be lost to the gatherer. The Karuk gathered ponderosa pine roots after the tree bloomed, because the roots were tougher at this time.[27]

The physiological and behavioral changes that occurred in animals during seasonal and even daily cycles were also crucial to Indian cultures. An-

TABLE 1

Gathering, Fishing, and Hunting Activities by Season

Spring	Pick basketry materials such as willow.
	Collect the unfolding new leaves of alum root for greens.
	Pick mushrooms after rains.
	Hunt geese as they are returning north.
	Spear salmon and hook eels as they are making their spring run.
	Dig edible bulbs before flowering.
	Build stone surrounds in the desert and hunt antelope and bighorn sheep.
	Harvest the larvae of the pandora moth around Jeffrey pines.
Summer	Net geese as they are molting.
	Prune tobacco to promote larger leaves.
	Gather juniper berries and process them into a sweet cake.
	Collect immature gray pine cones for roasting.
	Collect savory berries as they are ripening.
	Beat edible seeds and grains from the inflorescences of flowers and grasses into wide-mouthed baskets.
Fall	Hunt rabbits for their fur because their pelts are thick.
	Obtain deer in prime condition as there is a tendency for deer to group, especially when inclement weather forces their migration.
	Trap quail as they migrate from the mountains to the foothills.
	Gather ripe acorns from oaks and cones from many kinds of pines for their seeds.
Winter	Pick basketry materials from shrubs and trees because the bark adheres and the sap is down.
	Gather milkweed and dogbane stalks for their bast fibers for making cordage.
	Obtain shellfish easily at low tides.
	Gather wood for arrows before the shoots take up too much moisture and sprout new leaves.
	Hunt bear, wildcats, and otters.

imals were harvested at certain seasons, or when migration cycles caused natural congregations. Rabbits were taken when fattest, fish when spawning, birds when molting, crickets in the cool mornings when they bunched in the grass. (See Table 1.)

Seasonal changes in the environment such as a shift in wind direction, budding out of plants, different animal behavior, and swarming of particular insects were cues that certain culturally important plants would ripen soon or that particular animals would be migrating. Linking environmental cues with reproductive cycles required detailed knowledge. For example, Kashaya Pomo women watched for the first warm inland winds of summer as a sign that there would be only a few days to gather the seeds of wild oats (*Avena* spp.) before they fell to the ground. In summer, shellfish can become toxic to humans when they take in tiny, one-celled organisms called dinoflagellates *(Gonyaulax catanella)*. Coastal Pomo people stopped gathering clams and other shellfish as soon as elderberry shrubs flowered. When there were ripe elderberries, they knew it was time again to harvest shellfish. When the fruit of the California coffeeberry *(Rhamnus californica)* was ripe, the Tubatulabal people knew that the pinyon pines in the mountains were ready to harvest.[28]

RELATING ENVIRONMENTAL CONDITIONS TO DESIRABLE PLANT PROPERTIES

Indigenous groups were aware that plants occurring at different elevations and growing in different soil types, in certain plant associations, or subjected to different light intensities produced plant material of varying quality or desirability. For example, the bark of Pacific madrone *(Arbutus menziesii)*, used by the Tolowa in northern California for colds, is more potent if gathered in the high country. Hupa weavers preferred the strong black color of five-finger fern *(Adiantum aleuticum)* gathered at higher altitudes to the brownish stems gathered from the lowlands. The Karuk considered that the best yew *(Taxus brevifolia)* wood for making pipes and games was gathered in the wet, shaded areas along creek beds because it grows straight in the shade. The Yurok, who valued Oregon-grape (*Berberis* sp.) root as a medicinal, knew that its potency is heightened when it grows in red soil.[29]

According to the Nisenan, the best acorn trees were higher up in the mountains on the shady sides of canyons. Among the Pomo, the roots of *Angelica* sp. were most prized for medicine and as a talisman when collected from plants growing in cold places. The Hupa claimed that leaves of *mesteelen* *(Iris macrosiphon)* were better for making the twine and rope used in nets and snares when taken from plants growing under oaks rather than under pines. The Paiute preferred the roots of yucca plants growing above four thousand feet because their roots were a better color than those from plants at lower elevations.[30]

COMPETING SUCCESSFULLY WITH ANIMALS AND PATHOGENS

The same plants that are desirable for human food and material culture are often also palatable to birds, small and large mammals, and insects, and susceptible to attack by pathogens. In devising ways to cope with this competition from other species, native people had to understand these species' habits and life cycles. Often, harvest was timed to occur before insects could feed on the same resource. A Kashaya Pomo woman, for example, noted that the harvest of mountain dogwood *(Cornus nuttallii)* for making baby baskets was timed with insect life cycles. Sticks were harvested in early fall, right after the shrubs shed their leaves, because "[b]y late Fall, the twigs are infested with worms that weaken the wood." Many harvesting techniques were designed specifically to foil insects or herbivores. Black oak acorns, for example, were frequently harvested by knocking the trees with a long pole before the acorns dropped naturally to the ground. The anthropologist Helen McCarthy suggests that knocking acorns from the trees gave humans a competitive edge over other animals vying for the same resource, and it extended the available harvesting period, maximizing the quantity of acorns gathered during the acorn season.[31]

PROTECTING STORED FOOD

In many parts of California, the indigenous people built granaries to store small and large seeds. The structures were designed to keep out rain, snow, and hail. Waterproofing was often accomplished by shingling the structures with conifer boughs, brush, or grass thatch. If moisture penetrated the contents, certain kinds of bacteria and fungi could spread rapidly through the granary, destroying the entire contents. The cavity inside the granary also had to be insect-proofed by meticulously lining it with an insect repellent plant such as mugwort *(Artemisia* spp.). To thwart the many small mammals and birds—field mice, ground squirrels, gray squirrels, jays, woodpeckers— that would raid the stash if given the chance, the granaries were very tightly constructed and placed in special locations. The Cahuilla, for example, perched their granaries either on platforms made of poles or, in the mountains, on the flat tops of high boulders, out of the reach of field mice and kangaroo rats. The Sierra Miwok built their granaries on stumps or large rocks. All contents had to be carefully dried and thoroughly inspected for insect exit holes or molds *before* placing them in the granary. One rotting seed could contaminate the whole batch.[32]

The Cultural Fabric Surrounding Factual Knowledge

In native California societies, gathering and hunting knowledge was not just a set of facts to be memorized and mechanically followed in the daily harvesting rounds but rather was tied to a comprehensive cultural framework of values, beliefs, and behaviors that clearly defined the place of humans in the natural world. To an extent that it is difficult for a non-Indian to appreciate, nature and culture were closely entwined and interlocking. Knowledge of plants and animals informed culture, and culture shaped the way in which this knowledge was employed.

CULTURAL RULES

Nature was considered fully alive and sensate: every rock, hill, valley, wind, plant, and animal was inhabited by spirit forces. Many cultural rules existed to keep humans from offending these spirits, who might otherwise react emotionally and cause lightning, thunder, whirlwinds, or earthquakes. Taboos were rigidly observed concerning diet, menstrual cycles, and sexual activities before gathering or hunting expeditions.[33] While gathering or hunting, people all over California followed two overarching rules: *Leave some of what is gathered for the other animals* and *Do not waste what you have harvested.* Some acorns were left on oak trees, some berries on bushes, and some tubers in the ground for "the birds and squirrels and other animals," said Marie Potts, a Maidu elder. After picking two bucketfuls of blackberries in Trinity County sometime during the 1930s, Sam Young (Nor-Rel-Muk Wintu) told Edith Van Allen Murphey: "There will be someone here after we are gone. We have enough for our needs now. We Indians like to leave something for the one who comes after." This was a rule adhered to by most if not all California Indian cultures.[34]

The rituals that surrounded the act of harvesting, hunting, or fishing were as important as the act itself. How one approached a plant or animal—with what frame of mind and heart—was very significant. A personal connection was often made by saying a silent prayer, leaving an offering, and thanking the plant or animal for the gift of its life. Straying from this common practice might result in great diminishment of animal and plant numbers. A prominent Pomo basket weaver explained this personal connection to place and to the plants:

> When you come to dig these basket roots, you don't rush in there and run all over, you don't do that. My mother always approached

this grass very slowly. She'd come and stand and say a prayer. She also had a cane, and she'd touch this grass with it very slowly. She didn't go in there and just start digging. She'd come to a certain bed and try it; then she'd go on to another one and try there. Before she ever sat down, she'd do this three or four times. Then she'd sit down. She always asked the Spirit to give her plenty of roots. Then she'd say, "Thank you, Father," before she dug. And after she'd finished and had got what she wanted, she said a prayer which is like saying "That's good, you gave me enough. Amen Father."[35]

CEREMONIES, SONGS, DANCES, AND MYTHS

Legends, ceremonies, songs, dances, and artistic expressions suffused Indian life, forming the foundation of what Gary Snyder calls a "culture of place." They not only fed the human spirit but also instructed people in right and wrong behavior and how they were obligated to nature. Many of these social and religious observances were connected directly with the conservation of natural resources.[36]

Important cyclical events in the natural world—each the "return" of an important plant or animal food—were marked by songs, ceremonies, stories, and sacred events. The Concow Maidu celebrated the return of the geese in the goose dance; the Kashaya Pomo held festivals when the wild strawberries ripened; the Maidu honored the coming of the clover. Other rituals anticipated the time when a food crop would be ready to harvest. The Sierra Miwok, for example, held a feast to express the hope that the acorn crop might be abundant. Gashowu Yokuts shamans conducted special crop-prophesying dances to tell what kind of seed, clover, and acorn crops would come.[37] In these community rituals, native peoples humbled themselves to the more powerful forces that controlled the seasonal rhythms of plants and animals.

The plants and animals used by California Indian tribes figured prominently in their folklore, and this in turn influenced hunting and gathering behavior. In a Tubatulabal myth, for example, jimson weed was once a man, who, when he died, told the people to dig his roots if they were in need of help. With this story in mind, Tubatulabal people made a short speech to the plant before they dug its roots. In a Pomo myth, Marumda the Old Man instructed the first people about what was good to eat. "Here you will eat this kind of food," he said, plucking some clover and eating it. The people imitated him and ate. Then he dug up some potatoes. "These also you will eat," he said. He looked for a spot where acorns had drifted in a pocket in the creek: "These you will gather, and with them you will make mush!"[38]

Most tribes had legends that vividly told of the consequences that would befall humans if they took nature for granted or violated natural laws. According to a Southern Sierra Miwok legend, Half Dome in Yosemite Valley was at one time whole but was shattered in two long ago when the Creator, angry at the Miwok chief for failing to pray for timely rains to replenish the earth, made a powerful earthquake to punish him. In a Pomo legend, after the people had violated the marriage, hunting, and fishing laws, Marumda the Old Man sent a wildfire over the face of the earth. The Yurok believed that the greed for riches gained through trickery and dissention would ultimately cause the Creator to send a great conflagration that would consume all the world in flames. Legends about destructive fires reflect the almost universal belief among California Indian tribes that catastrophic fires were not a regular, natural occurrence but rather a rare punishment for a serious violation of religious and social rules.[39]

A KINCENTRIC VIEW OF NATURE

Although native ways of using and tending the earth were diverse, the people were nonetheless unified by a fundamental land use ethic: one must interact respectfully with nature and coexist with all life-forms. This ethic transcended cultural and political boundaries and enabled sustained relationships between human societies and California's environments over millennia. The spiritual dimension of this ethic is a cosmology that casts humans as part of the natural system, closely related to all life-forms. In this view, all nonhuman creatures are "kin" or "relatives," nature is the embodiment of the human community, and all of nature's denizens and elements—the plants, the animals, the rocks, and the water—are people. As "people," plants and animals possessed intelligence, which meant that they could serve in the role of teachers and help humans in countless ways—relaying messages, forecasting the weather, teaching what is good to eat and what will cure an ailment.

This view of other life as related, equal, and highly intelligent is what Enrique Salmón (Rarámuri) calls a "kincentric" view of the world.[40] In this vision of the world, nature is not to be treated as a separate entity "out there" that you do not touch or interact with, or labeled as a "scene" that is only to be viewed through a lens or shaped by the stroke of a paintbrush. *Homo sapiens* are full participants in nature, and they share mutual obligations and intricate interactions with many other forms of life.

Reflecting this belief, plants and animals throughout California were greeted with familiarity, as if one were visiting with old friends. The Mihilakawna Pomo elder Lucy Smith recalled what her mother used to say about taking care of her relatives, which she only later recognized as including her nonhuman relatives:

> [She said] we had many relatives and . . . we all had to live together; so we'd better learn how to get along with each other. She said it wasn't too hard to do. It was just like taking care of your younger brother or sister. You got to know them, find out what they liked and what made them cry, so you'd know what to do. If you took good care of them you didn't have to work as hard. When that baby gets to be a man or woman they're going to help you out. You know, I thought she was talking about us Indians and how we are supposed to get along. I found out later by my older sister that mother wasn't just talking about Indians, but the plants, animals, birds—everything on this earth. They are our relatives and we better know how to act around them or they'll get after us.[41]

Kinship with animals and plants was often fostered at birth. The anthropologist Lee Davis says of the Hupa in northwestern California: "The father of a newborn baby took the stump of its umbilical cord[,] . . . found a sapling of a Douglas fir tree and split its most upward pointing branch, placed the umbilicus inside the split and tied the branch back around the baby's navel stump. As the baby tree grew up, so the child's physical and spiritual growth was measured."[42] California Indians frequently named their children after a plant or animal—again demonstrating that humans were not superior to nature but a part of the natural system. Yuki names include *Aⁿsíumnának* ("woodpecker head") and *Musak' ćoⁿli* ("little cedar girl"). The Wappo name *Pipimeta* meant "quail woman," and *Awetulpi* meant "potato valley woman." The Sierra Miwok gave each other special names associated with trees, such as *Sumtciwe*, a female name referring to the fuzz on a young sugar pine cone. The Rumsen Ohlone held naming ceremonies under redwood trees.[43]

Jaime de Angulo, a physician who recorded the grammar and literature of the Pit River people in the Alturas region of California beginning in the 1920s, asked Bill, a Pit River man, how to say "animal" in his language. Bill replied, "Well . . . I guess I would say something like *teequaade-wade toolol aakaadzi* which means 'world-over, all living.'. . . I guess that means animals, Doc." Jaime then said, "I don't see how, Bill. That means people, also. People are living, aren't they?" Bill said, "Sure they are! that's what I am telling you. Everything is living, even the rocks, even that bench you are sitting on. . . . Everything is alive. That's what we Indians believe. White people think everything is dead."[44]

Trusting that respect and understanding would come through relationship, native people believed that animals—birds, toads, lizards, bears—could become familiar with, even grow accustomed to, the ways of *Homo sapiens*. Thus, we see many examples of indigenous groups living and interacting with potentially dangerous animals. When a man or a woman of the Pass Cahuilla met a California grizzly bear in the mountains, he or she called the latter *piwil* ("great-grandfather") and talked soothingly to him, saying, "I am only looking for my food, you are human and understand me, take my word and go away."[45]

As part of treating other life with respect, California Indian cultures shared an ethic that prohibited needless killing. One old Paiute man said, "I have never shot anything in my life but what is good to eat." Ishi attributed the cause of Saxton Pope's bad luck in deer hunting to his needlessly killing a rattlesnake on the trail. Pope said of Ishi, "He respected these reptiles, and always preferred to walk around a snake, wishing him well and leaving him unharmed."[46] In *Ecologies of the Heart*, the ethnobiologist Eugene Anderson writes, "[A] properly socialized individual had a powerful sense that the wild world was feeding him, and he ought to be as grateful and as anxious to act decently as he would to any human who fed him out of sheer kindness. Naturally wanton killing was virtually tantamount to murder, and ungrateful murder at that."[47]

Native people reconciled the killing of their plant and animal kin by following certain rules when hunting and gathering (e.g., do not waste or overharvest) and by sharing the bounty with needy relatives and friends. They believed that the necessary sacrifice of some creatures is made for the good of all creatures. "The killing and eating of other beings," writes the human ecologist Paul Shepard, "is understood by most tribal peoples as part of a larger gift of life rather than a victory over nature."[48]

Recognizing the impact of killing other creatures for food, California Indians in various tribes were careful to restore harmony through acts of reciprocity: offerings, prayers of thanks, special ceremonies, and management of habitats to benefit the plants and animals not taken. Thus, California Indians were predators of deer, antelope, and tule elk but also their benefactors. Clara Charlie, a Chukchansi/Choynumni (pers. comm. 1989), explained how animals were helped by regular burning: "My husband's family . . . burned to keep things clear. They also burned for the animals—the deer, bear, rabbits, and squirrels. The new growth the following spring gave them better and higher-quality foods such as buck brush."

Human Rhythms Tied to Natural Cycles

California Indians knew, unequivocally, that their lifelines as humans lay in the cycles of the land. Events were timed with the coming and going of animals and the ripening of culturally important plants. This synchrony was a constant reminder that human destiny and nonhuman life were intimately intertwined. Even the conception of a baby might be planned around specific seasons marked by animal births and the sprouting of plants. The most favorable time for a Miwok woman to have a child was between the months of March and June because, as a Southern Sierra Miwok man explained to a Belgian argonaut, "[i]t is when the Spirit gives existence to everything, that the *'Ohha* (woman) should also give existence." He further commented that during these months, "he [the sun] is superior in his turn, makes everything grow, the birds in the air, terrestrial animals, and plants."[49]

Time was not tied to one's wrist and mechanically broken down into tiny increments of seconds, minutes, and hours. The time of day, for instance, was estimated from the length of one's own shadow.[50] The round of the seasons was a circular continuum—solidly connected to complex biological phenomena such as the annual flights of waterfowl, the migration of whales, and the ripening of the acorn crop.

The continual renewal of life from the source—the salmon from their parents' birthplace, the rivers from their headwaters, the young sprouting redbuds from the ashes of their forebears—was intimately understood by indigenous people. Lunar calendars often reflected the importance of these cyclical events by devoting a lunation to the yearly arrival of a migrating animal or the coming into being of a food crop. The Wappo called one lunar cycle in late spring (our May) *Wa' 'ate-hin*, or "pinole moon." The Northern Maidu designated the same lunation as *Kon-moko*, the time to harvest the seeds, fish, and geese in the valley. A later lunation (September) was named *Se-meni*, or "seed," in the mountains. The Central Pomo calendar designated a time, *Umchachich-da*, when the seeds ripened. In the Cupeño calendar, one of the lunations is *Tausumbakmaiyil*, or the ripening of grass seeds. April, or *Kapchelam*, was the month for gathering yampah (*Perideridia* spp.) on the northern California Klamath calendar.[51]

In addition to dividing the year into a sequence of lunar cycles, the indigenous people of California recognized the cycle of the sun and its more direct connection to seasonal changes. Native people drew spiral pictographs—sets of connected concentric rings radiating out from a center—on cave walls and rock shelters in locations where they are illuminated by the rising sun on the winter solstice.[52] Solstice ceremonies, such as those practiced among the Chumash, acknowledged the seasonal changes of the sun, which in turn affected the availability of plants and animals for food and other needs. In *Stories and Legends of the Palm Springs Indians,* the Cahuilla chief Francisco Patencio wrote:

> In this tribe were some older people who put up signs to gauge how the sun shone. They found they had to keep moving the stick to the right for a long time, and then to the left, and so by this means they discovered what times the birds had their nests, and what times the animals had their young, also what time the plants grew, and the time the seeds were ripe. This they did year after year as they studied the signs of the sun.[53]

Native peoples' movements were also keyed to seasonal cycles. The territories of many tribes were for the most part laid out in an east–west orientation, covering a sequence of elevational zones. At each elevation was a seasonal camp, used when the annual cycles of ripening and population movements yielded abundant resources there. Early anthropologists mistook these movements for a nomadic life, but in actuality indigenous people migrated seasonally within a home territory, returning to the same summer camps and winter villages as had the generations before them. The areas around their villages were familiar, and each useful tree or patch was worked and reworked, thus creating a patterned, semisedentary life.

The cyclical departures and returns of wildlife were so predictable that California Indians, with their weirs, nets, and traps, could have extinguished large numbers of animals. Yet for the most part they did not, having learned that yearly abundance could be ensured by working *with* nature instead of taking advantage of it.[54] Similarly, they easily could have extirpated populations of annual wildflowers through overharvesting, but instead they were careful to leave seeds so that the plants were maintained in magnificent abundance.

According to various California Indian narratives, humans were given specific instructions through the spirit world to protect the earth's self-replenishing character. To carry out this directive, the people conducted world renewal ceremonies to drive away sickness, prevent natural disasters such as earthquakes, landslides, or floods, and tap the abundance of salmon, acorns, and other foods that lies dormant, waiting for humans to bring it forth. Some of these traditions are still enacted today.[55]

CHAPTER 3

The Collision of Worlds

White people want our land, want destroy us. Break and burn all
our basket, break our pounding rock. Destroy our ropes. No snares,
no deerskin, flint knife, nothing.

LUCY YOUNG, Lassik and Wailaki elder, in the year 1936 at age 90

Tales of a mythical island abounding in gold loomed large in the imagina-
tions of sixteenth-century Spaniards who read the novel *Las sergas de Es-
plandián (The Exploits of Esplandian)* by Garci Ordóñes de Montalvo. Montalvo's fabled realm, ruled by Queen Calafia, was populated by black women
of great strength whose arms, and the harnesses of the wild beasts on which
they rode, were made of gold. Griffins killed and ate any man who set foot
on the island. The name of this legendary place was California.[1]

Hernán Cortés and his followers, some of whom likely read this novel,
may have been responsible for fastening the name *California* on the Pacific
coast of North America in the 1530s. In any case, a name derived from a
mythical place became appropriate for a region onto which people of European extraction would continue to project their dreams, fears, illusions, and
ambitions for the next several centuries. The explorers, missionaries, settlers, and gold seekers entering California from the sixteenth through the
nineteenth centuries rarely saw it as it was—a land carefully tended by large
populations of people with remarkable and diverse cultures—but instead
saw the landscape and its inhabitants through lenses distorted by Western
ignorance, prejudice, and greed. As a result, the history of contact between
the indigenous people of California and outsiders of European heritage is
one of subjugation and devastation—of both the land and the people.

The colonizers variously saw California as a foreboding wilderness, a place
to do God's work, a giant untapped storehouse of wealth, and a place of raw,
unspoiled beauty. Although these conceptions varied, they were consistent
in two important ways: they ignored the essential humanity of the native
inhabitants, and they failed to account for the changes in the landscape these

people had wrought over millennia. When European explorers made their first contacts with indigenous peoples in the Americas, they saw them as wild, inferior beings, wandering aimlessly over the country, not subject to the order and routine of civilization. This was the only way Europeans could understand cultures vastly different from their own because their worldview was so resolutely dualistic. On one side of the divide was European civilization; on the other, everything it could be opposed to—nature, dark forests, disorder, and wild beasts and the humans who seemed to live like them.

In the European mind, the noncivilized side of existence was encapsulated by the concept of wilderness. Etymologically, the English word *wilderness* comes from roots that together mean "the place of wild beasts." According to historian Roderick Nash, the concept of wilderness gained common usage in Europe late in the fourteenth century when John Wycliffe inspired the first English translation of the Latin Bible. The notion of wilderness as uncivilized wasteland had already penetrated the consciousnesses of the earliest European explorers in North America in the fifteenth century,[2] ensuring that the wilderness concept would define the colonizers' mindset with regard to the New World.

California retained the wilderness image throughout the history of its colonization. The accounts of Gold Rush–era pioneers and settlers, in myriad journals and diaries, are rife with terms such as "frontier"[3] and "New World" that imply an uninhabited land, despite the fact that, as Rupert Costo and Jeannette Henry Costo remind us, California was "already populated, civilized, subject to authority and law, with a culture and religion of its own."[4]

Whether they were intent on Christianizing the Indians, extracting wealth from the land, extending territory, or making a livelihood, the Franciscan missionaries, Spanish soldiers, Mexican Californios, American miners, and American settlers who came to California wrought devastation both directly—through the subjugation and genocide of indigenous people—and indirectly—by developing economic enterprises that destroyed and vastly altered ecological systems and made it impossible or increasingly difficult for Indians to continue their traditional livelihoods. Growing alien food plants in extensive cropping systems, cutting down trees for fuel and construction materials, hunting native animals for the sale of their meat and fur, and grazing large numbers of non-native ungulates all impacted the supplies of salmon, deer, acorns, seeds and grains, greens, and edible bulbs and tubers on which the native people depended for food. Native gathering sites in the rich valley grasslands, coastal prairies, mountain meadows, oak savannas, and forests, kept fertile and open by burning and visited regularly for many generations, became rangelands, timberlands, and farms.

Unable to continue their gathering and vegetation management practices, Indians starved or were forced to depend on the whites for food.

As a whole, California Indians did not submissively accept their fate; rather, they practiced a wide array of resistance tactics, both active and passive. They fled to unsettled areas, set fire to buildings, stole cattle, engaged in open warfare, and practiced traditions in secret.[5] Ultimately, however, they were often forced to adopt the ways of the dominant culture to have any chance of survival. And a great many California Indians simply did not survive the colonization of their land: from 1769 to 1890 the population plummeted from approximately 310,000 to 17,000. This decline, which the historian Russell Thornton calls a virtual "holocaust," was caused by disease, starvation, warfare, and outright murder.[6] The subjugation of native peoples included the deliberate attempt to eradicate native cultural identities as expressed in material culture and in historical, religious, linguistic, and practical knowledge.[7] As a result, many cultural traditions and languages are extinct today.

California Indians provided much of the labor needed to build the new economies of the colonizers during each historical period, and many prominent non-Indian men in California's history built their fortunes on the backs of cheap Indian laborers.[8] But the California Indians gave the newcomers far more than their labor. The success of the mission economies and of the many fur trading, market hunting, gold mining, logging, ranching, and agricultural enterprises of early California rested on a land that was productive and healthy ecologically because of the careful stewardship of many generations of California Indians.

Early European Exploration (1542–1603)

Beginning in 1542, a string of European explorers reached California by sea from the western coast of New Spain (Mexico) or sailed the Atlantic around the tip of South America through the Strait of Magellan to the Pacific coast, each planting a flag or other marker and claiming the land for his host country. Hoping to discover "kingdoms and riches,"[9] Spanish and English explorers sailed the Alta (Upper) California coast searching for safe harbors and whatever the land could yield in the way of pearls, precious metals, spices, and medicinal herbs.

THE SEARCH FOR TRADE ROUTES

Sailing three ships under the flag of Spain, Juan Rodríguez Cabrillo left the port of Navidad (400 miles north of Acapulco) on June 27, 1542, in search of the legendary Strait of Anián. On September 28, he and his men became the first non-Indians to set foot on the shores of Alta California.[10] They described the Kumeyaay Indians they encountered as they came ashore at what is now known as San Diego as "well built and large" people who went about "covered with the skins of animals."[11] Farther up the coast, along what was probably Luiseño territory, Cabrillo found green valleys, broad savannas, and a great pall of smoke, suggesting Indian settlement. The expedition sailed to an island—Santa Catalina—and briefly encountered Tongva (Gabrielino) Indians. Sailing across toward the mainland, the chaparral fires, set by other Tongva Indians, indicated to Cabrillo and his men that the land around what is now San Pedro Bay was inhabited.[12]

As he sailed along the traditional Chumash territory, extending roughly to the Santa Clara River, Cabrillo noted that the land was thickly populated. Indians who came aboard the ships pointed out the many Chumash towns— Xuco, Bis, Sopono, Alloc—but Cabrillo renamed them in Spanish. On up the coast at present-day Carpinteria, Santa Barbara, Goleta Point, and Gaviota, Cabrillo and his band found the country "filled with people" and the villages well supplied with abundant acorns, grass grains, cattail seeds, fish, and maguey (probably *Yucca whipplei*).[13] Cabrillo made it as far north as the Russian River, 35–40 miles north of San Francisco Bay.

Approaching Point Conception on the voyage back south, Cabrillo's company was forced by winds to take shelter at a protected harbor—what is now Cuyler Harbor on San Miguel Island. The Island Chumash began a series of battles with the Spanish, and during one of these fights, Cabrillo fell on the rocky beach and shattered his shinbone; gangrene set in, and he died on January 3, 1543.[14]

Spain subsequently opened a regular trade route that enabled its galleons to move goods from its Philippines colonies back to Spain via Mexico. This route passed along the California coast, where many galleons stopped for repairs or to stock up on firewood, water, and food.[15] So long-lived and lucrative was the Spanish trade in Chinese silk and porcelain from the Philippines that by 1585 one or two galleons were sailing back and forth between Mexico and the Philippines annually, for a total of as many as two hundred trips between 1565 and 1822.[16]

SIR FRANCIS DRAKE

Unhappy with the Spanish monopoly on trade in the Pacific, Queen Elizabeth I sent Francis Drake (later knighted for his accomplishments) to plunder Spanish galleons and set up trade links with East Asia. Drake left Plymouth, England, in November 1577 and sailed south through the Strait of Magellan and up the South American coast, attacking Spanish ships along the Chilean and Peruvian shores and plundering their goods.[17] Drake continued northward and sought refuge on June 17, 1579, in an Alta California bay—possibly Drake's Bay—to administer major repairs to his leaky ship, the *Golden Hind*. Living on "mussels and sea-lions" and sleeping in "built huts" onshore, Drake and his men had many encounters with the natives during their five-week stay.

Six days after anchoring, there appeared "a great assembly of men, women, and children." If Drake was indeed at Drake's Bay, these may have been the peaceful Coast Miwok.[18] Drake claimed the land for the English crown and dubbed it Nova Albion (New England), despite the fact that it was the Indians' homeland. Drake and his men found "a goodly country, and fruitfull soyle, stored with many blessings fit for the use of man: infinite was the company of very large and fat Deere which there we sawe by thousands, as we supposed, in a heard; besides a multitude of strange kinde of Conies [pocket gopher]."[19]

Francis Fletcher, chaplain and diarist of the Drake expedition, recorded the bestowing of gifts and honor on Drake by the natives and wrote of "ceremonies of this resigning, and receiuing of the kingdome being thus performed." In his analysis of the Fletcher diary, the anthropologist Robert Heizer concludes, "Fletcher interpreted this ceremony as the giving up of the kingdom to Drake, a thought hardly ascribable to the Indians."[20]

FURTHER SPANISH EXPLORATION OF ALTA CALIFORNIA

Drake's voyage goaded the Spanish into focusing more attention on the still largely unknown area north of Mexico. At least three expeditions explored the California coast in the late sixteenth and very early seventeenth century and made contact with the indigenous people.

On July 12, 1587, the Spanish frigate *Nuestra Señora de Buena Esperanza* set sail from Macarera Island, off the southeastern coast of China, to locate two islands called Rica de Oro and Rica de Plata—perhaps part of the Hawaiian Island chain. After failing to find these islands, Captain Pedro de Unamuno sailed west to the California coast and on October 18, 1587, anchored in Morro Bay.[21] He and some of his men went ashore and took possession of the land in His Majesty's name. Several days later, following well-beaten paths and looking for signs of native people, they found more than thirty dugouts and "little cord bags, made like nets, in which were some pieces of rope made out of the bark of trees (very well made), and some old baskets in which they carry their luggage, and a trough made out of a tree trunk." The latter, the Spaniards inferred, was used to "gr[ind] roots or tree bark for some drink or dish of theirs."[22] Before getting back to the ship, they were attacked by Indians (possibly Salinan), who wounded several of the men and killed another. To avoid further conflict, the ship continued on its voyage south the next day and reached Acapulco on Sunday, November 22.[23]

In 1595 Luis de Velasco II, viceroy of New Spain, sent Sebastian Rodríguez Cermeño on a "voyage of discovery" to locate potential California ports on his return trip from Manila. On November 6, Cermeño anchored in Drake's Bay. The next day he went ashore and found nearby "many Indians—men, women and children—who had their dwellings there." Cermeño noted, "The soil will return any kind of seed that may be sown, as there are trees which bear hazelnuts, acorns and other fruits of the country, madrones and fragrant herbs like those in Castile. . . . There are also in the country a quantity of crabs and wild birds and deer, with which the people maintain their existence."[24]

On December 8 Cermeño and his men left Drake's Bay and sailed south, making note of Monterey Bay. Three days later they spotted "many people on top of some bluffs" near San Luis Obispo Bay. These were probably Chumash people. He anchored in front of these settlements and noted that the people are "well set up, of medium height, of a brown color, and like the rest go naked, not only men but women, although the women wear some skirts made of grass and of bird-feathers. They use the bow and arrow, and their food consists of bitter acorns and fish. They seemed to be about three hundred in number, counting men, women and children, some of them with long beards and with the hair cut round, and some were painted with stripes on the face and arms. The land seemed to be good, as it was covered with trees and verdure."[25]

On December 14 the ship arrived at what is now Santa Rosa Island. A canoe manned by two Indians, probably of Island Chumash descent, came alongside the ship. The Indians brought eighteen fish and a seal and gave them to the seamen, a welcome relief for the near-starving crew and passengers. This food was supplemented with thirty sea bass caught by the Spanish. On December 15 they reached Santa Monica Bay, where they saw on shore "many fires in Indian settlements."[26] After sailing along the coast of Baja California, on April 24, 1596, they arrived at the port of Chacala in New Spain. (See Map 2.)

In May 1602 Sebastián Vizcaíno, with three ships, the *San Diego*, the *Santo Tomás*, and the *Tres Reyes*, left the port of Acapulco for the California coast. He was instructed by the then viceroy of New Spain, Gaspar de Zúñiga, to

Map 2. Henry Briggs's *The North Part of America*, one of the earliest maps of North America, showing both Baja and Alta California as a massive island. From *Hakluytus posthumus or Purchas His Pilgrimes* (London, 1625). Courtesy of the Bancroft Library, University of California, Berkeley.

accomplish "the discovery and demarkation of the ports, bays, and inlets which exist from Cape San Lucas, situated at 22 degrees 15 minutes, to Cape Mendocino, at 42 degrees."[27] During the voyage, he gave many landmarks Spanish names, many of which are still in use today.

The fleet arrived at the Bay of San Miguel (named by Cabrillo) in November. The Carmelite father Antonio de la Ascensión, who accompanied the expedition, described the countryside in his diary as "very fertile and level" and noted that "near the beach there are very fine meadows" and that the soldiers recognized "the extensiveness, capacity and security of the port, its good depth and many fish."[28] On land, they encountered many Indians (probably Kumeyaay) painted black and white and with bows and arrows. The Spanish gave them colored glass, cords, and ribbons to appease them, and they remained peaceful. Vizcaíno renamed the bay San Diego de Alcalá.[29]

On November 20 they continued northward, and five days later they anchored near a large island, which they named Santa Catalina, after the martyr of the same name. As they sailed up the coast, they noted "many settlements of affable and inoffensive Indians" along the Santa Barbara coastline. North of Morro Bay they found themselves "near a very white high sierra, all reddish on the sides and covered with many trees[,] . . . named the Sierra de Santa Lucia."[30]

On December 16 they entered a "very good" port that was "well protected from all winds." Father Ascensión described the shore as having an immense number of great pine trees, smooth and straight, suitable for the masts and yards of ships, and many different kinds of animals—including birds the shape of turkeys with wingspans measuring more than a yard across, which must have been California condors. He said of the Indians: "They [probably Ohlone] brought us some of the skins of bears, lions, and deer. They use bows and arrows and have their form of government. They are naked. They would have much pleasure in seeing us make a settlement in their country." Vizcaíno named the bay Monterey in honor of the Count of Monterey. Vizcaíno reached Cabo Mendocino on Sunday, January 12, at which point there were "not more than six men on board all told who were well and up, the rest of the soldiers, sailors, cabin boys, and ship boys being sick in bed" with scurvy.[31] With the health of the men failing, Vizcaíno headed for Mexico. After entering the port of Acapulco on March 21, 1603, he sent an optimistic report to the king of Spain, urging that Spain colonize California.[32]

Despite Vizcaíno's report, much of California would remain terra incognita in official Spanish documents and maps for years to come. For 166 years after Vizcaíno's expedition, Spain launched no additional official expeditions into Alta California. It is likely, however, that there were some encounters between native people and the crews of the Manila galleons that continued to sail along the coast.

ENVIRONMENTAL AND CULTURAL IMPACTS

Very little is known about what impact early explorers may have had on the coastal flora and fauna. Susan Bicknell and Ellen Mackey hypothesize that when early explorers anchored near shore for repairs, they inadvertently deposited the seeds of species from other continents. It was standard practice for European explorers to take on fresh water, supplies, and beach sand as ballast along the coast of South Africa on the way to the New World. Bicknell and Mackey think it likely that early explorers transported propagules of the sea fig *(Carpobrotus chilensis)* from the beaches of South Africa to California dune habitat in this way.[33] The ecologist Harold Heady also theorizes that many alien species were introduced during this period:

> A brief land exploration, especially one with horses, from the first sailing vessel to reach the California shores probably left new plants. Many of the early ships carried a few live animals for meat; thus seeds in the manure thrown overboard could have resulted in alien plants reaching shore even though the sailors did not. Once a plant species produced seeds anywhere in the continent, birds could have carried them to other locations. California lies in the path of many bird migrations, especially north–south routes with one terminus in Mexico.[34]

For the most part, early explorers' contact with California Indians was brief and, as far as we know, limited to coastal California. Accounts in personal diaries, reports, letters, and ships' logs give us a far from complete picture of aboriginal coastal California, but they make it clear that parts of the coast were thickly populated with California Indians. It is also apparent that California Indian tribes greeted the explorers with a variety of attitudes: sometimes with trepidation, sometimes with friendliness, and other times with hostility. It is very likely that some European diseases were introduced to Indian populations during this period, but what these diseases were and what effects they caused are unknown.[35]

The Spanish Mission Period (1769–1823)

In the 1760s, with Russian fur traders advancing from the north and threatening Spanish claims on the Pacific coast, Visitador-General José de Gálvez of New Spain masterminded the expansion of Spanish territory into Alta California.[36] He hoped to thwart the territorial encroachment of other nations by establishing a series of missions and presidios along the coastline. He selected the president of the Baja California missions, Junípero Serra, to head a group of missionaries on an expedition to Alta California, accompanied by soldiers under the command of Captain Gaspar de Portolá.[37] On July 16, 1769, they founded the first California mission, which was the first permanent European colony in Alta California.[38]

Between 1769 and 1823 twenty additional missions were built in a territory stretching from San Diego to Sonoma. In addition to the missions, the Spanish established presidios, pueblos, and ranchos.[39] The presidios, which were made up of garrisons of soldiers and their families, were the political and social centers of early Hispanic California. Typically presidios were established close to the missions and were composed of a half dozen or more soldiers who protected the priests, helped them to enforce discipline on the California Indians, and led posses to recapture those who ran away. Pueblos, the secular villages of Spanish California, were established in San José (1777), Los Angeles (1781), and Branciforte (Santa Cruz; 1797). The ranchos, large grazing operations, developed in the 1830s (see below).

CREATING A NATIVE WORKFORCE

According to the historian Edward Castillo, "The mission was the most important institution used by the Spanish in the Americas to establish control of Indian territory and peoples." The typical California mission, according to the historian Jack Forbes, was "not erected in an already existing pueblo with sufficient population to support a church, but was utilized as a device for gathering together (congregating) natives who were dispersed in small villages, and for 'reducing' them from their 'free,' 'undisciplined' way of life to that of a disciplined subject of Spain."[40] Though initially some indigenous people willingly joined the mission system, by 1787, recruitment was sometimes forced.[41]

Socialization included converting the California Indians to Christianity and instructing them in European ways, which Father Serra and the other Franciscan priests thought of as inherently superior to those of California Indians. As Father José Señán at the San Buenaventura Mission wrote in a letter in 1796, "It is no small part of the missionary fathers' task to maintain their neophytes congregated in the missions, as members of society, so as to subdue the savage nature which impels them, like wild beasts, to live in the deserts and forests."[42]

Indians were seen as primitive because they lived at the mercy of nature in perishable, flimsy shelters, with their skin exposed to the elements. They tolerated lice, fleas, mosquitoes, heat, fires, floods, and unpredictable food supplies. They used crude stone and plant technologies such as chert and obsidian blades and digging sticks and had not yet figured out how to improve the land with agriculture.[43] Father Geronimo Boscana, who served at the San Juan Capistrano Mission from 1812 to 1826, wrote condescendingly of the Acâgchemem (Ajachmem) Indians who made up the mission congregation: "No doubt these Indians passed a miserable life, ever idle, and more like the brutes than rational beings. They neither cultivated the ground nor planted any kind of grain, but lived upon the wild seeds of the field, the fruits of the forest, and upon the abundance of game."[44]

Indians were taught the "civilized" skills of textile weaving, farming, cleaning and processing domesticated grains, cowhide tanning, sheep shearing, candle making, adobe brick making, wine making, and a hundred other tasks. These skills enabled Indians, who outnumbered the Spanish by up to 40 to 1, to provide the labor that was essential in the day-to-day operations of the missions. Each mission encompassed thousands of acres that needed to be worked—grazing lands, croplands, vineyards, orchards, and vegetable gardens. There were also many labor-intensive nonagricultural activities such as tanning hides, rendering tallow, and managing cows and horses.[45] Over time, the agricultural fields and cattle herds tended by Indians became vast economic enterprises. Altogether, the enterprises at each mission brought in an income ranging from $10,000 to $50,000 a year.[46] In a letter to Fray José Guilez dated April 18, 1812, Father José Señán, at the San Buenaventura Mission, wrote of the "very substantial profit" from the sale of products manufactured through Indian labor.[47] The clergy also loaned Indians as manual laborers and domestic servants to the garrisons and their families at the nearby presidios. Though it was initially understood that all such service should be paid for, after 1790 the work was done under "unmitigated compulsion."[48]

Ironically, although the Spanish considered the Indians culturally inferior, they did not hesitate to use Indians' traditional ecological knowledge when it suited their needs. When agricultural crops failed, Indians often provided themselves and mission staff with seeds, berries, greens, and bulbs that they gathered from surrounding wildlands, saving all from starvation.[49] The Indians taught the padres which plants would heal skin rashes, alleviate stomach- and headaches, check diarrhea, and soothe coughs. The Spanish names given to these plants by the padres—yerba santa, yerba mansa, yerba buena—appear today as common names in California's major flora—the

Jepson Manual.[50] California Indian medicinal knowledge was so effective and important that it was incorporated, over and over again, into Spanish mission pharmacopoeias. At Mission San Antonio de Padua in San Luis Obispo County, small vials of plants with their names and medicinal uses are on display, providing evidence of the ready adoption of indigenous knowledge by mission padres.[51] Native plants such as cascara sagrada (sacred bark),[52] originally used by the Indians and then adopted by the Spanish padres, have found their way into modern medicine chests and can be purchased in health food stores. Many of these plants, such as yerba mansa, which were cultivated by the Indians for medicinal purposes, are still found growing thickly on mission grounds.

ATROCITIES OF THE MISSION SYSTEM

English sea captains, Spanish alcaldes, and visitors from other countries who observed the mission system seemed to think that it bettered the conditions of the Indians, kept them from idleness with long hours of productive work, and improved the land by growing crops, grazing cattle, and other civilized endeavors.[53] About the Santa Clara Mission, George Vancouver wrote: "It is not possible to conceive how much these excellent men [padres] will feel rewarded in having been the cause of [a]meliorating the comfortless condition of these wretched humble creatures." G. H. Von Langsdorff noted in 1812 that the Indians' "attachment to a wandering life, their love of alternate exercise in fishing and hunting and entire indolence, seem in their eyes to overbalance all the advantage they enjoy at the mission, which to us appears very great."[54]

However, behind the outward appearance of productivity and order was a grim reality. Mission Indians labored long and hard, often lived under miserable conditions, had poor diets, suffered from epidemics, experienced physical abuse and intimidation, and died in huge numbers. After the establishment of the missions, diseases such as syphilis, tuberculosis, dysentery, diphtheria, and measles spread rapidly throughout native populations lacking any immunity, aided by crowded living conditions in drafty adobe dormitories with poor sanitation. Deficient diets consisting of a starchy cereal soup and a little meat predisposed Indians to infectious disease and malnutrition. Medical treatment was rare.[55]

In an 1813 report on Mission San Antonio de Padua, Father Señán noted that "for each two [Indians] that are born, three die."[56] Mission records show an even more morbid relationship: between 1779 and 1833 the padres recorded 62,000 deaths and only 29,000 births. At its height, the mission system boasted a total of 72,000 Indian converts, or neophytes, but by 1830 only 18,000 neophytes remained. The total loss of the mission Indian population has been estimated, on the basis of mission records, at 72 percent.[57] The anthropologist Alfred Kroeber wrote, "It must have caused many of the Fathers a severe pang to realize, as they could not but do daily, that they were saving souls only at the inevitable cost of lives. And yet such was the overwhelming fact. The brute upshot of missionization, in spite of its kindly flavor and humanitarian root, was only one thing, death."[58]

Disobedient Indians were whipped with a barbed lash, subjected to solitary confinement, mutilated, locked in stocks and hobbles, branded, and sometimes executed.[59] According to the historian James Rawls, "Once an Indian became a neophyte, he was considered no longer free to reject his vows and return to his former lands or way of life."[60]

Many Indians attempted to escape from their virtual captivity in the missions. Often they ran away into the interior, hiding among sympathetic tribes as refugees. Running away from the mission became more and more common over time, and by the second half of the Mission era, one out of ten mission Indians became fugitives. They escaped out of homesickness, out of a desire for sovereignty over their life and work, and because they could no longer endure hunger, cruelty, confinement, and poor living conditions. In 1797 Indians who ran away from Mission Dolores (San Francisco) and were forcibly returned gave some of the following reasons for having absconded:

> When he wept over the death of his wife and children, he was ordered whipped five times by Father Antonio Danti.
>
> He was put in the stocks while sick.
>
> Because of a blow with a club.
>
> When his son was sick, they would give the boy no food, and he died of hunger.[61]

Sherburne Cook notes that "forced translocation of large numbers of Indians could not fail to engender in many of them a conscious antipathy to their new environment, an antipathy which found an outlet in apostasy, fugitivism, and physical resistance."[62] But the padres did not tolerate any of these responses. From the earliest days of Spanish occupation, a garrison of from two to twelve men resided at each mission to enforce the rules and regulations.[63] If Indians ran away, the garrison would often pursue them far into the interior. Vassilli Petrovitch Tarakanoff, a Russian otter hunter, vividly described the cruel punishment of a group of Indians being brought back by soldiers and priests:

They were all bound with rawhide ropes and some were bleeding from wounds and some children were tied to their mothers. The next day we saw some terrible things. Some of the run-away men were tied on sticks and beaten with straps. One chief was taken out to the open field and a young calf which had just died was skinned and the chief was sewed into the skin while it was yet warm. He was kept tied to a stake all day, but he died soon and they kept his corpse tied up.[64]

If a padre inside the mission establishment criticized the treatment of the Indians, he was removed. In 1799 Fray Antonio de la Concepción Horra of Mission San Miguel angered his colleagues by reporting to the viceroy in Mexico: "The treatment shown to the Indians is the most cruel I have ever read in history. For the slightest things, they receive heavy flogging, are shackled and put in the stocks, and treated with so much cruelty that they are kept whole days without water." Horra was isolated, declared insane, and escorted under armed guard out of California.[65]

TRANSFORMATION OF THE ENVIRONMENT

During the Mission era, impacts on the environment were largely restricted to the marine-influenced strip west of the inner Coast Ranges from San Diego to Sonoma, but the impacts in this area were extensive. Grazing was among the activities that caused the greatest damage. Coastal prairies, oak savannas, prairie patches in coastal redwood forests, and riparian habitats, all rich in plant species diversity and kept open and fertile through centuries of Indian burning, became grazing land for vast herds of cattle, sheep, goats, hogs, and horses owned by the Spanish missions and rancheros. By 1832 the California missions had more than 420,000 head of cattle, 320,000 sheep, goats, and hogs, and 60,000 horses and mules.[66] At Mission San Antonio de Padua alone, the livestock in 1832 comprised 6,000 head of cattle, 10,500 sheep, 65 goats, 70 pigs, 82 mules, and 779 horses. Overgrazing eliminated native plant populations, favored alien annuals, and caused erosion. In addition, the high cattle and sheep densities reduced populations of native ungulates or pushed them farther inland. Though livestock grazing during the Mission period caused less environmental damage than that of later eras, it nevertheless brought about extensive and irreversible changes in many ecosystems.[67]

Agricultural activities also wrought major changes in the environment. At their height, the missions cultivated approximately 10,000 acres, producing 340,000 bushels of wheat, beans, and corn annually.[68] Agricultural fields replaced native ecosystems, but some important ecological impacts were less direct. The priests at Mission San Buenaventura near the mouth of the Santa Clara River, for example, used strychnine to poison the geese, swans, and other waterfowl that had begun to feed in the fields, overgrazing having destroyed their natural food sources in the coastal prairies.[69] Agriculture also resulted in the introduction of alien plant species. Cuttings and seeds of plants from Europe were planted for food, medicine, and ornamental uses on the grounds of the missions, presidios, and pueblos; they later spread to surrounding wildlands. The mission fig *(Ficus carica)*, for example, spread to riparian forests, and common mullein *(Verbascum thapsus)*, valued for its medicinal properties, spread to meadows, outcompeting natives.[70] The pepper tree, which the first great viceroy of Mexico had sent to Mexico from Peru, was planted at every mission.[71]

A great variety of alien species were introduced inadvertently during the Mission period. Research has shown that European forbs and grasses such as *Erodium cicutarium, Rumex crispus, Sonchus asper, Hordeum jubatum* subsp. *leporinum, Lolium multiflorum,* and *Poa annua* were brought into California at this time, contained in adobe bricks, livestock feed, livestock bedding, and other materials.[72] Soon these aliens overwhelmed the native species, markedly changing the character and diversity of grasslands and other habitats west of the inner Coast Ranges.

Hunting and fur trapping during the Mission period caused population declines in a number of mammal species. Marine mammals were among the first to be exploited, although more by the Russians than the Spanish. Sea otters, hunted extensively in the 1700s and 1800s by the Russian-American Fur Company, suffered the worst declines. According to Ray Dasmann, sea otters at the beginning of the Mission period were "abundant along the California coast, particularly around San Francisco and Monterey bays and the Channel Islands. Perhaps 300,000 or more swam in the offshore waters. Unfortunately for the otters, they had a dense, warm brown coat with a silvered frosting of guard hairs. This came to be regarded as highly desirable among fur wearers in Moscow, Peking (Beijing), and elsewhere among the world's elite."[73] Among terrestrial mammals, considerable numbers of elk were killed for their hides and tallow (to make candles) by the Spanish rancheros.

The Mexican Era (1822–1848)

In 1808 Napoleon I invaded Spain with a hundred thousand French troops, initiating a long struggle between Spain and France. With Spain's resources drained by the Napoleonic Wars, struggles for independence ignited in the Spanish colonies of Argentina, Chile, Paraguay, Uruguay, and Venezuela and soon spread northward to New Spain.[74] After eleven years of warfare, Mexicans won their independence in 1821, and the name "New Spain" was replaced with "Mexico." In November 1822 the first governor of California under Mexican rule, Luis Arguello, took office.

During the Mexican era, California had only a few thousand people of Spanish and Mexican descent, who called themselves "Californios."[75] Mexico intervened little in the affairs of its distant province, except for sending governors and some soldiers,[76] and California Indians experienced little change. The new Mexican government adopted the Plan of Iguala, which guaranteed citizenship to Indians and protected their rights and property, but it was good only on paper; in practice military forays continued to capture fugitive Indian neophytes, and Indians continued to be forced to labor for the missions.[77]

During this period, two major epidemics spread through the Indian population and left many dead. In 1833 an unidentified disease—possibly introduced by fur trappers—reached epidemic proportions in the San Joaquin and Sacramento Valleys. The "pandemic of 1833" annihilated whole Indian villages. Duflot de Mofras wrote of 12,000 indigenous deaths in the San Joaquin Valley and 8,000 in the Sacramento Valley. The epidemic also spread to the American, Feather, Yuba, Tuolumne, and Merced Rivers, and that year Wilkes discovered large numbers of Indian skeletons at the fork of the Feather and Sacramento. In 1837 a smallpox epidemic spread through the valleys of Sonoma, Petaluma, Santa Rosa, Russian River, Clear Lake, Suisun, and Sacramento, killing thousands of Indians.[78] Throughout the Mexican era, measles, pneumonia, diphtheria, and venereal disease, among other diseases, continued to spread from Californios to native populations.[79] These epidemics, together with armed conflict and destruction of the food supply, caused the population of indigenous people to plummet from 245,000 to 150,000.[80]

INFLUX OF AMERICANS AND OTHER FOREIGNERS

In the 1820s foreigners began to come overland into California to hunt and trap. Peter Skene Ogden, a Canadian-born explorer and fur trapper traveling south from Oregon, began to trap beaver along the Klamath River in 1826. Jedediah Smith and his party entered California from the Southwest late that same year, reaching the San Joaquin Valley in 1827 and trapping beaver along all of the tributaries of the San Joaquin River. Fur trapper A. R. McLeod of the Hudson's Bay Company reached Bonaventura (Sacramento) Valley in 1829 and found numerous well-populated Indian villages.[81]

Increasing numbers of trappers plied their trade along California rivers, lakes, sloughs, and marshes. In 1830 Suisun Bay yielded 4,000 beavers in just six months of trapping. From 1839 through 1841 Captain John Sutter saw 3,000 beavers taken annually from the Sacramento and San Joaquin Rivers by Hudson's Bay Company trappers.[82] He also noted that 1,500 beaver skins had been traded by the natives to Mission San José "at a trifling value" and were sold to ships at three dollars apiece.[83]

More and more Americans arrived in California during the 1830s, with Indians sometimes acting as guides for the newcomers. In 1834 two Indians guided Joseph Walker and his men over the southern Sierra Nevada through a pass that now bears Walker's name; and an Indian guide by the name of Truckee led Frémont and his men across the east side of the Sierra.[84]

Shipping traffic also increased during the Mexican era. Ships from Boston traded along the California coast, buying sea otter pelts and cattle hides and selling manufactured and exotic goods. The historian J. S. Holliday describes the diverse assortment of products sold to the Californios: "The New England merchant ships brought all kinds of goods: fish hooks, cotton cloth, woolen blankets, shoes, nutmeg, pepper, and more exotic temptations (from China and Europe) such as camel's-hair shawls, painted water pitchers, porcelain plates, brittany linens, even pianos."[85] In return, the Boston ships brought back sea otter skins, which in the 1830s were worth from $40 to $60 apiece.[86] They also bought whale oil, to fuel lamps and light streets at night.

Whaling companies operating out of Monterey and San Francisco bays had been established in the 1820s. Sperm, blue, finback, and right whales were slaughtered for their blubber, which was rendered into oil. Shore-based whalers targeted the migrating California gray whales, which each season faced "an increasingly well-armed and dangerous gauntlet . . . while en route to their calving grounds along Baja California, Mexico, where still other whalers awaited them."[87]

SECULARIZATION OF THE MISSIONS AND THE RISE OF THE RANCHOS

By the end of the Spanish Mission period, mission lands encompassed about one-sixth of present-day California. The pueblo and presidio land base had remained small under Spanish rule, and only a few individuals owned ranchos outside the military or town boundaries. This changed in 1834, when California governor José Figueroa drafted regulations to secularize the missions, following the decree of 1813 passed by the Spanish Parliament that ordered all missions in America that had been operating for at least ten years to be given up to the bishop in accordance with the laws.[88] After secularization, "the control and restraint of the church fathers was removed and the entire mission system went to pieces with terrific rapidity."[89] While the original decree for the secularization of the missions specified that the Indians were to receive land, livestock, and other goods, in fact they received little of the missions' wealth.[90]

Mission Indians scattered. Some had learned new skills and become acculturated into Spanish colonial society. In 1833 William Heath Davis, an Englishman who had married into a prominent Spanish rancho family, described the approximately two thousand "more or less civilized, well clothed" Indians he observed at Mission Dolores in San Francisco: "Among them were blacksmiths, shipwrights, carpenters, tailors, shoemakers and masons, all of whom had learned these trades at the Mission under the superintendence of the Padres."[91] Others, such as the Encinales family of Salinan descent, formerly at Mission San Antonio de Padua, were able to combine what they remembered of the "Old Ways" with the knowledge the mission padres had taught them to put together new lives and livelihoods. The Encinales occupied and made additions to an adobe brick house, and they harvested acorns and pine nuts in addition to growing corn, watermelon, and grapes.

However, many other California Indians were destitute and lived in dire poverty under conditions far worse than those of their ancestors. Many sought work in the pueblos, or went to work for the private ranchos. Although these individuals had employment of a sort, "life in the settlements led to an almost immediate breakdown of tribal organization and loss of cultural identity for the individual."[92]

The rancho movement grew following the secularization of the missions in the 1830s. By 1846 there were more than five hundred ranchos. The owners, called rancheros, presided over lands that were often enormous in extent. The extensive, fertile, and productive grasslands supported vast herds of cattle and sheep. Walter Colton, the alcalde of Monterey, was amazed at the number of stock: "Two thousand horses, fifteen thousand head of cattle, and twenty thousand sheep, are only what a thrifty farmer should have before he thinks of killing or selling."[93] Many of the rancheros were descendants of Spanish and Mexican soldiers and settlers, but some were American or from European countries other than Spain. Many men important in California history, such as George Yount, Abel Stearns, Hugo Reid, William Dana, John August Sutter, Jonathan Turnbull Warner, Peter Lassen, and John Bidwell, became naturalized as Mexican citizens to qualify for landownership.[94] Mission Indians also held a number of ranchos, but they were very few compared to the number of Indians who lost title to mission lands originally promised to them.

INDIAN LABOR ON THE RANCHOS

In early Mexican California, rancheros and their families enjoyed what was in many ways a pleasant and easy life filled with music, song, and dance. Colton characterizes California life during this period: "There are no people that I have ever been among who enjoy life so thoroughly as the Californians. Their habits are simple; their wants few; nature rolls almost every thing spontaneously into their lap. Their cattle, horses, and sheep roam at large—not a blade of grass is cut, and none is required. The harvest waves where ever the plough and harrow have been; and the grain which the wind scatters this year, serves as seed for the next."[95] However, this lifestyle was possible only because much of the labor on the rancho rested on the shoulders of California Indians.

William Garner, an Englishman who came to California in 1824, wrote the following in a letter dated November 19, 1846: "There are many persons who have tremendous large tracts of beautiful and fertile lands, containing from three to eleven square leagues, and the man who cultivates twenty acres of it, without taking the trouble to fence it, is considered among themselves an extraordinarily industrious man, and at the same time, were it not for the Indians who work about the farms for little or nothing (and generally get cheated out of that), there would be no land cultivated in California."[96]

Most of the five hundred ranchos were carved out of former mission-controlled lands.[97] The Californios were eager to have the Indians stay on the rancho lands as a workforce for the new owners. This was also true of the Americans who came to California, acquired ranchos, and married into Spanish or Mexican families. "It is indeed significant," writes Cook, "that well into the fifties nearly every wealthy American adopted without question the existing labor system. Bidwell and Reading, for instance, maintained serflike bands of Indian retainers until the Civil War period."[98]

Indians worked for the large landowners under a peonage system imported from Mexico: in exchange for their labor, Indians received homes, food, and commodities. But these benefits were given largely at the discretion of the employer, and the ranchero always had the upper hand. Indians did every kind of chore: they sheared the sheep, herded cattle, cut the lumber, harvested the crops, pounded the grain into flour, built the houses, tanned the hides, cleaned the houses, served the meals, and made tiles and adobe bricks.

Rancheros often resorted to coercion of indigenous people. "This might take the form of innocent persuasion," writes Cook, "or economic pressure through control of food reserves, or out-and-out kidnapping and slavery."[99] Sometimes, when labor was needed, ranchers organized expeditions against the Indian rancherias (from the Spanish word for Indian village) and brought in the required number as captive slaves. Active native resistance to these brutalities included warfare and the stealing of livestock. Edward Castillo describes these organized efforts: "Adopting guerilla warfare tactics perfected earlier by native resistance leaders like Estanislao, tribesmen underwent considerable physical and military adaptation. With the acquisition of horses from the colonists, these Indians changed from peaceful, sedentary, localized groups to semiwarlike, seminomadic groups. They began to take the offensive, making widespread cattle raids to supplement their diminishing native food supply."[100] Many California Indians lost their lives in the numerous military campaigns organized against them, accounting for a population decline of about 6 percent. For example, in an 1834 battle more than 200 Wappo people were killed and 300 taken prisoner by General Mariano Guadalupe Vallejo. In 1839 Mexican soldiers attacked a band of Indians near Los Gatos who had been responsible for numerous cattle raids. In the battle, the band's leader, Yozcolo, formerly a neophyte at Mission Santa Clara, was killed and 100 of his men captured.[101]

IMPACTS ON THE ENVIRONMENT

Environmental destruction and ecological change were more pervasive during the Mexican period than during the Spanish Mission era. Californio and American families overgrazed existing rangelands and expanded their herds into the Central Valley and the Coast Ranges. Sea otters were overharvested, and by the 1850s the valuable sperm and right whales had almost disappeared from the Central Coast of California. Beavers were trapped so relentlessly in the 1830s and 1840s that their numbers declined precipitously. In the mountains, wolverine, fisher, marten, mink, and river otter were reduced through persistent trapping.[102] Oak trees were logged for fuel and to make ox carts and saddle stirrups.[103] Native plants such as California bromes, blue wild rye, and clovers, which produced seeds, grains, and greens used as food by the Indians, were eaten by the cattle and sheep before they could set seed, and their formerly extensive populations shrank dramatically.[104]

American Control and the Gold Rush (1846–1860)

The United States Congress declared war on Mexico on May 13, 1846, and less than a month later American settlers began a revolt against Mexican rule in California. That summer, American naval and army forces encountered little resistance at Monterey, Sonoma, Yerba Buena, Los Angeles, and San

Diego. Although the Californios began to fight back in September and won several battles decisively, they faltered later that fall, weakened by internal dissension and poor-quality gunpowder. The last battle in California was fought on January 9, 1847, and hostilities ended on January 13, when the Mexican general surrendered to John Frémont. The war formally ended a year later, and a peace treaty was signed at Guadalupe Hidalgo on February 2, 1848.

Under the terms of the treaty, Mexico ceded 1.2 million square miles to the United States, including California, and the United States agreed to recognize Indians as citizens, with their freedom and property rights given full protection under the laws of their new government.[105] Unfortunately, Indians did not fare better under American rule. Between 1845 and 1855 the huge influx of miners and other settlers created what Cook calls an "overwhelming assault upon the subsistence, life, and culture of all California natives," the violence of which "ha[d] seldom been duplicated in modern times by an invading race." In only ten years, the California Indian population dwindled from about 150,000 to 50,000. According to Cook, "The direct causes of death were disease, the bullet, exposure, and acute starvation. The more remote causes were insane passion for gold, abiding hatred for the Red man, and complete lack of any legal control."[106]

Significant changes in the environment had occurred during Spanish and Mexican rule, but these were small and restricted compared to the changes wrought in the land during and immediately following the Gold Rush. Mining, logging, and agriculture caused tremendous damage to many of California's rivers and creeks, especially in the foothills of the Sierra Nevada and in the watersheds of the Salmon, Trinity, and Klamath Rivers of northern California, altering their ecologies and forever changing their hydrological functioning. The numbers of cattle, sheep, and pigs escalated, and exotic grasses and forbs spread to the farthest reaches of the state, drastically transforming the species compositions of grasslands and woodlands.

THE RUSH FOR THE GLITTERING METAL

Just nine days before the Treaty of Guadalupe Hidalgo, James Marshall discovered gold on the south fork of the American River near Coloma in the Sierra Nevada foothills. Discovering gold had been the dream of countless early European explorers; now it had become a reality. To control the gold-bearing land and keep future miners from stealing it, Marshall and his business partner, John Sutter, signed a treaty with the Yalesummy, a division of the Nisenan tribe, in 1848. "The tenor of this was that we were to pay them two hundred dollars yearly in goods, at Yerba Buena prices, for the joint possession and occupation of the land with them; they agreeing not to kill our stock, viz.: horses, cattle, hogs or sheep, nor burn the grass within the limits fixed by the treaty."[107]

The Nisenan chief warned Sutter that the gold "belonged to a demon, who devoured all who search for it."[108] Sutter ignored this warning, but it would turn out to be somewhat prophetic. Sutter sent the treaty to California governor Col. R. B. Mason for approval. Governor Mason wrote to Sutter on March 5, 1848: "I last evening received your letter of February 22, together with the lease to certain lands on the waters of the American fork, a tributary of the Sacramento, made by certain Indians of the Yalesummy tribe to yourself and Mr. James W. Marshall. The United States do not recognize the right of Indians to sell or lease lands on which they reside . . . after the war the United States might well claim all Indian lands as part of the public domain."[109]

As word of the gold discovery spread, the "world rushed in," and men from all over the United States, Europe, and South America and as far away as the Sandwich Islands, China, and Australia traveled to California by land and sea to try to strike it rich. They trespassed on numerous tribal lands, as well as on the vast ranchos of the Californios and Americans. With no courthouses or jails, except in San Francisco and Sacramento, a new civil government with few laws, and a distant federal government, miners and merchants often took the law into their own hands.[110]

About $400 million in gold (worth over $6 billion today) was recovered by miners between 1849 and 1855,[111] but for most eager newcomers, searching for gold was not as fruitful as imagined. Jean-Nicolas Perlot, a Belgian miner, recounted a common experience: "At the end of the week . . . it was found that we had each earned three dollars a day and consumed a dollar and a half in supplies, net a dollar and a half."[112]

TRESPASSERS ON THEIR OWN LANDS

At the end of the Mexican era, most of the tribes living in the Sierra foothills and the mountains of far northern California had largely escaped the fates of their cousins to the west and south. Though they had encountered disease and violence at the hands of whites, contact with settlers had been minimal, and their traditional lifeways had not been completely disrupted. This changed with the onset of the Gold Rush.

Many of the permanent Indian villages were in the Sierra foothills at or below the snowline between 1,000 and 3,000 feet in flat terrain and close to major rivers or streams. This was the prime elevational zone for what were termed the Northern and Southern Mines, located along the Mother Lode. The Northern Mines encompassed parts of the Feather, Yuba, Bear, and American Rivers and various creeks. These were the traditional territories of the Northern Maidu, Concow, and Nisenan. The Southern Mines encompassed the Consumnes, Mokelumne, Calaveras, Stanislaus, and Tuolumne Rivers and smaller creeks such as Mariposa. These were the traditional territories of the Northern, Central, and Southern Sierra Miwok. Other major mining activities occurred farther north along the Salmon, Trinity, and Klamath Rivers in the traditional territories of the Shasta, Tolowa, Chimariko, Hupa, Chilula, Yurok, and Wintu. The territories of other tribes, such as the Northern Valley Yokuts, while not gold country, were along the various routes to the goldfields. With more than 300,000 hopefuls coming into the goldfields between 1848 and 1857, one can only imagine what it must have been like for the native people. The miners overran native homelands, altering stream courses, destroying salmon runs, scaring away game with pistols and rifles, chopping down oaks and sugar pines, and grazing cattle, hogs, and horses on the grasslands.

Indians found themselves trespassers on the lands where their ancestors had lived for centuries. The best Indian lands—open, fertile, and flat or gently rolling and dotted with villages—were taken by Anglo settlers for farms, ranches, and mining settlements.

Indians were forced to retreat to surrounding areas where the terrain was less suitable for habitation, seed gathering, and hunting. Yet here too the burgeoning Anglo settlements interfered with the Indians' subsistence activities and vied with them for natural resources. Cattle fences and No Trespassing signs prohibited grasshopper, quail, rabbit, and deer drives. Wetlands rich in resources were drained. Cattle, pigs, and horses competed with Indians for acorns and grains of grasses. Lieutenant Livingston, in a letter dated 1856, wrote, "The acts of the whites so far seem to me to be utterly lawless. Those owning stock on King's River allow the stock to feed upon the Indian's acorns, and some even say, assist them to them."[113] A typical farmer or rancher in the Central Valley in the 1850s might let his hogs loose on Tulare Lake to feed on the tule roots and mussels formerly harvested by the Valley Yokuts and then take the hogs to the Sierra foothills to fatten them up on the acorns the Miwoks would have otherwise harvested.[114]

The practice of gold mining was particularly destructive. Nonrenewable gold was uncovered at the expense of the renewable fisheries on which many Indians depended. Gold panning, cradling, and sluicing muddied the rivers, devastating the major salmon runs. Brewer noted of the Sacramento River in 1851: "Previous to 1848 the river was noted for the purity of its waters, flowing from the mountains as clear as crystal but, since the discovery of gold, the 'washings' render it as muddy and turbid as is the Ohio at spring flood."[115] Moving the courses of whole rivers and streams to get at the gold in the stream bottoms destroyed not only many of the fish runs but also the freshwater mussels, other aquatic life, and the streamside vegetation from which animal and plant food had been gathered. Hydraulic mining released millions of tons of debris, silt, and gravel into streams, choking salmon-spawning beds.

Joaquin Miller describes the failed attempts of Indians to fish on the Klamath River after gold mining: "The trout turned on their sides and died; the salmon from the sea came in but rarely . . . . What few did come were pretty safe from the spears of the Indians, because of the coloured water; so that supply, which was more than all others their bread and their meat, was entirely cut off."[116] The situation was the same on the American, the Yuba, the Feather, and the other salmon streams along which gold mining was practiced.

In June 1860 C. A. Kirkpatrick wrote, "Salmon fish are fast disappearing from our waters—that is, upon all the streams upon which mining is carried on to any extent. . . ." He reminisced that in 1849 they had had no trouble whatever procuring all the salmon they wished and that by just constructing a rude barb or spear and wading out a few steps, they could literally pick up all they desired.[117]

With their lands overrun, their fisheries depleted, and their plant food resources destroyed, some tribes retaliated by killing whites and setting fire to trading posts. They also resorted to stealing cattle, horses, mules, and hogs, often to stave off starvation. Cook concludes that "the Indian could feel no ethical objection to appropriating the white man's property when the white man had already forcibly dispossessed him of his original means of subsistence."[118] The anthropologist Robert Heizer makes a similar point:

> It is a simple fact that in 1850 when California became the thirty-first state of the Union, the one hundred thousand Indians that still survived were not of the slightest interest or concern to most white Californians in terms of their human needs. If they ventured, as they occasionally did, to attack whites, this was not because they were an aggressive and unscrupulous lot of savages, but rather for the reason that they were human beings who had been pressed too hard—had seen their tribesmen and families shot down without mercy and without cause, and they finally understood that they were a people without hope and without a chance.[119]

White settlers responded to Indian depredations and thievery with violence. Numerous local and regional militias were organized to put a stop to livestock raids. Some settlers wanted Indians removed from the state altogether, and a series of Indian Wars ensued, some of which involved outright massacres of Indian people. In 1850, for example, 60 Pomo were killed on what has come to be known as Bloody Island in Clear Lake, and 75 Pomo were killed on the Russian River nearby. In 1851, in the town of Old Shasta, miners burned down the Wintu council meetinghouse and massacred about 300 of the people.[120] Edward Castillo labels such actions "subsidized murder," because the state legislature passed acts in 1851–52 authorizing payment of more than $1.1 million for suppression of Indian uprisings.[121] Altogether, more than 4,000 California Indians were killed in physical conflict with the Americans following the discovery of gold, mostly before 1865; many others suffered violent deaths at the hands of whites during quarrels and brawls over liquor or women, or as revenge for injury.[122]

In some counties, laws legalizing the murder and scalping of Indians were passed. In the Mariposa region, for example, miners drew up a petition demanding that Indians be outlawed; then the sheriff, sole civil and judiciary authority in Mariposa, rendered this decree: "I pronounce the Indian outlawed. Consequently, everyone is permitted to kill the Indians he encounters anywhere in the county of Mariposa, on the sole condition of burying them and of letting the sheriff know where and how many of them he has killed."[123] Even in California counties where homicide was illegal, few whites were brought to court to answer for killing Indians. Robert Heizer and Alan Almquist, in *The Other Californians*, conclude that "no more sorry record exists in the Union of inhuman and uncivil treatment toward minority groups than in California."[124]

EARLY INDIAN RESERVATIONS

Long before California was admitted to the Union, the U.S. Constitution empowered Congress to regulate commerce with Indian tribes. In 1793 Congress invalidated any title to Indian lands not acquired specifically through treaties under the Constitution. Indian relations were formally put in the hands of the president in conjunction with the treaty-making power. The treaty was the result of a federal policy of "removing the Indians from the advancing frontier" and of "gradually extinguishing their title to the public domain." Heizer and Almquist note, "Usually the basis of the Indian cession was the exchange of the occupied territory for lands further west, a policy which could no longer be followed when the Pacific coast was reached."[125]

Following California's admittance to the Union on September 9, 1850, three Indian agents, Redick McKee, George Barbour, and O. M. Wozencraft, were appointed by President Millard Fillmore and authorized by the U.S. Senate to go to California to negotiate treaties with the natives. In 1851 and 1852 they negotiated eighteen treaties involving about twenty-five thousand California Indians. In each of the treaties, the tribes were acknowledged as sovereign nations and the Indians promised provisions, cattle, and extensive tracts of valuable land—7,466,000 acres in all—to be set apart as reservations. Not all California Indian tribes were consulted. In June 1852 the Senate, yielding to the vigorous opposition of the California congressional delegation, rejected the California treaties, and the vast reservations proposed were never created.[126]

Congress assembled a land commission in 1851 to decide all cases of lands claimed by mission Indians under provisions of the Treaty of Guadalupe Hidalgo. The commission was legally responsible for informing the Indians about how to file claims for their lands, but it failed to notify Indians, and no claims were ever submitted. The swift result was the complete dispossession of the Indians from their lands, which were made part of the public domain and either opened to settlement or kept by the federal government.[127]

In 1852 Congress appointed Edward Beale the first Superintendent of Indian Affairs in California and appropriated $250,000 for the establishment of five reservations. In 1853 Beale focused on establishing the 50,000-acre Tejon Reserve at the southern end of the San Joaquin Valley near Tejon Pass, into which he gathered some two thousand Indians, neglecting other needy tribes.[128]

Beale was removed from office in 1854 and replaced by Col. Thomas J. Henley. Henley established several reservations: the Klamath reservation at the mouth of the Klamath River near the Oregon border, the Nome Lackee in western Tehama County, the Nome Cult Indian Farm in Mendocino County, the Mendocino reservation along the coast about fifty miles south of Cape Mendocino, the Fresno Indian Farm in the San Joaquin Valley, and the Kings River Indian Farm in the southern San Joaquin Valley.[129] Of course, having strong ties to their homelands, many Indians did not want to relocate to established reservations. A letter to Henley dated 1855 reads: "One of the Chiefs from high up on the McCloud River made mention of the fact that many of his men had perished the past winter while afraid to come into the lower hills for roots and clover, fearing the whites would compel them to remove onto the Reserve below."[130]

Furthermore, the reservations were completely inadequate for making a living. Several had sparse water supplies and little wild game for hunting. The administrators' purchases of food supplies often did not find their way to the Indians. In many reservations, white squatters took the best land and began to log trees, operate sawmills, set up stores, and raise livestock. According to Edward Castillo, "Most of these squatters were business partners or relatives of Henley and therefore impossible to remove."[131] To compete with squatters' livestock for food might cost an Indian her life: one witness reported seeing "a man drowning squaws from a cloverfield inside the reservation, they were picking clover or digging roots; he said he would be damned if he would allow them to dig roots or pick clover, as he wanted it for hay."[132]

Indians on reservations feared kidnapping and enslavement by outsiders. In 1856 E. A. Stevenson, agent for the Nome Lackee reservation, wrote in his report: "Of the Indians residing in this neighborhood, a large number are on the ranches or farms of private individuals, who are using them as working hands, and who seem to have adopted the principle that the Indians belong to them as much as an African slave does to his master, and that they have the right to control them entirely. Many of these Indians have left their places and come to the reservation, and have been followed and demanded by the persons claiming them as private property."[133]

Corruption was widespread during the establishment and administration of the reservations: "Subagents gerrymandered reservation boundaries to buy developed land, beef was seldom delivered to the Indians for whom it was intended, private businesses were allowed to operate on the reservation, books were incomplete and vouchers irregular. It is little wonder that few Indians ever stayed on the reservations. Native people found themselves again used as forced labor to enrich their overseers."[134] Except for Nome Cult (which became the Round Valley reservation), eventually all of these early reservations were closed or abandoned by the late 1860s.[135]

BUILDING A NEW ECONOMY WITH INDIAN LABOR

Disenfranchised on their own lands and unable to support themselves on scarce resources, many Indians had no recourse but to work for Euro-American settlers. The lucky ones got jobs at low wages. The rest were forced into virtual or actual slavery. Once again, Native Americans in California became a major labor force in the building of a new economy. California Indians sheared sheep, grew crops, threshed domesticated grains, rowed boats, dug irrigation ditches, cut timber, maintained railroad tracks, washed clothes, rounded up cattle, built structures, ferried travelers across rivers, panned for gold, and served meals for white settlers.[136]

Many California Indians tried to survive by learning to mine. Baskets used for sifting seeds began to be used for panning gold. Indians operated independently in the mines, but, again, they also served as laborers for white miners. Mine operators preferred to hire Indian laborers, for they could pay them less than whites. Some well-known figures in California history—the Murphy brothers (Murphys), Andrew Kelsey (Kelseyville), James Savage (Yosemite region), John Bidwell (Chico Creek), Peter Lassen (Deer Creek), Pierson Reading (Cottonwood Creek)—made great financial gains in gold mining by using Indian labor. In any event, by August 1848 Indians made up more than half of the four thousand miners in the goldfields.[137]

Merchants took advantage of Indian miners whenever they could. With little idea of the true value of gold, Indians were easily cheated. Merchants vastly overcharged for goods, and traders grossly understated the value of gold in trading it for cash. Native people became more cautious after the old mission Indians told them that whites sold to each other by ounces and pesos and that they could get more for their gold if it was weighed. But the merchants soon began to rig the weights and scales.[138] One such method was a special two-ounce counterweight, called a "Digger ounce," which produced a sum that was half what the gold was actually worth.[139] Prospector Chester Lyman jotted in his journal: "From the Indians all sorts of prices are taken & much deception is practiced."[140] Whites found numerous ways to extract labor from California Indians without fair compensation. A legislative act passed in 1850 authorized the indenture of Indians. Heizer calls it "a thinly disguised substitute for slavery."[141] But many whites simply kidnapped, bought, and sold Indians. The *Alta California* newspaper ran an article in October 1854 that read in part: "Abducting Indian children has become quite a common practice. Nearly all of the children belonging to some of the Indian tribes in the northern part of the state have been stolen. They are taken to the southern part of the state and there sold."[142]

Major Horace Bell described a slavery system in the vineyards surrounding Los Angeles:

> The cultivators of vineyards commenced paying their Indian peons with *aguardiente,* a veritable fire-water. . . . About sundown [on the second day] the pompous marshal . . . would drive and drag the herd to a big corral in the rear of Downey Block, where they would sleep away their intoxication, and in the morning they would be exposed for sale, as slaves for the week. . . . [T]he slave at Los Angeles was sold fifty-two times a year as long as he lived, which did not generally exceed one, two, or three years, under the new dispensation.[143]

Like African Americans during the same time period, California Indians had no legally recognized rights. If you were a California Indian in the 1850s you could not gain U.S. citizenship, and hence you could not vote. You could not testify against a white person in a court of law on your own behalf, or own a gun to protect yourself. If you were accused of a crime, your case would be heard before an all-white jury—if it was not first resolved through vigilante justice. You had to yield your land to government or military authorities, or non-Indian squatters. If you were a woman, you could be forced into prostitution against your will or raped. If found loitering or strolling, you could be arrested for vagrancy and auctioned off to the highest Anglo bidder. Your children could be stolen and sold as slaves.[144]

THE COMMODIFICATION OF CALIFORNIA

As James Rawls and other historians have pointed out, the Gold Rush was an instrumental event in the economic history of California, setting the tone, mind-set, fervor, and conditions for the exploitation of other resources and the mistreatment of minorities.[145] Gold Rush immigrants were quick to realize that gold was just one of the resources that could make one wealthy; the state was full of animal, plant, mineral, and hydrological resources waiting to be tapped, brought to market, and transformed into commercial products. Thus the elements of the landscape that native Californians had so carefully integrated into their ways of life were transformed into commodities.

Miners' diaries and other primary sources from the Gold Rush era show that the newcomers tended to make three closely related assumptions about California's natural environment. First, they assumed that its diverse natural resources lay idle, untapped, and uncultivated by lazy Indians and Californios. Second, they thought nature's abundance and diversity were going to waste, and they had a God-given right to use them for profit. Third, they viewed the resources of California as inexhaustible.[146] Reflecting this last assumption, a German visiting the redwoods of Mendocino in 1852 commented, "As far as the eye could reach, a seemingly interminable forest lay at our feet . . . and yet only a part—a very small part—of the Mendocino Coast Range lay before us. It was sufficient to convince me that California will for centuries have virgin forests, perhaps to the end of Time!"[147]

Euro-Americans milked the land and sea for every product imaginable. Market hunters hunted and trapped the wildlife, loggers harvested the virgin timber, farmers plowed the fertile valley soil, and ranchers loosed their stock on the verdant fields and woodlands. Early entrepreneurs even commercialized hot springs and caves, collected natural asphaltum, and harvested salt from the Salton Sea.[148] J. D. Borthwick, a miner, wrote of the intensity of California life in 1851: "The every-day jog-trot of ordinary human existence was not fast enough pace for Californians in their impetuous pursuit of wealth."[149]

To support the appetites of the many thousands of miners, farmers, and other settlers, hunting became a growing and lucrative business. By 1851 the markets of California were well supplied with nearly every variety of flesh, fish, or fowl. Sandhill cranes *(Grus canadensis)* made a favorite Thanksgiving or Christmas dinner. Grizzly bear meat was on the menu at most California restaurants. The cottontail rabbit's flesh was reported to be "white and very good eating," and the jack rabbit's meat was pronounced by Englishmen to be "fully equal to the English hare." While taking a train from the Midwest to California in the 1870s one might feast on elk, antelope, bison, or grouse as commonly as beef.[150]

The greater white-fronted goose *(Anser albifrons)* was highly prized for food, and until its numbers declined it was a common bird on the market. Using a large-bore gun aimed at a thick flock, a hunter could bring down hundreds of birds with four shots. The lesser and cackling Canada geese, two subspecies of the Canada goose *(Branta canadensis)*, were often to be found in the market, sold under the name "brant." The canvasback duck *(Aythya valisineria)*, reputed to be the best tasting of ducks, could be found by the tens of thousands in the lakes and lagoons of southern California. The lucrative nature of hunting was demonstrated by a report by Borthwick that in 1851 the wild geese and ducks were so numerous around the shores of San Francisco Bay that a Frenchman had been able to kill $1,500 worth of game in two weeks.[151]

Even the islands off California were surveyed for commodities. The Farallons, a favored nesting site of gulls, murres, and, to a lesser extent, tufted puffins, became the harvesting grounds of the Farallon Egg Company. The preferred egg was that of the common murre *(Uria aalge)*. The company sold between 3 million and 4 million of these eggs to the San Francisco market between 1850 and 1856. According to James Hutchings, their eggs "[are] unaccountably large, for the size of the bird, and afford excellent food, being highly nutritive and palatable—whether boiled, roasted, poached, or in omelets." In the 1850s the sale of eggs of common murres from the Farallon Islands reached the value annually of between $100,000 and $200,000. Over the span of fifty years these birds furnished the market with "some twelve million dozen fresh eggs."[152]

Shore-based whaling, instituted in the 1820s, expanded to encompass gray, finback, humpback, and blue whales, as well as the increasingly rare sperm whale. The baleen from toothless whales was used to make umbrella ribs and to add body to the corsets made for women. The blubber of a blue whale might yield over 20 tons of oil—a finback, 10 tons. Sperm oil, obtained from an organ in the whale's head, was valued as a high-grade lubricant for machinery and as a dressing for leather. The oil also yielded a

waxy substance called spermaceti, which was used in making smokeless candles, ointments, and cosmetics such as skin creams. Ambergris, a substance
found in the intestines of sperm whales, was used to stabilize perfume.[153]

Just as the adjacent ocean was perceived as an unlimited supplier of com-
modifiable resources, the land was seen as an endlessly malleable source of
food and other products. This sentiment is apparent in the 1848 writings of
James Carson, a gold prospector:

> The tule marshes, about which much has been written[,] invite the
> planter to convert them into rice fields; they can be drained or flooded
> at pleasure for that purpose. Along the rivers and in the drained tule
> beds, hemp, flax and tobacco can be raised to an extent and perfection
> that would stand unparalleled. A gentleman from the southern states
> informed me that he had closely examined the soil of the Tulare Valley,
> and that from his observations, he felt assured that cotton and the sugar
> cane could be brought to high perfection any place within the plain. For
> the cultivation of the grape, California will contend with sunny France
> or Italy; and the whole of this valley could be made one vast vineyard
> and orchard. . . . The soil is rich and deep, and the bottoms are heavily
> timbered with oak of the best quality, and sufficient for all purposes
> of fencing, etc. In cultivating the lands on the east side of the valley,
> between the rivers, an apparent obstacle may arise from the want of
> timber. This scarcity can be easily remedied, from the inexhaustible
> supplies of the finest timber from the adjacent Sierra Nevada moun-
> tain, not only for agricultural purposes, but for plank or railroads.[154]

During the Gold Rush, trees became extremely valuable as fuel and construction material. Huge sugar and ponderosa pine trees that had provided
edible pine nuts for Indians were cut down for lumber to build the cabins
in the quickly assembled mining towns. Borthwick described the decima-
tion surrounding Placerville: "The number of bare stumps of what had once
been gigantic pine trees, dotted over the naked hill-sides surrounding the
town, showed how freely the ax had been used."[155] Bayard Taylor described
the clear-cutting in the hills surrounding Grass Valley: "The hills were
stripped of wood, except here and there a single pine, which stood like a mon-
umental obelisk amid the stump head-stones of its departed brethren."[156]
Within nineteen years most of the coast redwoods surrounding Eureka and
Arcata were gone—and Humboldt Bay was deforested.[157]

The Late Nineteenth Century (1860–1900)

Following the Gold Rush, the influx of white settlers steadily increased, especially with the coming of the four railroads that shortened the overland

trip West to three or four days.[158] Cities were built, the economy diversified, and the resource-extracting and agricultural activities that had marked the Gold Rush era—mining, logging, market hunting, ranching, and farming— continued to expand. These activities became both more extensive and more intensive, spreading over larger parts of the state and harnessing new technologies to more efficiently convert resources into more products for a rapidly expanding population.

DESTRUCTION OF REMAINING WILDLIFE POPULATIONS

In the late nineteenth century, trappers, market hunters, trophy hunters, and egg collectors, as well as farmers and ranchers protecting their crops or stock, continued their relentless exploitation of bird and mammal populations across the state. Because they had already caused precipitous declines in many species, such as grizzlies, pronghorn antelopes, and common murres, they shifted their attention to those species that remained relatively abundant. By the end of the century, many of these animals had also succumbed. Over and over, the story was the same: an animal population was overharvested until the numbers were too small to bring an economic return. Restraint, conservation, and sustainable yield never entered into the picture, as they had for the native people of California.

By the early 1900s, the numbers of marine mammals, wildfowl, elk, deer, bear, and other birds and mammals had been so drastically reduced that Joseph Grinnell would write: "Throughout California we had been forcibly impressed with the rapid depletion everywhere evident among the game birds and mammals."[159] In *The Destruction of California*, Ray Dasmann identifies the period between 1850 and 1910 as one of "massive faunal change."[160]

Snow geese, Ross's geese, and greater white-fronted geese, once numbering in the millions, were stalked with horses, and hunters mowed them down by the thousands. Farmers also set a price on the heads of these grass-eating birds, and they were slaughtered beyond any possible demand of the market. Great egrets were killed by the tens of thousands in a single season, and of the former abundance of the egret in California, Dawson said, "We have no clear conception." They were shot during the nesting season in order to secure the aigrettes (a magnificent train of feathers), which were used to adorn ladies' bonnets. Western grebes were killed by the tens of thousands in northern California and Oregon, shot or clubbed for their glistening white, watertight feathers, which made excellent muffs, stoles, and capes. Those that remained at the beginning of the twentieth century were a "mere centesimal of the former hosts."[161]

In the 1880s and 1890s, millions of quail were shot or trapped for the markets in various cities. During the open quail-shooting season of 1895–96, for example, 177,366 quail were sold in the open markets of Los Angeles and San Francisco alone.[162] Valley quail populations were also diminished by overshooting, trapping, and poisoning from grain impregnated with thallium, a metallic poison that was scattered far and wide to poison ground squirrels.[163]

Eggs from the common murre dwindled from 41,000 dozen in 1854 to 25,000 dozen per year in the period 1864–73 to only 7,645 dozen per year by 1896. In 1897 a law was passed to prohibit gathering eggs for sale. William Dawson said in 1923: "The infamous egg traffic is a thing of the past, but the Farallon rookeries have never recovered."[164]

The bighorn sheep, thought by John Muir to be safe from man's harm in its high rocky outposts, was wiped out over much of its range. The sheep were sought by stockmen and miners for meat and trophies and succumbed to epidemics brought in by domestic sheep as early as the 1870s.[165] In 1878 the California legislature placed a four-year moratorium on hunting bighorn sheep and in 1883 extended the moratorium indefinitely. The majority of bighorn sheep disappeared by 1900, felled by diseases transmitted by domestic livestock and illegal hunting.[166] Elk in turn were wantonly destroyed: "Elk were here in great numbers. . . . They were mercilessly killed by hunters, killed not for their flesh, but for the fun of the killing." William Brewer wrote of elk, "All are now exterminated, but we find their horns by the hundred." By 1904 only three bands of tule elk, a mere 145 individuals, were left, living within the confines of the Miller and Lux Ranch near Bakersfield. Pronghorn antelope, described by Brewer in 1851 as "very rapidly disappearing," declined to almost none.[167]

Abalone populations were intensely harvested by the Chinese, and by 1879 annual catches were in excess of four million pounds. As stocks of black and green abalone were depleted in southern California, fishermen moved north to San Luis Obispo County. In 1900, due to population declines, county ordinances were passed that made it illegal to gather abalone from less than twenty feet of water.[168]

Both northern and Guadaloupe fur seals continued to be hunted for their luxuriant fur by commercial sealers on islands off western Mexico and California. The numbers of both species dropped precipitously, and the Guadaloupe species was presumed extinct until its rediscovery in 1954. Sea lions were killed by the thousands in the 1860s and 1870s for their oil. Sea otters were hunted so persistently that they were nearly extirpated, the California population being about fifty otters in 1911.[169]

Predators that posed a threat to ranchers' stock and birds that ate farmers' crops continued to be the targets of extermination campaigns. Shepherds in the Sierra Nevada carried strychnine to kill coyotes, bears, and mountain lions. Henry Henshaw wrote of his field experiences in 1875: "Perhaps few animals have suffered more from persistent and relentless warfare waged by man than this formidable bear [grizzly]. . . . A supply of strychnine is part of the outfit of every shepherd, and by means of this the number of bears is each year diminished, till in many sections where formerly they were very abundant they have entirely disappeared."[170]

To protect their apricot and fig crops, orchardists sometimes put out drinking fountains filled with arsenic-tainted water to kill finches, grosbeaks, and orioles. Poisoned grain was also widely distributed to curtail the huge flocks of blackbirds that frequented rice fields.

EXPANSION OF FARMING AND RANCHING

During the latter part of the nineteenth century, farming and ranching expanded, and by the 1870s "more men made their living in the broader geography and economy of farming—48,000—than in all the mines of the Sierra foothills—36,000."[171] To accommodate the acreage devoted to growing crops, marshes were drained, underground water was tapped by artesian wells, streams and rivers were dammed and diverted for irrigation, and lands were fenced. In the process, huge tracts of former native grasslands, riparian corridors, and vernal pools were converted to artificial, human-managed agricultural systems.[172]

The historian Alfred Crosby emphasizes that the successful exploitation of the New World, particularly California, depended on the settlers' ability to "Europeanize the flora and fauna of the New World," and successful they were. By the mid- to late 1800s, much of the Central Valley was devoted to mechanized, monocultural farming operations with ties to worldwide markets. More than 11 million of California's 100 million acres were in cultivation by 1870; over 28 million by 1900.[173] As agriculture became industrialized, machines were brought in to plow the land and harvest the wheat and other crops. Steam plows were introduced about 1871, and by 1881 they could turn a quarter section of land in a day. Steam-powered harvesters could move quickly across vast expanses of ground.[174] (See Figure 8.)

Van Dyke in 1886 explained this industrialization of agriculture:

> The labor, money, and anxiety of the whole year were staked upon the success of some one thing, such as wool, wheat, honey, oranges, or something else. Not only this, but as it was done to get rich, it, of course, had

Figure 8. *Harvest Time,* painted by William Hahn in 1875, depicts the harvesting and threshing of wheat. Taming the wilderness of California involved the large-scale transformation of fields of native grasses and wildflowers into grain fields. The soils were so fertile that a wheat farmer could become rich with eight good harvests. Courtesy of the Fine Arts Museums of San Francisco, gift of Mrs. Harold R. McKinnon and Mrs. Harry L. Brown, #1962.21.

> to be carried on upon the largest possible scale. And this, of course, could not be done without hiring labor and buying machinery; both of which were very expensive at this distance from manufacturing centers.[175]

Monocultures of wheat and other crops were hard on the land's ecology. Conservation-minded farmers such as Peter Shields of Davisville (now Davis) and John Bidwell of Chico realized early on that monoculture was depleting the rich soils of the Central Valley. Farming the modern way—to satisfy global rather than simply local markets—exhausted the soil, lowered the water table, and salinized the land, in great contrast to both Native Californian and early European methods of farming.[176]

Dasmann estimated the grasslands of California to be about 22 million acres at the point of Euro-American contact—about one-fifth of the total landmass of the state. Where the grasslands were not plowed under for crops, they were grazed. Domestic livestock—hogs, cattle, sheep, horses, and goats—replaced the native herbivores—tule elk, pronghorn antelope, and mule deer.

By 1862 cattle boomed to more than 3 million head. The number of sheep increased to a peak of 5.5 million head by 1875.[177]

There were productive, expansive grasslands not only along the coasts, in the Central Valley, and in montane meadows but also in the open understories of many woodlands and forests throughout the state that had been perpetuated by Indian burning. Many of the nutritious native grasses and forbs originally used by the Indians for food (e.g., *Bromus carinatus*, *Elymus glaucus*, many *Trifolium* spp.) provided substantial forage value to domestic livestock and sheep in the herbaceous layers under trees. The U.S. Forest Service estimated in 1936 that over half of the original 21.5 million acres of conifer forest land in California was grazeable (11 million acres), as were 8 million of the 10.6 million acres of original oak woodlands.[178]

With the grazing of livestock, exotic grasses and wildflowers were brought in both deliberately, through direct seeding for livestock feed, and unintentionally, through seeds carried in hay bales, folds of textiles, hooves of livestock, and a thousand other means.[179] The resulting competition from non-native plants, along with overstocking, periodic drought, and the lack of any kind of regulation, resulted in the rapid deterioration of many of the rangelands.[180] In an 1876 U.S. Geographical Survey report, J. T. Rothrock documented overgrazing in many parts of southern California, including the area around Cassitas Pass, the western edge of the desert near Lake Elizabeth, the area near Fort Tejon, and even Santa Cruz Island: "On the grounds most visited by the herds of sheep, all vegetation, save sage-brush, cactees and the erodium or storksbill, had been entirely swept away. . . . It is impossible to conceive a more dreary waste than was here produced as the result of over-pasturage."[181] L. T. Burcham described the results of overstocking the open range as "drastic weakening of the plant cover, trampling and compaction of the soil, and increase in species of lesser value for forage— depletion of the grazing capacity and inestimable damage to the forage resources."[182]

Many prominent ranchers admitted to an overstocked range. John Bidwell said in 1866, "It cannot have escaped the observation of those engaged in rearing stock in California that the indigenous grasses, once so abundant as to pasture thousands of animals where only hundreds are able to subsist now, are fast disappearing from the plains. This is attributable no doubt to excessive grazing, especially by sheep and horses."[183] Israel Russell, a geologist who traveled to Mono Lake in 1881, also recorded the effects of overgrazing: "There was formerly sufficient wild grass in many portions of the basin to support considerable numbers of cattle and sheep; but, owing to overstocking, these natural pastures are now nearly ruined."[184] By the end of the nineteenth century, the conversion of much of the state's perennial bunchgrass grassland to annual, exotic grasses was complete.

HYDRAULIC MINING

Hydraulic mining began during the Gold Rush, in about 1853, and rapidly became the preferred method for recovering gold. It involved directing forceful jets of water at hillsides containing gold-bearing alluvium, deposited long ago when the streams draining the early Sierra Nevada had different courses. To provide the high-pressure water required to wash away mountainsides, vast networks of reservoirs, flumes, ditches, and iron pipes were built to carry billions of gallons of Sierran water to the giant nozzles. The crumbled masses of earth were washed in sluice boxes, where a series of riffles caught the gold. The waste mud and gravel flowed into nearby creeks or ravines. Whereas gold panning, cradling, and sluicing had earlier undermined the major salmon runs, hydraulic mining had even more devastating environmental effects.

Most of the hydraulic mining occurred in the watersheds of the Feather, Yuba, Bear, and American Rivers. The discharge of debris was so great each year that it forced rivers out of their banks, and major flooding occurred in the lower elevations in wintertime. These floods swept away farm structures, drowned cattle, and wiped out orchards—driving hundreds of farm families to other regions. Large acreages of bottomland became covered with rocks, gravel, and mud. In 1878, farmers rallying in defense of their properties organized the Anti-Debris Association of the Sacramento Valley to act against the ravages of the mining debris. In 1884 the Sawyer decision gave them an important legal victory. In the 1890s, with new injunctions and tougher enforcement, the last of the hydraulic mining operations folded.[185]

LOGGING

Between 1860 and 1900 more trees were cut in California than had been cut in the one hundred years preceding 1860. William Alverson and colleagues have labeled the logging in this era "rapacious," "dominated by a 'cut-and-take' ethos that afforded little concern for environmental values or the sustainability of forest growth. Lands often were abandoned after cutting, slash fires raged, and no reseeding or other active efforts at regeneration were made. This stage extended into the early twentieth century in many areas."[186]

Logging occurred everywhere in the state. In the Coast Ranges, tan oaks (*Lithocarpus densiflorus*), growing in association with coast redwoods, were used for firewood and to fuel the kilns in which quarried limestone was heated to extract the lime used to construct buildings and foundations. The bark of these trees was also highly valued for tanning heavy leathers. Trees were girdled and killed as their bark was stripped off to supply regional and even overseas tanneries. Coast redwoods *(Sequoia sempervirens)* were cut for cradles, pencils, packing sawdust, tanning vats, railway cars, railroad ties, fences, shingles, bridges, and buildings.[187]

Valley and blue oaks *(Quercus lobata* and *Q. douglasii)* were harvested to fuel the state's steam-powered locomotives. Canyon live oaks *(Quercus chrysolepis)* were used for ships' knees, wagon wheels and axles, and wedges for splitting redwood. Interior and coast live oaks *(Quercus wislizenii* and *Q. agrifolia)* and valley oaks fueled blacksmiths' ovens, smelters that extracted minerals from the earth, and ovens in Los Angeles bakeries. Western sycamores *(Platanus racemosa)* were highly regarded for making furniture.[188]

Giant sequoias *(Sequoiadendron giganteum)* were cut for fence posts and palings, shingle bolts, roof shakes, split boards, and grape stakes. The oldest, broadest, and tallest trees were singled out and cut. During the late nineteenth and early twentieth century, thousands of giant sequoias were felled with axes in what Hank Johnston described as "the greatest orgy of destructive lumbering in the history of the world." The challenge of cutting the grandest giant sequoias was irresistible—and frequently the trees were so massive that they shattered into fragments as they hit the ground. Less than half the timber cut was actually used. Sugar pines *(Pinus lambertiana)* were felled for such diverse uses as construction, fruit boxes, door blinds, piano keys, and apiary supplies.[189] By 1900, 40 percent of California's 31 million acres of old-growth forest had been logged.

"CIVILIZING" THE INDIANS

The Indian population continued to decline as a result of disease, malnutrition, murder, and war. With whites still seeing Indians as a menace, the latter were subjected to three overlapping strategies: assimilation into white society, forced removal to reservations, and calculated genocide. (See Figure 9.)

During the late nineteenth century, it became increasingly difficult for Indians to maintain an existence outside of white society, and at the same time the need for cheap labor increased in all areas of economic activity. These two factors resulted in many Indians becoming relatively assimilated into white society and losing much of their native identity and culture. This process occurred with varying degrees of compulsion, but the result was

1. *Wahla*, chief of the Yuba tribe,—civilized and employed by Mr. S. Brannan. 2. A partly civilized Indian. 3. A wild Indian.—From daguerreotypes by Mr. W. Shew.

Figure 9. The supposed stages of transformation from wild to civilized were vividly portrayed in this lithograph that appeared in *Annals of San Francisco* (1855) by Frank Soule, John Gihon, and James Nisbet. Such views of Indians justified both their assimilation and their murder.

that many Indians dressed in European fashion, ate domesticated foods, practiced Western religion, took Christian names, learned English, lost their own languages, labored at Western jobs according to Western time schedules, and attended Western schools or oppressive Indian boarding schools.[190] Assimilation was an extremely tough road for California Indians because of rampant prejudice against them. Thus cooperation with whites often meant a harsh life, full of uncertainties, poverty, and violations of their human rights.

Many Indians were hired by whites to work as general farm laborers, herdsmen, grain harvesters, fruit pickers, and domestic servants.[191] Many did seasonal farm and fruit work, moving from crop to crop and returning to a winter home near where they had lived before whites came. Frequently, they were not paid fairly for their work.[192] In many employment situations, the household arrangements discouraged Indian marriage and child rearing. The historian Albert Hurtado has noted that "the scarcity of potential mothers was a severe problem for California's Indian population, already under stress and in rapid, prolonged decline." The Indians working on Bidwell's Rancho Chico in 1860, for example, were segregated by sex: thirty-nine male herders, gardeners, and farm laborers lived in three households; eleven women day laborers lived in another dwelling. "[T]he Indians at Rancho Chico," writes Hurtado, "were there to work, not to raise families or sustain tribal populations."[193]

Since the need for Indian labor was seasonal, many people were forced to depend on hunting, gathering, and fishing for a food supply during the winter.[194] J. Ross Browne, confidential government agent for the U.S. Treasury Department, describes the tenuous and unfair working conditions of California Indians in northern California in 1877:

> In general, they [the settlers] engaged them [the Indians] at a fixed rate of wages to cultivate the ground, and during the season of labor fed them on beans and gave them a blanket or a shirt each; after which, when the harvest was secured, the account was considered squared, and the Indians were driven off to forage in the woods for themselves and families during the winter. Starvation usually wound up [killing] a considerable number of the old and decrepit ones every season; and of those that failed to perish from hunger or exposure, some were killed on the general principle that they must have subsisted by stealing cattle, for it was well known that cattle ranged in the vicinity; while others were not unfrequently slaughtered by their employers for helping themselves to the refuse portions of the crop which had been left in the ground. It may be said that these were exceptions to the general rule; but if ever an Indian was fully and honestly paid for his labor by a white settler, it was not my luck to hear of it. Certainly, it could not have been of frequent occurrence.[195]

Although many Indians entered employment relationships with whites voluntarily (if only out of desperation), others were compelled to labor as indentured servants. In 1860 an amendment to an 1850 law legalized long periods of forced indenture to whites. A young Indian child could be held by a white family until he or she reached the age of 25; a teenager of 14 could be indentured until the age of 30.[196]

The amendment was lauded in an article in the 1861 *San Francisco Bulletin:* "This law works beautifully. A few days ago V. E. Geiger, formerly Indian Agent, had some eighty apprenticed to him, and proposed to emigrate to Washoe with them as soon as he can cross the mountains. We hear of many others who are having them bound in numbers to suit. What a pity the provisions of the law are not extended to greasers, Kanakas, and Asiatics. It would be so convenient to carry on a farm or mine, when all the hard and dirty work is performed by apprentices!"[197]

With such laws on the books, Indian children and women were openly kidnapped and sold into service in different parts of California. Children could bring as much as $300 each in some communities. Such tragedies are documented in dozens of newspaper accounts and hundreds of letters and reports, such as this letter from Second Lieutenant of the United States Army Edward Dillon to his captain in 1861:

> I have the honor to report that there are several parties of citizens now engaged in stealing or taking by force Indian children from the district in which I have been ordered to operate against the Indians. I am reliably informed that as many as forty or fifty Indian children have been taken through Long Valley within the last few months and sold both in and out of the county. . . . It is needless to say that this brutal trade is calculated to produce retaliatory depredations on the part of the Indians and exasperate them to a high degree.[198]

In 1893 Helen Carpenter wrote in an article in the *Overland Monthly* magazine that the kidnapping and selling of Indian children was "so prevalent, that at one time there were few families in town [Ukiah] that did not have from one to three Indian children." Children would be put to work washing dishes, cooking, cleaning, rocking white babies to sleep. Many whites believed that these Indian children were "much better off" than they would have been with their own biological parents.[199]

Another fate of California Indian children was to be taken from their families by government officials and sent to Indian boarding schools, where they were forcibly assimilated into mainstream society. Indian boarding schools were established at the Tule River Reservation in 1881, Round Valley in 1883, Middletown in Lake County in 1885, the Hoopa Valley Reservation and Perris in 1893, and Fort Bidwell in 1898.[200] The philosophy behind this policy was to place "the savage-born infant into the surroundings of civilization" so that the Indian would "grow to possess a civilized language and habit."[201] Charles Lummis characterized government-run Indian boarding schools as aiming to "destroy the family ties" among the indigenous populations.[202]

Margaret Archuleta and colleagues sum up the devastating effect of these schools on native cultures: "Indian boarding schools were key components in the process of cultural genocide against Native cultures, and were designed to physically, ideologically, and emotionally remove Indian children from their families, homes, and tribal affiliations. From the moment students arrived at school, they could not 'be Indian' in any way—culturally, artistically, spiritually, or linguistically."[203]

REMOVING INDIANS TO RESERVATIONS

As white settlers pressed into all corners of the state, all but eliminating Indians' ability to subsist in the traditional manner and increasing the possibilities of conflict, removing Indians to reservations became a preferred method of dealing with the "Indian menace." Indians continued to be forced by volunteer militias or the U.S. Army to move to a handful of military reservations—Tejon, Fresno Farm, Nome Lackee, Mendocino, and Klamath— where they joined Indians from a variety of other tribes. "Those lucky enough to have reservations established in the aboriginal territories," writes Ed Castillo, "were understandably reluctant to share the same advantages they enjoyed with newly arrived emigré tribes."[204]

Tribes from other regions, for their part, were passionately against relocating to a strange land and leaving their homelands behind.[205] The Modoc of northeastern California, for example, were forced onto the Klamath Reservation in southern Oregon. Klamath men, who resented the fact that the Modoc were living on Klamath land, harassed them, stole their fence rails and logs, confiscated their fish, and beat the Modoc women gathering wocas seed. One Indian agent, a man named Knapp, did nothing after several complaints, and the Modoc moved back down to near Tule Lake in 1870. Another attempt by troops from Fort Klamath to deport the Modocs back to Klamath resulted in the Modoc War in 1872, the final armed resistance by California Indians.[206]

Corrupt agents continued to administer the reservations. Reservation horse teams, farming implements, and food supplies were often seized by white settlers. Ranchers continued to let their livestock onto the reservations to feed on the Indians' crops. In 1862 an agent wrote from one reservation: "The settlers have succeeded in destroying a large portion of the small grain, and the corn crop entirely. The corners of the fence had been raised and chunks of wood put in, so that the largest hogs could walk in. Where they had destroyed the crops, the Indians were told that there was nothing for them to eat, and that they would have to starve or steal, and that if they did not leave they (the settlers) would kill them."[207]

Once again unable to make a living on the land, the Indians were forced to steal hogs and cattle for food outside the reservation boundaries. More and more white settlers surrounding the reservations wanted the Indians removed and the land opened to white settlement. In 1859, for example,

Tehama residents signed a petition demanding that the secretary of the interior move the Nome Lackee reservation "beyond the pale of [their] thickly settled districts."[208]

As a way to appease the white settlers, Congress passed the "Act to Provide for the Better Organization of Indian Affairs in California" in 1864. This act established three tracts of land for new reservations that were more remote from the white settlements. The reservations were created over the next decade and included Hoopa Valley (1864), Round Valley (1864), and Tule River (1874). These civilian reservations, established by federal statute and executive order, replaced the five military reservations set aside for California Indians.[209] The Indians of Fresno Farm and Tejon Valley were forced to relocate to the Tule River Reservation; those at Nome Lackee and Mendocino went to Round Valley; and some northern California tribes were forced to relocate to the Hoopa Reservation. To a greater extent than before, diverse tribes speaking very different languages were forced to live together.[210]

Finally, Indians whose homelands had been in the areas colonized by the Spanish missionaries were recognized by the U.S. government. An act passed in 1891 and amended in 1898 allotted reservations "for each band or village of Mission Indians residing within said State, which reservation shall include, as far as practicable, the lands and villages that have been in the actual occupation and possession of said Indians."[211] This led to the establishment of fourteen southern California Indian reservations. However, nine-tenths of the land on these reservations was not arable, and white settlers adjacent to reservation lands often diverted streams for their own use.[212]

Outside of these reservations, very few Indians had land of their own. Until 1875 no Indian in California was allowed to homestead, and after 1875 Indians could homestead on the condition that they renounce their tribal affiliation. There were also efforts to break up Indian landholdings. In 1887 Congress passed the General Allotment Act, also known as the Dawes Act, which enabled reservation lands to be divided into individual parcels of 160 acres for an Indian family and 80 acres for each single Indian man. The surface intent of the act was to teach Indians the white man's ideas about owning property and farming. But the actual result was to allow whites to buy Indian land.[213]

EXTERMINATING INDIANS

An increasingly desperate Indian population and an expanding white population created a volatile combination in the second half of the nineteenth century. The common but isolated violence against Indians during the Gold Rush escalated into what Rawls has called a "program of genocide," or "extermination," which resulted in the deaths of thousands of California Indians.[214]

The Euro-American view of Indians as "a different order of man" not only justified their exploitation and annihilation but also led whites to manufacture various self-serving explanations for any Indian behavior that seemed threatening.[215] The self-defense measures undertaken by Indians, which escalated to include the theft of livestock and attacks on white settlements, were seen by whites as evil and uncivilized, as demonstrating a lack of respect for property or even jealousy. For example, when white settlers chopped down large quantities of pinyon pine near Mono Lake in 1877 for the mining companies of Bodie, depriving the Paiute of an extremely important food resource, the settlers could not understand why the local Indians made death threats: "What has given rise to this sentiment, people are at a loss to conceive, unless it be the chronic jealousy of the redskins on beholding the rapid growth of the white population, and contrasting the difference between their own debased condition with the continually increasing comforts and luxuries of the white settlers."[216]

Instances of Indian thievery and attacks on whites became an excuse for many settlers to wantonly murder any Indians anywhere. Massacres of California Indians—who were clearly outnumbered, outarmed by sophisticated weapons, and unprotected by county, state, or federal laws—became common events. The *San Francisco Bulletin* reported in 1860 that settlers in Mendocino county had killed 32 Indians, because a Mr. Woodman had lost 100 head of horses, 74 of which were found dead not far from his home, and upon which the Indians were having a feast.[217] Major G. J. Raines at Fort Humboldt wrote a letter to the assistant adjutant general in 1860 describing a massacre of about 188 Indians at Indian Island:

> Volunteers, calling themselves such, from Eel River, had employed the earlier part of the day in murdering all the women and children of the above Island and I repaired to the place, but the villains—some 5 in number—had gone, and midst the bitter grief of parents and fathers—many of whom had returned—I beheld a spectacle of horror, of unexampled description—babes, with brains oozing out of their skulls, cut and hacked with axes, and squaws exhibiting the most frightful wounds in death which imagination can paint—and this done . . . without cause, otherwise, as far as I can learn, as I have not heard of any of them losing life or cattle by the Indians. Certainly not these Indians, for they lived on an Island and nobody accuses them.[218]

A majority report to the California legislature in 1860 stated: "Accounts are daily coming in from the counties on the Coast Range, of sickening atroc-

ities and wholesale slaughters of great numbers of defenseless Indians in
that region of country. Within the last four months, more Indians have been
killed by our people than during the century of Spanish and Mexican dom-
ination."[219] D. R. Brown summarized the situation well when he wrote in
1876:

> They [Indians] were driven from their hunting-grounds and fishing-
> places; the result was that they stole cattle for food, and the whites
> punished them for this by the sharp law of the rifle. The end of this
> is, that at this day the Indians throughout California, with a few ex-
> ceptions, are used in the most unjustifiable and brutal manner by
> the whites—buffeted, robbed, and ill-used on any or no provocation,
> butchered, often with the most abominable cruelty, by men hardly
> worthy of the name, and even without the excuse of self defence, the
> Indians being under their protection at the time.[220]

INDIAN RESISTANCE AND CREATIVE ADAPTATION

Many native Californians did not passively stand by but protested both vi-
olently and nonviolently against their treatment. Indigenous people showed
their discontent with government schools by destroying the day school at
Potrero in 1888 and burning down the school at Tule River in 1890. The
Luiseño burned down the school and killed the teacher in 1895. Some groups
actively resisted the allotment policies of the government, such as the Kumeyaay at Mesa Grande, who warned allotment agents not to set foot on
the reservations or death might be the consequence.[221]

The Yurok, Karuk, Hupa, and other tribes resisted relocation to reservations and retaliated for the destruction of their villages and the kidnapping
of their children. They organized violent uprisings that led to a number of
Indian wars in northwestern California between 1860 and 1872.[222] Other
Indians, such as the Cupeños from Warner Springs, chose to seek legal title
to their lands through the courts. With the aid of the Indian Rights Association, they began a suit to stop their eviction from their ancestral village
of Cupa, renamed Warner Hot Springs. The case went to the U.S. Supreme
Court. Although they lost the case in 1903, they were able to gain lands on
the nearby Pala Indian Reservation in San Diego County.

Still other Indians chose to purchase land that was once theirs and reside
on it. The Yokayo Pomo collected almost $1,000 from their people and pur-
chased a 120-acre site near the Russian River, on which they thrived.[223]
However, not every transaction was just. On March 7, 1904, the *San Francisco Chronicle* reported that Indians who bought land from whites were
being dispossessed by the heirs of the granters, who gave no valid titles. "The

Northern California Indian Association reported that about 10,000 Indians lived on land to which whites hold title. They were subject to eviction 'at any time.' "

ROMANTICIZING NATURE AND THE NOBLE SAVAGE

The perception that California was a primeval wilderness free of human influence has its roots in this period of California history. By the eighteenth century, wilderness areas in Europe had come to be viewed as places for self-renewal, where one could escape the hectic, burdensome life of the cities for the tranquility and purity of nature. The splendor and nobility of nature had become linked with God's creative energies and omnipotence. Coupled with this favorable view of wilderness was the idea of the noble savage—a kind of wild man uncorrupted by the vices of civilized life—who lived a simple, harmonious, unfettered existence in nature.[224]

During the nineteenth century, this view of nature, a central part of the Romantic movement, began to receive attention from American writers and philosophers such as Henry David Thoreau, Ralph Waldo Emerson, and Walt Whitman. Many of the late-nineteenth-century Americans who would figure prominently in the state's history, including John Muir, were strongly influenced by Romanticism and its proponents. Muir and those with similar views responded to the destruction and exploitation of California's natural resources with a preservationist ethic that valued nature above all else but which defined nature as that which was free of human influence. Thus while he championed the setting aside of parks as public land, Muir also contributed to the modern notion that the indigenous inhabitants of the state had no role in shaping its natural attributes.

Muir was clearly troubled by the Indians he encountered, unable to fit them into his worldview. (See Figure 10.) He wanted them to be natural, like animals, but they disappointed him by showing some of the qualities he disliked in his fellow whites. He wrote the following account of Miwok Indians in the Sierra Nevada in 1869:

> We had another visitor from Brown's Flat to-day, an old Indian woman
> with a basket on her back. Her dress was calico rags, far from clean.
> In every way she seemed sadly unlike Nature's neat well-dressed ani-
> mals, though living like them on the bounty of wilderness. Strange
> that mankind alone is dirty. Had she been clad in fur, or cloth woven
> of grass or shreddy bark, like the juniper or libocedrus mats, she might
> have seemed a rightful part of wilderness; like a good wolf at least, or
> bear. But from no point of view that I have found are such debased

Figure 10. John Muir among the Northern Paiute, 1903. Muir found the California Indians a disappointment because they did not meet his standards of nobility but instead showed signs of being a "fallen," poverty-stricken people. The idealized image of the "noble Red Man" living freely on the bounty of the wilderness belied the much more complex truth that California Indians were skilled harvesters and sophisticated wildland managers. Muir and other preservationists deemed much of California a virgin, uninhabited wilderness, thus missing the connection between nature's abundance and the stewardship role of indigenous peoples. Courtesy of the Bancroft Library, University of California, Berkeley, #032080.

fellow beings a whit more natural than the glaring tailored tourists we saw that frightened the birds and the squirrels.[225]

Muir longed to experience the wildness of raw nature; the utility of the biota did not interest him. "Just bread and water and delightful toil is all I need," he said during his first summer in the Sierra Nevada, "not unreasonably much, yet one ought to be trained and tempered to enjoy life in these brave wilds in full independence of any particular kind of nourishment." Yet more than once, he faced starvation when food supplies from civilization did not arrive on time to his shepherd outpost in the Sierra Nevada wilderness. A sentiment that would trouble him constantly was the oddness of feeling "food-poor in so rich a wilderness."[226] Muir showed admiration of the native peoples' knowledge of ethnobotany, but he preferred the society of woodchucks and squirrels to that of Indians and never seriously considered the possibilities for wilderness living offered by California Indian ways.[227]

Muir's view of California nature was a necessary counterweight to the view that had prevailed before—that nature was there to be used, exploited, and commodified—but it left us with a schizophrenic approach to the natural world: humans either conquer nature and destroy its integrity, or they visit it as an outsider, idealizing its beauty and largely leaving it alone. These seemingly contradictory attitudes—to idealize nature or commodify it—are really two sides of the same coin, what the restoration ecologist William Jordan terms the "coin of alienation" (pers. comm. 2002). Both positions treat nature as an abstraction—separate from humans and not understood, not real.

Early Twentieth Century (1900–1930s)

For California's native people and natural environment, the turn of the century brought grim realities as well as glimmers of hope. The California Indian population reached its low point around 1900. Then through the early years of the twentieth century it began a gradual rebound, as the plight of Indians was recognized and they were able to regain some of their human rights. The population declines of many wildlife species also bottomed out at about this time—some neared extinction—and conservation laws and bag limits began to be enacted to save them. With destructive logging, overgrazing, and conversion of land to agriculture continuing unabated, growing concern about the loss of resources important to the public welfare spawned a conservation movement that for the first time put the federal and state governments in the role of land stewards. Although the conservation movement resulted in significant areas of the state being set aside as public land, it also fostered the concept of wilderness and the policy of wildfire suppression, both of which overlooked the historical role of Indians in managing the landscape and ignored the needs of contemporary Indians.

INDIAN LIVING CONDITIONS

The California Indian population in 1900, according to estimates, was a mere 15,500.[228] Castillo notes that this figure "represents a gut wrenching descent from over 300,000 into a vortex of massive death in just 131 years of colonization." "Every Indian who survived to see the dawn of the 20th century," writes Castillo, "had witnessed great suffering and the irreplaceable loss of numerous grandparents, mothers, fathers, and children. Some lineages disappeared altogether. The nadir had been reached."[229]

Many of the 15,500 survivors of genocide and persecution lived a miserable existence. Landless and poor, lacking health care, treated as less than human, deprived of their native culture but not accepted into mainstream white culture, they suffered malnutrition, disease, and unemployment. In 1919 the Indian Board of Co-operation wrote about the conditions of elderly and sick California Indians:

> Generally throughout California the indigent sick or aged Indian receives no aid from the State; there are no Indians admitted to the alms houses or to the county hospitals, although there are many instances of pitiable poverty among the aged and cases of curable diseases which go untreated. Children die of tubercular spines and hips and many become blind from trachoma and conjunctivitis. The aged eke out a miserable existence, half clad and half starved, dependent upon the precarious help of their own poverty-stricken race or the intermittent assistance of charitable white people.[230]

Many California tribes, landless and lacking federal recognition as "tribes," were denied health care at the federal level. At the same time, they were denied health care by the state because they were Indians. Only those tribes that were federally recognized received services, assistance, and funds through the Bureau of Indian Affairs (BIA) and the Indian Health Service, and this was long in coming. It was not until 1924 that the Indian Service established a special division for health,[231] and although there were a few Indian hospitals in California (seven by 1930), the care they offered often was substandard. Moreover, the treatments did not address the root of many of the illnesses: dire poverty and lack of adequate nutrition. The federal government admitted that "even if they [government doctors] provided minimally satisfactory medical treatment they took too little interest in related aspects of Indian community life."[232]

The BIA, established in 1824, was charged with protecting American Indians, acting as trustee for lands and money belonging to Indians, and providing public services such as education and welfare aid. But between the turn of the century and the Great Depression, it interfered more and more with native life, and its policy of prohibiting California Indians from practicing their religions and other customs on reservations contributed to an effacement of Indian values and forced assimilation into Western society. In 1902 the BIA advised its reservation agents: "You are therefore directed to induce your male Indians to cut their hair, and both sexes to stop painting. . . . Noncompliance with this order may be made a reason for discharge or for witholding rations and supplies. . . . Indian dances and so-called Indian feasts should be prohibited. . . . Feasts are simple subterfuges to cover degrading acts and to disguise immoral purposes. . . . The government has a right to expect a proper observance of rules established for their good."[233] (See Figure 11.)

When the United States entered World War I in 1917, the federal government cut back Indian health, education, and welfare services on the reservations. Many California Indians were drafted or enlisted, though they were not U.S. citizens. In 1924 the Indian Citizenship Act was passed by Congress, granting citizenship to all noncitizen Indians born within the territorial limits of the United States. Tribal members did not lose their tribal citizenship by becoming American citizens.[234]

In 1926 the Institute for Government Research launched a study of Indian economic and social conditions that culminated in the Meriam Report of 1928. This report exposed the tremendous poverty, suffering, and discontent throughout Indian country. The authors found disease and malnutrition on reservations and documented an average life expectancy for Indians of only forty-four years. The average annual per capita income was $100. The report drew two conclusions: "(1) The BIA was inadequately meeting the needs of Indians, especially in the areas of health and education; and (2) Indians were being excluded from the management of their own affairs."[235]

Robert Spott, a Yurok man from the Klamath River, made a moving speech to the Commonwealth Club of California in June 1926:

> I am here to tell you that we are almost at the end of the road. My English is broken, but I will explain to you as near as I can. In the old time, away back, we had a place where we used to go and pick berries for our winter supply. Then, again, we had a hunting ground where we killed the game for our winter supply. And again, we had a place where we used to go to gather acorns for our winter supply. Then, again, we could go up along the river to where a fishing place was left for us. But today, when we go back to where we used to go for our berries, there is the sign, "Keep out." Then, again, we go to where we used to go to hunt. You see the sign, "Keep out. No shooting allowed." All right. We go away. Then, again, we go down to where we used to fish. That is taken up by white man. What are we going to do? We cannot do anything.[236]

INDIAN LAND POLICIES

In 1905 Congress passed the Indian Appropriation Act, which authorized a survey to evaluate the living conditions of the Indians of northern California and make recommendations for improvement. The results of this

Figure 11. At a time when most surviving California Indians had been forced to abandon their traditional lifeways, a small band of Yahis managed to remain hidden in the rugged foothills west of Mount Lassen. In 1911, Ishi, the last survivor of this band, close to death from exposure, hunger, and fright, came down to the outskirts of Oroville and was discovered by non-Indians. Ishi was brought to the University of California to live. There he demonstrated a wealth of skills to anthropologists and the public, including how to fashion arrowheads, manufacture string, make fire, build houses, and track game. His hunting techniques included imitating a rabbit call by pressing two fingers against his lips and making a kissing sound, as shown here. These examples of the rich body of knowledge about nature had been incorporated in the cultures of California Indians, but were disappearing. Courtesy of the Phoebe Apperson Hearst Museum of Anthropology and the Regents of the University of California. Photograph by Saxton Pope, 1914, #15–5814.

survey, along with the urging of citizens and government officials, convinced Congress to pass a series of appropriation acts that funded the purchase of small parcels of land for landless Indians in central and northern California.[237] By 1930 thirty-six rancherias spanning sixteen northern California counties had been set aside for California Indians. However, most of the rancherias were small (from five to several hundred acres) and unsuitable for agriculture and therefore could not support a significant number of Indians. Thus many Indians moved to towns or to farms to secure employment, supplementing their small incomes with hunting, gathering, fishing, trapping, and woodcutting.[238]

Meanwhile, non-Indians continued to steal reservation resources and dispossess California Indians from their lands. For example, in 1904 white men claimed that the valuable timber resources on the Tule River Reservation were under their ownership and began cutting timber illegally.[239] In 1916 El Tejon Land Company of Kern County filed suit to evict the El Tejon Indians from the El Tejon ranch, which the Indians owned under Spanish and Mexican laws that the United States had agreed to uphold under the Treaty of Guadalupe Hidalgo. The company had been renting the land from the Indians and employing Indians to work on it. They began to withhold rent payments, and the Indians, fearful of losing their jobs, dared not demand the rent be paid. After waiting the requisite number of years, the company claimed ownership of the land on the basis of peaceful possession.[240]

The allotment program authorized under the Dawes Act continued during the early twentieth century, and by 1930 about 2,300 allotments had been made throughout the state. Although a fair number of the allotments were cancelled because Congress never appropriated the money, those that were implemented probably did more harm than good. The allotment system caused divisive conflicts on the reservations, sometimes pitting family against family because it shifted tribal lands from communal holdings that benefited everyone to a checkerboard of land parcels owned by different individuals, some of whom were non-Indian.[241]

PLOWS, DAMS, SAWS, AND "HOOVED LOCUSTS"

During the early twentieth century, California's natural resources continued to be harnessed for economic growth, with predictable results for native plant and animal populations. Clearcut logging expanded into previously untouched forests, large areas of rich bottomland were converted into agricultural systems, and the numbers of livestock grazing the grasslands, woodlands, and deserts increased.

Figure 12. A day's catch of salmon on the Eel River by an early-twentieth-century commercial seining operation. Other streams that were affected by gold mining and diversions no longer supported such numbers of fish. Courtesy of Peter E. Palmquist.

Agriculture boomed. More and more acres of former native plant habitats were put to the plow, and landownership became increasingly concentrated, with huge operations worked by groups of seasonal laborers living in company towns replacing family farms. By the mid-1920s California had surpassed Iowa to become the nation's leading agricultural state. The boom was fueled by innovations in agronomy, new crop varieties, the development of more efficient farm machinery—and in particular the complete reshaping of California's hydrological systems to suit the needs of agriculture. Streams were diverted and dammed and aqueducts built to create extensive irrigation systems.[242] The dams and diversions "tamed" the waterways, but they also destroyed many remaining salmon runs and drastically reduced the habitat for waterfowl and riparian vegetation. (See Figure 12.)

The number of livestock continued to exceed the capacity of the range well into the early 1900s. One sign of further deterioration was the shift in vegetation from grasses and forbs to shrubs and the accompanying disappearance of culturally important plants. In 1932 the Surprise Valley Paiute in northeastern California told the anthropologist Isabel Kelly that many of the plants with edible roots could no longer be found. Kelly reported that the meadowland covering the valley floor in aboriginal times had been almost completely replaced by sagebrush and cultivated land.[243] Another sign of deterioration was the reduction of plant cover on the rangelands, which exposed the soil to erosion. Some areas were almost completely denuded of vegetation, accept for the occasional non-native weed.

To protect livestock in California and other western states, an extensive government campaign was begun in 1916 to trap and kill predatory animals. The gray and red wolves were entirely exterminated. Mountain lion populations declined as the cats were poisoned and hunted, practices encouraged by a state bounty put into place in 1907. Between 1907 and 1933, 6,990 mountain lions were killed in California.[244]

THE CONSERVATION MOVEMENT

In the early 1900s a "widespread turning to nature" had begun to grow among the urban middle class, as its members became aware that the natural resources of the state, thought limitless during the Gold Rush, were in fact very limited and even seriously threatened.[245] Many people joined the growing movement, whose roots were in the writings of Thoreau and Emerson, to protect and preserve nature for future generations.

For the most part, conservationists did not understand how California Indians, over centuries, had influenced the animals, plants, and landscapes and maintained the land's fertility. Some, such as John Marshall, referred to Native Americans as wise stewards and conservationists but failed to defend the Indians and insist that they be treated humanely. Stephen Fox believes that conservationists' interest in Indians was "manipulative and vicarious" because "few conservationists dealt with actual living Indians, or listened to their problems in a white man's world, or helped them maintain the old ways of which conservationists spoke so reverently."[246]

In contrast, Galen Clark, guardian of the Yosemite Grant for many years, had considerable firsthand experience with the Sierra Miwok and their ways. He saw the link between Yosemite Valley's beauty and its stewardship by its longtime residents. His was a lone voice in pointing out that the classic landscape of Yosemite Valley, with large open meadows, California black oak groves, and open mixed conifer forests of ponderosa and sugar pine, was under the care and management of the Indians, probably for centuries, and he argued that non-Indians could learn much from the indigenous people. (See Figure 13.)

During this period, the conservation movement's advocacy for protection of remaining wildlands and wildlife populations helped to solidify a shift in public opinion and government policy that had begun in the 1890s. Increasingly, state and federal governments stepped in to manage the use of natural resources, enacting harvesting quotas, limits on grazing, and environmental laws, and setting aside lands for the public welfare.

Following up the passage of the National Park Protective Act in 1894, the National Park Service was set up in 1916, and the national parks were removed from military guardianship.[247] Congress had authorized the president in 1891 to set aside forest reserves in the public domain, and by 1905 the reserves had increased in size to 63 million acres. In 1907 Congress renamed the forest reserves the "national forests" and set up the U.S. Forest Service to manage them.[248]

The national forests were still subjected to commercial use, but that use was regulated. For example, in the 1930s the Forest Service recommended selective management to industry through a series of bulletins. Selective management curtailed clear-cutting and emphasized cutting in small openings to retain most of the forest cover.[249] Starting in the 1920s, the Forest Service also set aside areas within the national forests that were to be protected from road development, timber cutting, and other uses. They could be used for limited grazing and public hunting but not developed in any way; these were Wild Areas and Wilderness Areas. In 1927 the California Division of Beaches and Parks was established to administer areas set aside for recreational purposes, preservation of historical resources, and preservation of native vegetation.[250]

Livestock grazing began to be regulated on national forest land. Not only were stockmen no longer at liberty to graze the mountains without restriction, they had to pay for grazing permits. Will Barnes, in an article published in *Out West* magazine in 1908, wrote that the grazing regulations would entail meticulous calculation of the forests' stock-carrying capacity that would specify number of animal units per unit of land and thus prevent overgrazing and exclude from grazing special areas, such as fragile watershed areas.[251] The Taylor Grazing Act was passed in 1934 in order to "stop injury to the public lands by preventing overgrazing and soil deterioration, to provide for their orderly use, improvement, and development, and to stabilize the livestock industry, dependent on the public range."[252]

Many of the fur-bearing animals—beaver, marten, gray fox, sea otter,

Figure 13. Galen Clark, guardian of the Yosemite Grant, in front of Yosemite Falls. Clark was one of the few conservationists who tied preservation of wildlands to Indian management. Courtesy of the Yosemite Museum, Yosemite National Park. Photograph by George Fiske, #YM187.

mink, fisher—had become so reduced by the end of the century that laws began to be passed in an effort to save them from extinction. In 1911 a law was passed that protected the beaver, and in 1913 a ban was placed on the killing of sea otters. General fur-trapping regulations were enacted in 1917 that required a license to trap fur-bearing animals and instituted an open season during which an individual could legally trap.[253] In 1901 a law established a bag limit for quail and prohibited their sale. Had it not been for this law, wrote A. Starker Leopold, "the supply of quail in this state would have been totally exterminated." Greater sandhill cranes were nearly decimated by 1900, but with the protection received under the Federal Migratory Bird Treaty Act of 1918, they were able to hang on and recover.[254]

THE BEGINNINGS OF FIRE SUPPRESSION

Sheepherders and cattlemen learned about burning from the Indians or from their fathers and grandfathers, and up through the early 1900s they lit fires to keep meadows open and keep down the brush in the forest understories. These fires increased the numbers of palatable grasses and forbs for grazing animals. Timbermen also set fires in forest understories to reduce hazardous fuels and brush. Around the turn of the century, however, many Forest Service officials began to oppose deliberate burning and to argue for the suppression of all fires. They viewed lightning-caused and Indian-set fires as inherently destructive, with no ecological role in forest or grassland development and maintenance. It was thought that all fires threatened valuable tree resources, protective watershed cover, and wildlife.

A "light burning" versus "no burning" controversy began, pitting timbermen and ranchers against those pushing for fire suppression. The timbermen and ranchers argued that light, frequent burning was necessary to prevent catastrophic fires and was beneficial to timber and grazing habitat.[255] This view was supported by some government officials. For example, the acting superintendent of Yosemite National Park in 1903 advocated "a return to the old Indian custom of systematically burning over portions of the forests of the park each year in the autumn."[256]

The no-burn faction eventually prevailed. Conflagrations in Idaho and Montana in 1910, coupled with a light burn in California that got out of control and destroyed 33,000 acres that same year, were instrumental in spurring new legislation aimed at suppressing wildfires. The Weeks Act of 1911 provided for matching funds and agreements between the Forest Service and state foresters to expand fire protection on public and private lands. The Clarke-McNary Act of 1924 greatly broadened the cooperative fire protection program by giving major funding to the states for fire protection, with oversight by the U.S. Forest Service.[257] In 1935 the state and federal governments began a paramilitary-like program to quickly stamp out wildfires through the rapid mobilization of firefighters, equipment, and technology. This program was fairly successful for many years, because there was as yet only a modest buildup of fuels in most western forests.[258]

An Unsettled California

In the short span of sixty years, from 1850 to 1910, relatively few Europeans and Asians (about 379,000 in 1860, 864,000 in 1880, and 1,485,000 by 1900)[259] caused major declines in dozens of bird and mammal populations, denuded entire landscapes, caused major changes in the state's hydrological processes, vastly accelerated rates of erosion, destroyed countless acres of productive wildlife and plant habitats, radically altered much of the state's vegetation, and decimated Indian populations and destroyed their cultures. So successful were Europeans at transforming California into a version of Mediterranean Europe that the historian Alfred Crosby Jr. calls this period "the greatest biological revolution in the Americas since the end of the Pleistocene era."[260]

The rapid decline of animal and plant life spurred the enactment of numerous government laws, ordinances, and quotas designed to enforce restraint. Legally discouraging humans from overharvesting specific areas or species and making certain lands off-limits to most human activities were necessary compromises—lest every woodland be deforested, every stream dammed, every grassland denuded, every wetland drained. In more recent years, with the passing of the Wilderness Act, more and more land has been set aside and protected in areas designated as "wilderness."

Wilderness preservation, a concept perhaps unique to Western culture, is necessary given our burgeoning population, economic policies that do not recognize the finiteness of resources, and shrinking natural areas. But setting aside wilderness is only a reaction to the plundering of natural resources, and both spring from a mind-set of alienation from nature. Moreover, the wilderness concept tends to compartmentalize nature and culture, giving humans the illusion that activities done outside of protected areas will not affect what is within. The setting aside of wilderness areas has also been extremely divisive, setting many people against conservation initiatives in general.

Even with our wilderness areas, Californians today are separated from nature. California Indians would say that we are homeless. Seeing the lack of authentic attachment to place in the first non-Indian explorers, trappers, and prospectors, the Pit River people called them *enellaaduwi,* which literally means "wanderers."[261]

Perhaps indigenous people can offer us a way to really find our home in California by helping us to locate a middle ground between exploiting and degrading the environment on one hand and insulating it from all human influences on the other. By understanding the ways that native people established long-term relationships with the land and its creatures, we may be able to forge a new kind of relationship with nature, based on connection, that carries with it both obligations and privileges—what Californian Indians would call reciprocity.

Part II

INDIGENOUS LAND MANAGEMENT AND ITS ECOLOGICAL BASIS

CHAPTER 4

Methods of Caring for the Land

Take care of nature and it will take care of you.

WILLARD RHOADES, Pit River (1995)

Indigenous peoples have been pigeonholed by social scientists into one of two categories, "hunter-gatherer" or "agriculturist," obscuring the ancient role of many indigenous peoples as wildland managers and limiting their use of and impacts on nature to the two extremes of human intervention. The image evoked by the term *hunter-gatherer* is of a wanderer or nomad, plucking berries and pinching greens and living a hand-to-mouth existence; *agriculturist*, at the other extreme, refers to one who completely transforms wildland environments, saves and sows seed, and clears engulfing vegetation by means of fire and hand weeding. This dichotomous view of nature–human interactions has shut out the fact that Indian groups across California practiced many diverse approaches to land use, and it has led to a focus on domestication as the only way humans can influence plants and animals and shape natural environments.

Recently anthropologists have learned about the complexity of traditional ecological knowledge and the extent of indigenous peoples' management of wildlands by going to other parts of the world to study more intact cultures (e.g., Darrell Posey's work with the Kayapó Indians in the Brazilian Amazon and Henry Lewis's fire management work with Australian Aborigines).[1] But a reassessment of the record in California reveals that land management systems have been in place here for at least twelve thousand years— ample time to affect the evolutionary course of plant species and plant communities. These systems extend beyond the manipulation of plant populations for food. Traditional management systems have influenced the size, extent, pattern, structure, and composition of the flora and fauna within a multitude of vegetation types throughout the state. When the first Europeans visited California, therefore, they did not find in many places a pristine, virtually uninhabited wilderness but rather a carefully tended "garden" that was the result of thousands of years of selective harvesting, tilling, burning, pruning, sowing, weeding, and transplanting.

Much of the rich material disclosing the ancient management of wilderness lies in the dusty diaries and handwritten notes of anthropologists and the eyewitness accounts of early European settlers. For example, Kroeber's 1939 field notes, housed at the Bancroft Library at the University of California, Berkeley, record that the Yurok of northwestern California practiced burning at a frequency that was appropriate for each cultural purpose: burning of hazelnut for basketry occurred every two years; burning under the tan oaks to keep the brush down took place every three years; burning for elk feed occurred every fourth or fifth year; burning in the redwoods for brush and downed fuel control occurred every three to five years. These observations did not change his thinking about "hunter-gatherers," nor did he publish them. Other early anthropologists found examples of hunter-gatherers saving and sowing wild seeds, pruning wild crops, and managing wildlife and vegetation with fire.

Other rich sources of information about California's human past are tribal elders. Mono women still save jars and bags of tiny, edible seeds—collected long ago by their mothers—which give clues as to what plants grew in the understory of the lower montane forest when it was much more open. Sierra Miwok women can still pry edible tubers of sanicle *(Sanicula tuberosa)* and remember that long ago these plants were common when the lower montane forests in the central Sierra Nevada were regularly burned. Sons of North Fork Mono basket weavers can take one to former deergrass *(Muhlenbergia rigens)* gathering sites in chaparral and lower montane forests that are dying out from the lack of burning and from the discontinuance of the practice of gathering flower stalks and tillers. Wukchumni Yokuts men still gather medicinal plants, though exotic plants are crowding out this native pharmacy.

Did Indians, through millennia, demonstrate a measure of ecological harmony with some vegetation types of California? Does the statement "Indians lived in harmony with nature" bear any truth? If so, native people in California must have developed a sophisticated understanding of the inner workings of nature and acted deliberately to sustain, not degrade, ecological systems. While it is hard to prove or disprove native peoples' role in sustaining ecosystems, there are still many kinds of questions we can ask, and information we can seek, to get closer to the truth. If, for example, California Indian interactions with nature shaped ecosystems in ways that supported these systems and their biodiversity, we would expect widespread rules calling for moderation in harvest levels. We would also expect to see harvesting strategies that were least detrimental to the continued existence of target species. We would anticipate harvesting and management that assisted, rather than destroyed, an organism's cycle of renewal—what the ethnobiologists Nancy Turner and Doug Deur describe as actions designed to "keep it living."[2] In other words, Indian peoples' interactions with nature should demonstrate true relationship, which implies reciprocity, continuity, familiarity, and continual learning.

We would also expect to see practices that increased an area's primary productivity, maintained and recycled an ecosystem's nutrient capital, and maintained the fertility of the soils. And we would expect the maintenance of diverse successional stages and stand structures in different vegetation types that provided habitat for many native species. With ecological harmony, production activities would be balanced with the maintenance of resiliency, diversity, and critical functions of natural ecosystems.

In thousands of pages of unpublished field notes, in dozens of published ethnographies, and in the statements of contemporary California Indian gatherers, the same harvesting rules can be found over and over again. We also find evidence of diverse resource management techniques—burning, pruning, sowing, weeding, tilling—that Indians assert were practiced to *help nature along*. These techniques, especially fire, were integral to the health and vigor of myriad species of plants and animals and were the basis of successful subsistence economies throughout the state.

Behind the seemingly simple act of digging leopard lilies *(Lilium pardalinum* subsp. *pardalinum)*, pinching clover greens, or capturing quail lay intimate, complex knowledge. It is this knowledge, based on the keen observation of plants' physical requirements and responses to fire and animals' reproductive biology, nutritional needs, and migration cycles, that provided the foundation for figuring out how to both harvest and sustain plants and animals. In other words, human cultures that established intimate relationships with plants and animals time and again informed and dictated the elements of harvest and management, and these practices eventually led to some measure of ecological harmony.

The Elements of Plant Harvesting

At the beginning of Euro-American contact, the population density of California Indians in some areas was about one person for each two square miles, a density that ranks among the highest in the world among native people.[3]

Consequently, California Indians could easily have overexploited wild plant resources, even using simple, small-scale technologies (digging stick, seedbeater, obsidian and chert tools, human hands). It is possible to over-harvest resources if they are gathered at the wrong season, too frequently, at an inappropriate intensity, or without regard for replacement. To determine whether native people had negative, benign, or beneficial impacts on the natural environment requires a deep understanding of their total gathering system.

To both use and conserve nature requires complex knowledge and practices, far more complex than leaving nature alone. Hidden in the simple act of gathering lay sophisticated rules that safeguarded the plant stock from overharvesting. An overarching gathering rule was, Spare plants or plant parts; do not harvest everything. Anthropologists recorded the existence of this rule from the northern extremity of the state, among the Tolowa, to the southern border, among the Quechan (Yuma). Indeed, rules commanding the user to take no more of a resource than is absolutely necessary are common throughout North America and existed among groups as diverse as the Cree, Beaver, Koyukon, Kaska, Inland Tlingit, and Tutchone of Alaska and Canada.[4] There were many more specific rules in California, such as, Take roots from one side of the tree only. When digging ponderosa pine roots for baskets, the Maidu cut roots from one side of the tree one year and from the opposite side the next year, so as not to kill the tree.[5]

There were at least five elements of the harvest, described below, that when combined in various ways formed a regime or regimen—a total gathering system—that the indigenous people of California repeatedly applied to a given plant population or populations. Ultimately, these regimes led to plant conservation or extirpation.

HARVESTING TECHNOLOGIES

Harvesting technologies, that is, the tools used to retrieve plants and plant parts, included digging sticks for harvesting rhizomes, roots, and tubers; obsidian or chert knives for harvesting young basketry branches and the leaves and stalks of some plants (the Sierra Miwok, for example, used a stone knife made of a naturally fractured stone to cut hazel branches or maple shoots for basketry);[6] poles for knocking down the cones of various pines and the acorns of different kinds of oaks; shorter staves to knock off the flower buds of tree yuccas and the center stalks of agaves; sharpened bones or antlers from deer, elk, and other wildlife for prying seeds or other edible parts out of holes; oak wood prys with fire-hardened tips for harvesting the spiny leaves of agave; sharply pointed sticks for spearing the prickly buds of some cacti; seedbeaters for removing the seeds from herbaceous plants; and, of course, human hands for pinching, plucking, shaking, or tearing plant parts. Whenever possible, native people used tools appropriate to the resource. The "means" were purposefully designed *not* to destroy the "ends"—the source of production.

Although the tools used in California appear primitive and unlikely to affect vast areas, investigations show that their power to transform landscapes has generally been underestimated. The digging stick, for example, in the hands of thousands of women, could turn over and aerate large areas of soil in meadows, coastal prairies, or valley grasslands—greatly affecting the composition and densities of the species found there. Neither should the seed beater be underestimated in terms of its potential effects on California vegetation. The seed beater is a shallow basket with a handle, made of the branches of sourberry *(Rhus trilobata)*, buck brush *(Ceanothus cuneatus)*, or other shrubs. The basket is thrust over the spike or inflorescence of a grass or wildflower to knock off the grains and seeds into a burden basket or wide-mouth basket. Seed beating was a common technique throughout California.[7] Rarely were whole plants uprooted or the fruit heads broken off, which would have been a much more destructive harvesting strategy. Seed beating kept annual and perennial plants in place while ensuring that a certain proportion of the seeds fell to the ground, perpetuating the seed stock on the site. Frank Latta affirms that this occurred among the Yokuts: "In beating seeds like wild oats or wild rice into a basket with a flat beater, such as was used by some Indians, the seed was scattered all over the place."[8] Seed beating also allowed for repeated harvests of indeterminate inflorescences, maximizing numbers of ripe seeds gathered and minimizing the extent of the vegetative part detached with the seed.

Seed beating may have enhanced the herb population in a number of ways. In addition to scattering seeds around the collection area, the method ensured that only ripe seeds were collected, leaving immature seeds to develop fully and replenish the stock. Breaking off fruit heads or uprooting whole plants and removing them from the site, in contrast, easily could have extirpated wild populations of annual plants in a matter of years or decades. Leaving most of the plant behind was important in terms of lightening the load a woman carried in her burden basket. But it also ensured that the biomass would eventually be recycled in *that* place. This fact becomes important in the application of another management practice—the periodic setting of fires. The seed-bearing grasses and wildflowers, as they dry and wither, become the fine fuels that will carry frequent, light surface fires.

SEASON OF HARVEST

The time of year that a plant is harvested affects both quantity and quality of production and, in the longer term, the productivity of a population. The harvesting of the tender, immature flower stalks of *Agave* spp. and *Yucca* spp. before flowering may have stimulated vegetative reproduction in the form of pups.[9] The digging of the bulbs and corms of wildflowers such as camas and blue dicks for foods after they had gone to seed helped to prepare the seedbed and perpetuate plant populations by reseeding the site.

Young shoots or branches of many shrub species (e.g., dogwood, buck brush, redbud, sourberry, mock orange, and California button willow) used for arrows, baskets, and other cultural items were often harvested in fall or winter after the leaves had dropped. Many groups referred to this time of year as "when the sap's down"; the Western Shoshone weaver Florence Brocchini called it "when the energy of the plant is in its roots." Cheme-huevi, Washoe, and Paiute weavers, for example, harvested young willow shoots from fall until early spring. Paiute weavers told Isabel Kelly, an an-thropologist, that if gathered in summer, the willow was "brittle and not easily worked." Pomo women harvested hazel switches for fish traps and surf fish baskets after leaf drop in fall. This period after leaf drop is called the dormant period, which according to Western horticulturists is a resting stage. Cutting or pruning at this time of year is least detrimental to the con-tinued existence of the target species.[10]

Some herbaceous perennials were harvested by California Indian tribes after the plants had already died back for the year. This practice was used for important cordage plants such as milkweeds (*Asclepias* spp.) and Indian hemp (*Apocynum* spp.). Cutting and gathering this dead material for mak-ing string and other cultural products probably had no impact on the plant population. As long as the underground organs were not disturbed, the non-woody aerial parts that died away in fall and were replaced the following spring by new growths could be harvested with no effect on the plant. Some studies published in the ecological literature demonstrate that removal of dead, aboveground material has no negative impact on the plant and some-times benefits it.[11]

FREQUENCY OF HARVEST

"Frequency of harvest" refers to how often a plant (or a plant part) is gath-ered from the same area. Populations of different plant species were har-vested by California tribes at different frequencies, sometimes allowing for rest periods to rejuvenate plant populations. Pomo Indian basket weavers collected sedge (*Carex* spp.) rhizomes along California rivers for use as lacing. They allowed a two- or three-year rest period between harvests. Such conservation practices were not unique to California. Klikitat basket makers of southern Washington dug the roots of western red cedar *(Thuja plicata)* every three years, giving the trees time to regrow and replenish their roots. Navajo medicine men still refrain from harvesting medicinal plants from the same stand two years running, granting periods of rest and regrowth between those of tillage and extraction. Thousands of other examples exist around the world.[12]

Those native plants that indigenous people knew would not be harmed thereby were harvested annually. Indeed, they frequently asserted that many plants and mushrooms *did better* if they were picked. The picking of some tule, cattail, dogbane, and milkweed stems removes parts that might shade out new growth the next year.

Deergrass basketry sites were (and still are) harvested annually for flower stalks, and many elders say that gathering flower stalks from deergrass increases culm production the following year. For example, a North Fork Mono elder told me, "I gather *monop* [deergrass] in the fall—first part of September on. That's when they're most strong. It comes back plentiful when you pick it. I pick from the same plants each year. By clipping them, they grow better" (pers. comm. 1991). Barbara Bill, a Mono/Yokuts, said, "Picking the flower stalks is just like clipping the sourberry; you remove them and the new shoots come out better" (pers. comm. 1991). And Norma Turner Behill, a Mono/Dumna elder, said, "The more you gather it the more it grows. If you don't gather, it doesn't produce" (pers. comm. 1991).

INTENSITY OF HARVEST

Intensity of harvest, or the proportion of a plant's parts or the percentage of a plant population that is removed, appears to have been considered by California Indian tribes with regard to its effects. Harvesting of many kinds of plants, algae, and fungi did not exceed the population's biological capacity to recover. Those plant parts that were renewable—fruits, branches, leaves, flowers, corms, stems—could be gathered year after year without causing harm. Seaweeds, such as laver (*Porphyra perforata, P. lanceolata*), were harvested by lightly pulling them from rocks by hand, leaving the holdfast and a mat that regrew blades. (See Figure 14). *Porphyra* is usually an annual and will also be replaced via spore settlement each year. Grapestone *(Mastocarpus papillatus)* is a long-lived perennial, and it is very likely that it was

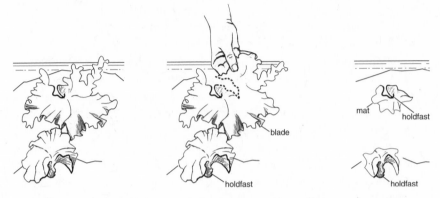

mat   holdfast

blade

holdfast

holdfast

Figure 14. Annual harvest cycle of the seaweed laver *(Porphyra perforata)* by California Indians. Laver is torn gently from the rocks by hand, leaving a holdfast with a ruffled mat that will regrow new blades. The Kashaya Pomo and other indigenous groups have harvested laver annually from the same rocks for centuries.

harvested similarly so that the alga would grow back. There is evidence for ecologically sustainable harvesting of seaweeds by indigenous people in Hawaii and elsewhere. The Southern Sierra Miwok still harvest mushrooms by leaving the stipe behind and putting earth over it. Miwok grandmothers instruct their grandchildren that the fine threads under the ground, known to mycologists as mycelia, must not be disturbed, to ensure future abundance; in addition, some mushrooms must be spared.[13]

Elsie Allen, a superb basket weaver, noted in her book *Pomo Basketmaking,* that when gathering large quantities of white root (rhizomes of *Carex* spp.) it is necessary to leave about half of the rhizomes to reproduce. The Washoe still tear the odoriferous leaves of aspen onion *(Allium bisceptrum)* with their fingers one or two inches from the ground, leaving some of the plants to go to seed. Yokuts harvest brodiaeas, yampah *(Perideridia* spp., also known as wild carrots), and lilies, leaving parent plants or bulblets, cormlets, and tuber fragments purposely behind in the loosened earth to grow the following year. In harvesting juniper wood for Paiute bows or bark from alders for Maidu dyes, trees were not felled or girdled, allowing for repeat harvesting over several generations.[14]

In historic times, every late summer or fall, agile Plains Miwok and Sierra Miwok men climbed gray pines *(Pinus sabiniana)* in the foothills, instead of felling the trees with firebrands, for a onetime harvest of nuts. Colonel Z. A. Rice described the process of harvest among the Indians of the Sacramento basin in 1850: "As the trunks of these trees are frequently without branches to a height of thirty or forty feet from the ground, the Indians ascended them by means of spliced poles long enough to reach to the first limbs. The pole was held in place by Indians on the ground, while an expert climber ascended and beat off the pine cones with a short pole."[15]

Cultural rules required the gatherer to spare or pass up some plants or plant parts of a certain size, for example, mother bulbs that are too big or baby bulbs that are too small. Various kinds of plants are still highly valued and remembered by gatherers from Sierran cultural groups for their edible bulbs and corms, and recent interviews substantiate that plant parts are purposely left behind by Indian harvesters to ensure future abundance. Clara Jones Sargosa, a Chukchansi elder, said, "In digging wild potatoes, we never take the mother plant. We just select the babies that have no flowers, just leaves. We are thinning the area out so that more will grow there next year. Also, when harvesting wild onions with a little stick, we would leave some of the young ones behind. They don't have a taste yet and will be ready the following year" (pers. comm. 1989).

LONG-TERM PATTERNING OF HARVEST

The regularity with which a gathering site is used over a span of time measured in decades or generations is significant for California Indians because it again points to the limitation of the artificial dichotomy hunter-gatherers/ agriculturists. It has been widely assumed that nonagriculturists were casual inhabitants who drifted from place to place on a nomadic food quest, visiting widely dispersed and variable food resources in a specified tribal territory. In this view, gathering sites for the most part were not visited over many generations, and therefore the effects of human occupancy were not cumulative and were soon erased, like the ephemeral marks of a storm or the tracks of birds and squirrels. Early anthropologists adopted this view in part because California Indians always said "No" when asked if land was owned in their cultures.[16] They mistakenly inferred that this denial of landownership meant that California Indians never stayed anywhere long enough to leave lasting influences and had no interest in repeated use of areas.

But California Indians embraced a different concept of ownership, one based on usufruct rights. Under this conception, if an area is used and tended, it becomes the domain of the gatherer. For example, throughout California, individuals or families repeatedly gathered from and cared for specific oak trees and groves, giving them usufruct rights to those resources. Under the usufruct system, each family had a combination of exclusive rights to certain resources and communal rights to other resources. The Pomo distinction between individual and communal rights, as reported by the anthropologist Fred Kniffen, was typical: "Like larger manzanitas, all the great oaks of the valley flat were privately owned; those of the hills were owned by the village as a whole."[17]

Usufruct rights were often extended through ties of kinship and marriage. Extended families commonly frequented the same stands of different kinds of oaks for many generations, thus passing down to each successive generation valuable ethnobotanical knowledge and management techniques. Eleanor Beemer said of the Luiseño, "Certain acorn gathering places have belonged from time beyond memory of the living people, to their families."[18]

California Indians used areas over and over because their shrubs had sweeter fruit, their trees bore larger and less bitter acorns, or their sandy soils grew the straightest and longest rhizomes. Special areas were often designated for basketry materials, bulb gathering, seed collecting, cordage harvesting, or greens picking—designations shaped by continual long-term use and management. This was also true of special fishing sites.[19] Tracts of narrow-leaved willow *(Salix exigua)* and sedge *(Carex barbarae)* were used repeatedly for basketry by the Pomo. Agave, oak, and mesquite gathering areas were owned by lineages of the Cahuilla in southern California.[20]

The anthropologist Anna Gayton noted of the Chukchansi Yokuts and Sierra Miwok, "When manzanita berries are nearly ripe, women specify certain bushes as theirs . . . and no one else can take the berries from them." She stated further of the Chukchansi Yokuts, "They went up in the mountains to get berries or seeds, always going to the same place on which they had a traditional claim."[21]

Family sites were frequently arranged in discrete patches, the boundaries of which were marked. Among the Pomo, for example, an oak tree or manzanita bush might be marked by tying a piece of brush to it, blazing the trunk, or setting up a stake.[22] Proprietorship of a patch was carried on for many successive generations, which is clearly evident from numerous references in both the historical literature and interviews with elders today. Family sites contained the best plants for the intended purposes: "When acorn gathering, there was a family area that you go to year after year because you know that the trees will produce. These were picking areas. You didn't really own the trees but you knew what the trees were going to give you" (Gladys McKinney, Mono, pers. comm. 1992).

Techniques of Traditional Resource Management

Any landscaper, gardener, or horticulturist knows what happens to untended plants. The apple tree gets leggy; the daffodils overcrowd themselves. Sowing a seed or putting a plant in the soil is just the beginning of a long relationship between the person and the plant. California Indians had a similar working relationship with the plants in their environment: culturally significant plants, populations, and habitats were directly manipulated to maintain and heighten their abundance, productivity, and diversity so that they could meet an array of cultural needs. This culturally mediated relationship with the natural world, in which humans intervene in the life cycle of native plants and animals to direct their growth and reproduction, is often termed "traditional resource management."[23]

In *Almost Ancestors*, Theodora Kroeber and Robert Heizer wrote, "The Indians' preservation of the land and its products for the ten thousand or more years of their undisputed occupancy was such that the white invaders wrested from them a garden, not the wilderness it salved their conscience to call it."[24] But Kroeber and Heizer never fully understood the gravity of these words, or the importance of their research and that of dozens of other academic scholars that recorded the ethnobiological uses of plants and animals and the land management techniques designed to augment populations. These historical records provide a precious road map for restoring many native plants and animals in California.

California Indians practiced resource management at four levels of biological organization: the organism, the population, the plant community, and the landscape. They used resource management techniques at each of these levels, or scales, to promote the persistence of individual plants, plant populations, animal populations, plant associations, and habitat relationships in many different vegetation types in California. The techniques in the Indians' repertoire included burning, irrigating, coppicing, pruning, sowing, tilling, transplanting, and weeding. All of these techniques, especially burning, represented a disturbance; by applying them in various ecosystems, Indians became agents of controlled, culturally mediated disturbance, using it to maintain plant populations of special importance and habitat diversity.

BURNING

Burning is the application of fire to particular vegetation areas under specified conditions to achieve select cultural purposes. The use of fire entails a number of important considerations, such as seasonality, the frequency of burning in a particular area, and the aerial extent of the fire. Tribes in California made fire either by drilling or by percussion. "Drilling" refers to the rotating of a slender wooden shaft in a hole in a stationary board called the hearth to create intense heat by friction; "percussion" refers to striking two objects together, such as two stones, to create sparks. (See Table 2.)[25]

Most tribes used a slow match or torch, which consisted of a tightly packed flammable material that smoldered at one end for a considerable period. Various tribes used fire fans to control the intensity or direction of the fire. The Pomo fitted the tail feathers of herons with long wooden handles to create fire fans for augmenting a blaze. The Foothill Yokuts and Western Mono used a fire fan made of hawk or buzzard tail feathers to smoke out ground squirrels.[26]

Fire was the most significant, effective, efficient, and widely employed vegetation management tool of California Indian tribes. The slow match gave them the technological capability to burn both small patches and extensive tracts of vegetation in a systematic fashion. When they set fire to vegetation types occurring as continuous fuel beds, such as grasslands, the fire conceivably could burn uninterrupted for miles. The adventurer Frank Marryat witnessed such a fire near the Russian River: "The rainy season was now approaching, and the heat became occasionally intense. At times the Indians would fire the surrounding plains, the long oat-straw of which would ignite for miles. The flames would advance with great rapidity, leaving everything behind them black and charred." In addition, most tribes had the ability to fell trees and large shrubs with fire for meeting cultural needs. They used this tool to create village sites and to convert riparian habitat and floodplain into farming areas in southeastern California. The acreage that was burned by California's earliest humans may have been significant; the fire scientists Robert Martin and David Sapsis estimate that between 5.6 million and 13 million acres of California burned annually under both lightning and indigenous people's fire regimes.[27]

Deliberate burning increased the abundance and density of edible tubers, greens, fruits, seeds, and mushrooms (see Table 3); enhanced feed for wildlife; controlled the insects and diseases that could damage wild foods and basketry material; increased the quantity and quality of material used for basketry and cordage; and encouraged the sprouts used for making household items, granaries, fish weirs, clothing, games, hunting and fishing traps, and weapons. It also removed dead material and promoted growth through the recycling of nutrients, decreased plant competition, and maintained specific plant community types such as coastal prairies and montane meadows.[28]

The Pomo burned bracken fern patches to enhance them; the new fronds were eaten and the rhizomes used to create basket designs. The Luiseño of

TABLE 2

Plants Used in Fire-making Kits

Fire drill	Young shoots of buckeye, elderberry, wild rose, willow, blue oak, manzanita, seep-willow *(Baccharis salicifolia)*, Basin sagebrush *(Artemisia tridentata)*, desert almond *(Prunus fasciculata)*, alder, juniper, and palm-fruit stems.
Hearth	Incense cedar wood, buckeye wood, dry palm-fruit stem, redwood root, Basin sagebrush, and cottonwood root *(Populus fremontii)*.
Tinder	Rotten pine wood, tree moss, shredded dry grass, cattail fluff, dry pine needles, shredded incense cedar bark, rotten buckeye wood, dry white punk from rotten hollow trees, white alder wood *(Alnus rhombifolia)*, shredded conttonwood roots, the dried and powdered leaves of California black oak, and oak galls.

southern California fired shrubs, such as sumac *(Rhus trilobata)*, to bring out the young shoots used for basket weaving. The Western Mono burned the understory of blue oak woodland and lower montane forest to stimulate the growth of Hall's wyethia *(Wyethia elata)* and other plant species with edible seeds. The anthropologist Harold Driver recorded that the Wiyot burned every two or three years to increase feed for deer. Fires were also set to aid in hunting. John Muir witnessed such fires among the Sierran tribes and noted in his diary, "Indians burn off the underbrush in certain localities to facilitate deer-hunting."[29]

IRRIGATING

Supplying selected land areas with water by means of artificial channels was an indigenous technique, practiced long before the Spanish and other Europeans introduced their agricultural knowledge. The Owens Valley Paiute on the east side of the Sierra Nevada, for instance, diverted water from tributaries of the Owens River to nurture "wild" stands of *nahavita*, or blue dicks *(Dichelostemma capitatum)*, perhaps re-creating the natural conditions that existed in the swampy lowlands of the Owens Valley. A host of other native and non-native plants were watered artificially to increase their productivity and abundance. These species included lovegrass *(Eragrostis* sp.*)*, crested wheatgrass *(Agropyron* sp.*)*, and Great Basin wild rye *(Elymus* sp.*)*.[30]

TABLE 3

Indigenous Burning to Cultivate Food Resources

Consultant and Tribe	Plant Species Burned	Time of Year Burned
Lydia Beecher, Mono	*Sanicula tuberosa* (sanicle, *tena*)	October–December
	Fragaria vesca (wood strawberry)	October–December
	Allium sp. (wild onion)	October–December
Rosalie Bethel, North Fork Mono	*Trifolium* spp. (clover)	fall
	Ribes roezlii (gooseberry)	fall
	Sanicula tuberosa (tena)	fall
	Sambucus mexicana (elderberry)	fall
	Arctostaphylos spp. (manzanita)	fall
	Quercus kelloggii (black oak)	fall
	Salvia columbariae (chia, *saat*)	fall
	Clarkia spp. (farewell-to-spring)	fall
	Plagiobothrys sp. (snowdrops)	fall
	Wyethia elata (Hall's wyethia)	fall
	Allium sp. (wild onion)	late summer, fall
Clara Charlie, Chukchansi/Choynumni	*Ribes roezlii* (gooseberry)	late August
	Prunus virginiana var. *demissa* (chokecherry)	late August
	*Rubus vitifolius (*blackberry)	late August
	Vitis californica (wild grape)	late August
	Quercus kelloggii (black oak)	late August
Ruby Cordero, Chukchansi/Miwok	*Perideridia* sp. (yampah, *homogi*)	mid-August to September
	Sanicula sp.? *(dana)*	mid-August to September
Lalo Franco, Wukchumni Yokuts	*Arctostaphylos* sp. (manzanita)	fall
	Astragalus bolanderi (rattleweed)	October, November
	Lathyrus sulphureus (scrub pea)	October, November
	Pickeringia montana (chaparral pea)	October, November
	Trifolium spp. (clover)	October, November
Bill Franklin, Miwok	*Fragaria californica* (wild strawberry)	October, November
	Sanicula sp.? (Indian potato)	October, November
Ron Goode, North Fork Mono	*Arctostaphylos* spp. (manzanita)	September, October
	Quercus kelloggii (black oak)	September, October
Avis Punkin, Mono	*Arctostaphylos* spp. (manzanita)	mid-July

Table 3 (continued)

Consultant and Tribe	Plant Species Burned	Time of Year Burned
Francys Sherman, North Fork Mono	*Calandrinia ciliata* (red maids, *kaseen*)	fall
	Madia sp. (tarweed)	fall
	Clarkia spp. (farewell-to-spring)	fall
	Plagiobothrys sp. (snowdrops)	fall
	Wyethia elata (Hall's wyethia, sunflower)	fall

PRUNING AND COPPICING

Pruning is the removal of dead and living parts from native plants to modify growth form or enhance fruit or seed production. Coppicing is a severe form of pruning that involves cutting down a shrub or small tree to a few inches above ground level to promote the growth of long, straight shoots. Many California Indian cultures pruned shrubs and trees repeatedly, in many cases prolonging the life spans of the plants by removing the dead twigs and branches that could harbor insects and diseases. In the 1950s, the Northeastern Maidu elder Lily Baker told James McMillin, an anthropology student, that Indians used to clean around specific elderberry bushes, removing the dead portions in the canopy to enhance the growth of the bushes for food.[31]

Individuals of many tribes harvested acorns by climbing the trees and cutting the limbs, a process Galen Clark recorded among the Yosemite Miwok: "In order to get the necessary supply [of acorns] early in the season, before ripe enough to fall, the ends of the branches of the oak trees were pruned off to get the acorns, thus keeping the branches well cut back and not subject to being broken down by heavy snows in the winter and the trees badly disfigured, as is the case since that practice has been stopped."[32] The Mono elder Lydia Beecher remembered the former pruning of oaks: "My grandpa Jack Littlefield would climb black oak trees and cut the branches off—just the tips so that many more acorns would grow the next year" (pers. comm. 1991).

The Timbisha Shoshone pruned the dead limbs and lower branches of honey mesquite *(Prosopis glandulosa* var. *torreyana)*, an important food resource, and kept areas around the trees clear of undergrowth. Special hooked sticks were used by the Miwok to retrieve dead limbs from the canopies of trees for firewood. Tribes in various parts of California pruned plants to induce the rapid elongation of young shoots for basketry. The branches of redbud were cut in winter or early spring by the Yuki and Pomo to ensure that there would be suitable material for baskets the next fall.[33]

The Pomo pruned (and continue to prune) narrow-leaved willow, which

Figure 15. Kawaiisu women pruning a stand of wild tobacco plants (*Nicotiana quadrivalvis*, formerly *N. bigelovii*) near Piute Rancheria in 1937. Patches of tobacco in open shrub lands and forests are one of the many unique vegetation associations maintained with burning and pruning that are disappearing throughout California as a result of the cessation of these practices. Courtesy of Maurice and Kathy Zigmond.

activated the growth of the underground lateral runners used in large baskets. David Peri and Scott Patterson observed that "the cultivation by the weavers helped the willow to spread, kept it from getting top heavy and falling down, removed diseased branches, and increased its effectiveness in controlling erosion along the stream banks."[34]

Many tribes pruned tobacco patches. (See Figure 15.) Charles Smith described the practice among the Tubatulabal: "Two species of tobacco (*so'ogont*) grow wild in Tubatulabal territory: *Nicotiana bigelovii* [now *Nicotiana quadrivalvis*] and *N. attenuata*. In early summer women stripped off the side shoots to encourage larger leaf growth. The plants were pruned twice more."[35]

Today California Indians assert that the way in which the expendable plant part is harvested is often good for the plants. Indians pinched back leaves of herbaceous plants or cut back stems in the harvesting of edible

Figure 16. Annual harvest cycle of prickly pear cactus (*Opuntia* spp.) fruit and young pads. Cutting off the new pad at the end of the native *Opuntia* does not damage the plant but increases its production. Where there was one pad before, two pads will grow, one from each side of the base of the cut. Periodic cutting keeps the plant from growing too tall, which would reduce its accessibility, and activates young pads.

greens, which may encourage side branching and new leaf production on the plant that is left behind. A common practice among different California Indian tribes was to knock the branches of pinyons and oaks with a wooden pole, causing the pinecones or acorns, both highly desirable foods, to fall to the ground. According to David Peri and Scott Patterson, knocking acted as a pruning process because "some branch tips and leaves and brittle and dead twigs were also removed in the process, while dead or diseased limbs were intentionally broken off. This autumnal 'pruning' increased the surface area of the canopy and fruit production by stimulating the growth of new branchlets and foliage the following year." Today elders of Miwok, Paiute, and Mono descent gather acorns from the ground, but many elders remember the former practice of knocking. Hazel Hutchins, a Mono elder, said, "Some people would knock acorns because they're hungry and they want to get some. The acorns are still a little green. They would hit them down with sticks in about September. They used to have the men climb the trees and shake the acorns down or hit them down" (pers. comm. 1992).[36]

Desert tribes harvested at least five kinds of prickly pear (*Opuntia* spp.) for their flat oval branches or stems, called pads (Spanish, *nopalitos*). Harvesting these pads, which made a tasty vegetable food when boiled, was essentially a pruning process. (See Figure 16.) Frederick Coville described how a Panamint woman harvested the pads: "When the fleshy joints, buds, blossoms and immature fruit are distended with sweet sap she breaks them off with sticks and gathers them in baskets. She then rubs each prickly joint with grass and dries the harvest in the sun. In this form the joints and fruit may be boiled and eaten with a little salt." The fleshy pads also have a juicy pulp that, along with the fruit, quenched the thirst of Native Americans while traveling through the desert.[37] Empirical observations on commercial plantations of a spineless variety of prickly pear *(Opuntia ficus-indica)* show that pruning is effective in promoting pad yields.[38] According to William Pink, Luiseño/Cupeño, cutting off the new pad at the end of the native *Opuntia* with a knife does not damage the plant but rather increases production. Where there was one pad before, two pads will grow, one from each side of the base of the cut. Periodic cutting prevents the plant from growing too tall, maintaining accessibility, and activates young pads. Older pads become lignified and thus too tough to eat. The new growth of *Opuntia* also was favored by bighorn sheep, who would bite on the pads and suck out all the moisture (pers. comm. 1997).

SOWING

Sowing, the broadcasting of seeds collected from native plants, usually in a recently burned area, was practiced by many tribes long before the invention of domesticated agriculture. To avoid extinguishing the populations of the annual plants from which seeds were gathered, many tribes broadcast some of the seeds at the gathering site. The Paiute sowed the seeds of *Mentzelia* and *Chenopodium* plants on burned-over ground. Western Shoshone women broadcast the edible seeds of lamb's quarters (*Chenopodium* sp.) and blazing star (*Mentzelia* sp.). The Quechan broadcast the seeds of native plants in southern California. The Cahuilla and probably other tribes of the Southwest planted the seeds of the desert fan palm (*Washingtonia filifera*) at oases.[39]

TILLING

The moving of soil to harvest underground perennial plant organs (e.g., roots, rhizomes, corms, tubers, bulbs) frequently involved dividing these underground organs and leaving individual fragments or smaller clumps in the soil. Underground swollen stems and other organs were gathered for five major purposes: food, medicine, dye, basketry, and ceremonies. Huge quantities of these plant parts were gathered with a digging stick, potentially causing considerable soil disturbance and ultimately affecting plant abundance, density, and composition. Digging bulbs, rhizomes, and tubers tended to increase the size of the tract, aerate the soil, lower weed competition, and sever the tiny bulblets and cormlets, allowing them a better opportunity to grow.[40]

Various tribes left tubers, bulblets, cormlets, and rhizomes in the soil and harvested after the plants had gone to seed, preparing the seedbed for new plants. The Yokuts, for example, harvested medicinal roots in this way:

> This root is called Lip'-trow, and the Indians there dug it in great quantities to trade with those who passed by. They made tea of the root and drank it. It was very bitter. They say that it grew up from the same root each year and was about 20 to 24 inches high, straight stemmed and with a small white flower. Mrs. Francisco spent some little time trying to locate a specimen of it for me to preserve, but was unable to find any of it. She said that she would find some of it on Tule river and save it so I could identify the plant. (From Woody toward Famosa about five miles below Woody on the road—not at Willow Springs yet.)[41]

On his way from San Diego to Monterey in 1791, the Spanish naturalist and surgeon José Longinos Martínez jotted in his diary: "The gentiles of the mountains cultivated a root called, in the language of the Channel Indians, chuchupaste, meaning 'plant of great virtue.'. . . . The greatest virtue they ascribe to it is that it will cure any kind of headache by inhalation, and any pain in the stomach by mastication and swallowing the saliva." This tended plant was most likely *Lomatium californicum*—called "chuchupate" by John P. Harrington in his fieldwork among the Chumash.[42]

Marie Potts said that her people, the Maidu, steadfastly limited the gathering of the tubers and bulbs of wild carrot and camas, always leaving some plants behind for seed. The Pomo cultivated the rhizomes of black root (*Schoenoplectus* spp.) for basketry and cultivated various edible wild potatoes such as yampah (*Perideridia kelloggii*), soaproot (*Chlorogalum pomeridianum*), and mariposa lilies (*Calochortus* spp.).[43]

TRANSPLANTING

Indians widened the ecological amplitude of native species by introducing them to new areas and transplanting or sowing seed at traditional collection sites away from and adjacent to village sites. Thus the ranges and distributions of favored plant species were expanded through human intervention. Today, individual plants or small colonies of the California walnut (*Juglans californica* var. *hindsii*) are often found growing in association with archaeological sites, where they may very well have been planted more than a century ago.[44] Deliberate movement of plants and animals to new areas is discussed more thoroughly in Chapter 5.

WEEDING

Certain wild plants in different plant communities were favored through removing the unwanted plants around them. Beds of sedge (*Carex* spp.) along lowland streams and rivers in California and bracken fern *(Pteridium aquilinum* var. *pubescens)* at 5,000 to 6,000 feet in the Sierra Nevada were weeded intensively to encourage the production of long, creeping rhizomes, which were used extensively by many tribes as the weft or lacing in basketry, or as a design material in coiled and twined baskets. Other tracts of plants prized for their delectable bulbs or medicinal properties also were weeded. David Peri and colleagues described the benefits of weeding by the Pomo: "The elimination of unwanted plants interrupts the natural process of plant succession and insures the continual presence of a particular plant by reducing competition for sun, nutrients, and moisture. Collecting areas were also raked to clear away dead branches, rocks, and excessive leaf litter from around the plants, so that the potential for the plant being crowded out was reduced."[45]

The Ecological Rationale for Using Fire as a Management Tool

Fire was the primary land management tool of California Indians because it had many significant ecological effects, five of which stand out as the most fundamental and compelling.

DECREASING DETRITUS AND RECYCLING NUTRIENTS

Many wild plant populations accumulate aging parts (dead branches and shoots, leaves, cones, and seed pods) that may reduce plant vigor and productivity over time. Fires set by California Indians consumed this biomass and released some of the plant nutrients it contained.

Scientific studies have recently shown that nutrient movement can take a long time, relative to human life spans. The turnover rates of many nutrients are slow. In some ecosystems the nutrient storage compartment (e.g., the litter on the forest floor) can become a vault, locked against internal cycling. Various ecosystems will not remain productive over time if dead material accumulates much faster than it decomposes. Like soil arthropods, bacteria, and fungi, fire is a mineralizing agent in forests and other vegetation types, but it works much faster than decay organisms and thus speeds up nutrient recycling and the return of sites to high productivity. Although fire can accelerate nitrification and thus loss of nutrients, research is demonstrating that the leguminous, nitrogen-fixing forbs (such as lupines and clovers) often promoted by fire can rapidly provide nutrient replacement.[46]

Freshwater marshes in the Central Valley of California were burned by the Wukchumni Yokuts, and in the Panamint Valley by the Timbisha Shoshone, to clear out old reeds, recycle nutrients, stimulate new plant growth, and provide open water for waterfowl. Lalo Franco of the Wukchumni Yokuts says, "Tules, the bulrushes—whole forests of tules—were burned. In the Kaweah Oak Preserve there are bulrushes and roots that we know they burned and you can see what I'm talking about—today without the burning there is congestion, you can see what it does to the plants. It does not let them grow healthy."[47]

Deergrass, an important native bunchgrass for coiled basketry in central and southern California, was burned in chaparral, lower montane forests, and oak woodland plant communities by the Cahuilla, Foothill Yokuts, Kumeyaay (Diegueño), Luiseño, Sierra Miwok, and Western Mono tribes to clear away accumulated dead material and increase flower stalk yields. Virgil Bishop, of the North Fork Mono, pointed to deergrass with very little new growth and said, "Some of that stuff just stays there and doesn't produce anything, so they would burn it and it would come back in again better than it was—full of vitality, with many more stalks to get. My family would burn it whenever they thought it was necessary—every two or three years" (pers. comm. 1991).[48]

CONTROLLING INSECTS AND PATHOGENS

Fire helped to control the pathogens and insects that would otherwise compete for the same resources used by native people. Ruby Cordero (Chukchansi Yokuts/Sierra Miwok, pers. comm. 1992) recalls burning to eliminate insects that attacked shrubs that were important for basketry: "Mother used to talk about burning buffalo grass [deergrass], sourberry sticks, and chaparral [Ceanothus cuneatus] sticks. The men would burn sourberry, redbud and chaparral in the fall after the women picked their sticks. The smoke would decrease the insects on the sourberry. They'd burn a little hillside where there was a group of redbuds or chaparral."

Many Indian tribes in California burned in oak (Quercus spp.) woodlands and tan oak (Lithocarpus densiflorus) stands to reduce insect pests that inhabit acorns and overwinter in the oak leaf duff. Harold Driver noted during his work among the Tolowa: "Burned under tree to make acorns drop off; also to kill parasites on or underneath tree." According to Kathy Heffner

(Wailaki, pers. comm. 1992), all of the tribes she interviewed in northern California (Hupa, Wailaki, Tolowa, Yurok, and Karuk) burned under the California black oaks and other oak species to destroy insect pests: "They needed to eliminate that duff that was underneath the oak trees because the oaks will drop their leaves and create a big pile of duff. As long as all that duff stayed there, when the acorns dropped, the acorns could only be on the ground just a little while, because that duff was the home of a lot of bugs. The minute they hit the ground, those bugs were into those acorns. So if they burned it, that eliminated the duff and the insects that would get into the acorn."[49]

In a 1916 letter to the California Fish and Game Commission, Klamath River Jack from Del Norte County makes the link between eliminating wormy acorns and setting fires: "Fire burn up old acorn that fall on ground. Old acorn on ground have lots worm; no burn old acorn, no burn old bark, old leaves, bugs and worms come more every year. . . . Indian burn every year just same, so keep all ground clean, no bark, no dead leaf, no old wood on ground, no old wood on brush, so no bug can stay to eat leaf and no worm can stay to eat berry and acorn. Not much on ground to make hot fire so never hurt big trees, where fire burn."[50]

Arthropods in two major genera feed on acorns during their larval stage, causing severe damage or destruction. These are the filbert worm (*Cydia latiferreana*) and the filbert weevils (*Curculio occidentalis, C. pardus,* and *C. aurvestis*). Studies of California oak species have shown that individual trees can exhibit up to 80 percent acorn damage by the filbert worm and 20 percent to 75 percent destruction by filbert weevils. Individual trees can exhibit up to 95 percent acorn damage from a combination of these pests. The larvae tunnel throughout the inside of the acorn, leaving frass, destroying the embryo, and rendering the acorn inedible.[51]

This phenomenon could not have gone unnoticed by any of the tribes that depended on the acorn as a major food source. Elders today say that infested acorns were never purposely gathered, or eaten if accidentally gathered, but were selectively discarded when gathering. Gladys McKinney (Mono) says, "We were instructed never to gather any acorns with little holes in them" (pers. comm. 1992). Consultants also comment that lighter than normal acorn weight is the best indicator of insect damage (Margaret Baty, Western Mono, pers. comm. 1992). Norma Turner Behill, Mono/Dumna, says, "I run my hand through the acorns to find the bad ones. If they have a hollow sound, worms are in there and I take them out" (pers. comm. 2004). Two consultants I interviewed remembered burning under the oaks for acorn pests. Lalo Franco (Wukchumni Yokuts, pers. comm. 1991) said, "The old people said there were a lot of bugs and that setting fires under the oaks would help get rid of these. Burning took place in the high country right after the harvest of acorn and other things. Ron Goode (North Fork Mono, pers. comm. 1992) said, "I learned about burning under the oaks mostly from my mom. The burning would create a higher quality acorn crop with less worms. The ash was necessary." Ted Swiecki, a plant pathologist who has studied California oak pests and diseases, told me the following about the habits, feeding, and life cycles of the filbert worm and filbert weevil:

> These insects invade acorns while on the tree, and the insects continue to develop as the acorns fall. In fact, insect-infested and diseased acorns tend to drop earlier than sound acorns. Eventually, the larvae exit the acorn and overwinter as pupae in the duff beneath oak trees. If you were to burn off the duff and old acorns in the fall, you would destroy most if not all of the infested acorns as well as pupae that are in the duff. This would greatly reduce the number of filbert worm and filbert weevil adults that emerge in the following year, which would reduce the level of infestation in the acorn crop. If you were to do this every year, or even every couple of years, I would think that you would end up with a pretty clean crop of acorns, with relatively low losses due to insects. However, you would have to keep burning pretty regularly, because there will commonly be a few insects that are missed by the burn, and adults could also move in from beyond the burned area. (Pers. comm. 2004)

He comments further:

> Measures like disposing of understory debris and discarding obviously bad acorns are good cultural practices that any good orchardists would follow if they were trying to minimize the types of pest and disease problems that acorns have. They were following integrated pest management for acorns. Also, burning off plant debris beneath the trees would make harvesting easier whether you are knocking acorns out of the tree or simply waiting for them to fall. It makes the acorns easier to find and pick up and eliminates any old acorns that would need to be sorted out. This is similar to modern nut orchards in California, where the debris and vegetation on the orchard floor is completely cleaned off before harvest, although it is done with equipment instead of burning. (Pers. comm. 2004)

Fungi and bacteria also have the potential to decrease substantially the mast crop of oaks. Fungi that commonly colonize acorns are *Penicillium* (Robert Raabe, pers. comm. 1992) and *Aspergillus*. Fire may have helped to curb these pathogens as well. After collecting acorns, the Lassik and Pitch Wailaki burned the area "so that disease was kept down and the ground was cleared of undergrowth for easier collecting in the future." Yearly fires set by the Karuk in northwestern California reduced the likelihood of disease attacking the forest trees. The Luiseño in southern California burned regularly as well to destroy insect pests and diseases that damaged native food crops: "Some ground, they [Luiseño] stated, was cleared by fire in order to later scatter seeds of desired food plants, grass seed was mentioned by the older coastal informants particularly. Regular burning, some informants commented, destroyed insect pests and parasites, such as dodder, which damaged food crops."[52]

MANAGING WILDLIFE

An extremely important reason for setting fires was to increase forage for wildlife. Alfred Kroeber reported of the Maidu in the northern Sierra Nevada: "Like most of the Californians who inhabited timbered tracts, the Maidu frequently burned over the country, often annually. . . . Travel was better, view farther, ambuscades more difficult, certain kinds of hunting more remunerative, and a crop of grasses and herbs was of more food value than most brush."[53] José Joaquin Moraga, a chronicler of the second Anza expedition, jotted down in 1776, "The heathen [probably the Ohlone] had burned many patches [southeast of the mission of San Francisco], which doubtless would produce an abundance of pasturage." John Muir noted that in the lower montane forests of the Sierra Nevada the Indians would burn off the underbrush in certain localities to facilitate deer hunting. Philip Drucker recorded in 1937 during his work among the Tolowa in northwestern California: "The densely wooded nature of the country renders stalking difficult. Informants maintain that near-by hills were kept clear of brush by annual burning; this also improved the grass, so that deer frequented such clearings and could be shot easily."[54]

It has been shown that pruning or burning vegetation increases the forage value for certain wildlife and that the number of larger game animals increases after fire.[55] The disturbance caused by fire changes plant architecture, often making plants more accessible to animals. Fire also encourages vegetative reproduction, and the resulting young, tender shoots of such plants as mountain misery (Chamaebatia foliolosa), deer brush (Ceanothus integerrimus), and whitethorn ceanothus (Ceanothus cordulatus) are much more palatable and nutritious to deer and other browsers (Harold Biswell, pers. comm. 1989).

Today elders from a number of tribes substantiate that the practice of burning is highly beneficial to wildlife. The Sierra Miwok elder Bill Franklin learned about burning from his father and grandfather: "They said the Indians used to burn in the fall—October and November. They set the fires from the bottom of the slope to decrease the snowpack, get rid of the debris so there's no fire danger and they burned in the hunting areas so there was more food for the deer. They burned every year and in the same areas" (pers. comm. 1990).

Fire was also used in hunting many kinds of animals. Often, burning carried out for the immediate purpose of securing food had secondary ecological effects noted above. The methods used for hunting were ingenious and numerous. Tribes employed fire to attract animals, to drive them in certain directions, to create smoke for killing rodents in their burrows, or, in the case of insects, to reveal their nests or disable or kill them so that they could be collected. Alfred Kroeber recorded among the Yokuts: "When the geese traveled, inflammable brush was piled up, and when the birds were heard approaching on dark, still nights these were suddenly lit. The birds swooped down to the flare, and in their bewilderment were easily killed."[56] The Owens Valley Paiute stalked deer in disguise, used a surround with people and trained dogs, and sometimes fired brush. The Sierra Miwok set small fires in the hills around a meadow, into which deer went. These men then kept building new fires. As the deer descended to the meadow, they approached the fires out of curiosity and concealed hunters shot them with bows and arrows. A Hupa man described how fire was used to capture deer: "Two fires set in canyon so as to burn toward bottom. Some hunters drove with the fire; others awaited game at bottom."[57]

Fire was a tool used often for driving rabbits. For example, brush was fired for driving rabbits on the floor of valleys by the Tubatulabal between July and the middle of August. Twenty or more men stood in a circle outside of the burning brush and as the rabbits fled, the men shot them with bows and arrows. The historian Thomas Fletcher said, "In late summer and early fall the Kuzedika [Mono Lake Paiute] held rabbit drives around the lake flats. The drives required the participation of many people, some to hold the long nets into which the rabbits were driven, others to light fires in the sagebrush and force the rabbits into the nets."[58]

Many tribes captured ground squirrels, a reliable food source, by smoking them out of their burrows with the aid of a fire fan or burning them out. For example, a Tubatulabal hunter would pick a dry, flat area next to a meadow where there were many squirrel burrows. Dry grass and weeds were piled by some of the openings, and the rest would be blocked with dirt. The dry grass was set on fire, and with flat sifter trays smoke was fanned or forced down the open holes. After five or ten minutes the remaining holes were also blocked or sealed with dirt, and the hunters left for an hour to give the smoke time to do its work. When the holes were opened again, there would be many squirrels dead where they had crowded in the opening trying to get to fresh air. The Foothill Yokuts plugged ground squirrel burrows with grass, which was fired and kept burning by drafts from a feather fire fan. As the animals attempted to escape or get air at openings they were dragged out with a hook, seized, and their necks broken.[59]

The larvae of wasps and yellowjackets were a delicacy eaten by many tribes, and fire was sometimes used to find them. Marryat noted, "The Digger Indians burn the grass to enable them to get at roots and wasps' nests."[60] Similarly, tribes throughout California used fire to capture grasshoppers. Stephen Powers wrote in his journal in the 1870s, "In the mountains they [Yokuts] used to fire the forests, and thereby catch great quantities of grasshoppers and caterpillars already roasted, which they devoured with relish, and this practice kept the underbrush burned out, and the woods much more open and park-like than at present. This was the case all along the Sierra. But since about 1862, for some reason or other, the yield of grasshoppers has been limited. This was perhaps due to the effects of overgrazing by cattle and sheep."[61]

In 1855, near the South Fork of the Cosumnes River, the gold miner Alfred Doten (1849–1903) wrote in his diary: "In the fall when the grass is dry and grasshoppers plenty, all hands turn out; men, women, boys and girls. Just as a lot of us would go blackberrying a party of them surround half an acre or so with large brushes or brooms of chaparral, driving the grasshoppers in towards the center. The grass is then fired on all sides and, it being very dry, it burns like a train of gunpowder and the poor grasshoppers catch a scorching."[62]

The anthropologist Cora Du Bois described this practice among the Wintu: "Grasshoppers were obtained by burning off large grass patches. Two or three villages might participate in a drive. . . . The grass was set with torches three to five feet long made of dry wormwood or of devils' stems tied into bundles. . . . The grassy area was encircled by people who sang and danced as they whipped the grass and drove the grasshoppers into a center ring. The grass within the narrowed circle was then fired. After the blaze had subsided, men and women combed the ground for the insects, now partly roasted and with wings singed off."[63]

Colonel Rice described game hunting in the forests by the Cosumnes Indians (Plains Miwok) in 1850: "A whole settlement would turn out and begin operations by starting a number of small fires at regular intervals in a circle through the woods, guiding the flame by raking up the pine needles, and stamping out the fire when it spread too far. When the fires burned out there was left a narrow strip of bare ground enclosing a circular area of several acres, within which the game was confined. A large fire was then kindled at a point inside of the circle, taking advantage of the direction of the wind, and allowed to spread unchecked. The men, armed with bows and arrows and accompanied by their dogs, kept to the windward in front of the fire and shot down the rabbits and other small animals as the heat drove them from cover, while the women, with their conical baskets on their backs, followed up the fire to gather up the grasshoppers, which merely had their wings singed by the fire, but were not killed."[64]

MODIFYING THE STRUCTURE OF FOREST AND WOODLAND VEGETATION

In forests and woodlands, thickets of shrubs and small trees tended to accumulate over time, creating a potential wildfire hazard. Aware of the danger uncontrolled fires would pose to villages and collecting sites, California Indians regularly fired the understory in forests and woodlands "to keep the brush down" and promote the growth of wildflowers and grasses. After repeated burning, the fires were of low intensity and crown fires uncommon in many areas. There are dozens of examples of this deliberate modification of the structure of forests and woodlands, a few of which are mentioned here. The Karuk in northwestern California deliberately set fires in tan oak stands to decrease the brush, which facilitated acorn collection and prevented the buildup of dangerous fuels. Roland Dixon, a curator at the American Museum of Natural History in New York, reported that the Maidu burned every year, eliminating the underbrush, keeping the forests open, and reducing the likelihood of destructive forest fires. The anthropologist Frank Essene wrote that the Lassik regularly used fire to keep their territory clear of underbrush and make it easier to hunt and to travel, particularly adjacent to the Eel and Mad Rivers. Merriam recorded in his 1902 journal that the Chukchansi Yokuts kept the "brush burnt out of the flat parts of the valley and wild oats were thick and tall" in the Sierra Nevada. The Tolowa in northwestern California kept the nearby hills clear of brush by annual burning.[65]

Many elders in the Sierra Nevada foothills recalled how their ancestors set fires to clear the brush. Several examples are included below.

> Mother and Grandmother used to talk about burning the brush. The fires were set up around Cascadel [North Fork, California] in the ponderosa pines. The fire would creep along because it was late in the fall when the winds have stopped that they would set the fires. They'd burn

coming out of the forest, usually late fall before the storms start, about September, October and November. The fires would just go out. It wasn't covered like it is now with pine needles, old limbs and stuff that is all dry. Wherever the brush or trash was, they would set fires. But there wasn't the brush like there is now. Every year fires were set and they never got hot enough to burn the big trees. In 1913 we could ride straight in the saddle and not knock the brush out of the way. It was open country. Then they came with their fire restrictions. (Nellie Lavell, North Fork Mono, pers. comm. 1991)

I'm going by what the elders told me happened in the 1800s. Burning was in the fall of the year when the plants were all dried up when it was going to rain. They'd burn areas when they would see it's in need. If the brush was too high and too brushy it gets out of control. If the shrubs got two to four feet in height it would be time to burn. They'd burn every two years. Both men and women would set the fires. The flames wouldn't get very high. It wouldn't burn the trees, only the shrubs. They burned around the camping grounds where they lived and around where they gathered. They also cleared pathways between camps. Burning brush helped to save water. They burned in the valleys and foothills. I never heard of the Indians setting fires in the higher mountains, but don't take my word for it. (Rosalie Bethel, North Fork Mono, pers. comm. 1989)

My parents talked about the old-time Indians burning. All the elders talked about the Indians burning as they came down the mountains in the fall to the lower elevations. It was common knowledge. They'd burn in October or the last of September. The fires didn't burn out of control. Nacomas Turner said they let Yosemite National Park go to heck because they let the trash stay on the ground for so many years. Walking in the forest is like walking on foam rubber. The litter must have been a foot deep. Everything our people did 50 years ago they don't dare do today. My dad and mother used to burn on their properties. (Sylvena Mayer, North Fork Mono, pers. comm. 1991)

MAINTAINING HABITAT FOR SHADE-INTOLERANT PLANT SPECIES

The majority of plant species that California Indians relied on for food and medicine and for making cordage, basketry, and tools thrive only in full sun or partial shade. They include certain mariposa lilies (*Calochortus* spp.), lilies (*Lillium* spp.), chia *(Salvia columbariae)*, brodiaeas (*Brodiaea, Dichelostemma, Triteleia* spp.), buttercups (*Ranunculus* spp.), farewell-to-spring (*Clarkia* spp.), clovers (*Trifolium* spp.), mule ears (*Wyethia* spp.), and many others. The areas where the favored plants occurred frequently were burned so as to keep them open and decrease competition from weeds. Ecologically, fire was used to maintain the earlier successional stages that these species require.

Heightened species diversity, abundance, and density have been associated with regular, intermediate-intensity, spatially heterogeneous disturbance. Based on this relationship, it can be hypothesized that the disturbance caused by California Indians' use of fire in a variety of ecosystems, occurring at intermediate intensities and frequencies, promoted a maximally heterogeneous mosaic of vegetation types and increased species diversity. California's coastal prairies provide a good, well-researched example of how native practices promoted vegetational heterogeneity and high biodiversity (see Chapter 5). According to the ecologist Mark Stromberg, "The coastal terrace prairies in California and Oregon are the most diverse grasslands in North America. If you count the number of species in a square meter of California's coastal terrace prairie you average 22.6 species—more than the inland prairies of the Midwest, which have between 8 and 12 species" (pers. comm. 2001).[66]

Taking Care of Nature

We have learned that native plants were tended with a variety of resource management techniques, including pruning diseased parts of favored plants, weeding around plants to decrease competition and aerate the soil, replanting the smaller bulblets of harvested plants, and scattering seed. Many different habitats were burned to heighten the amount of forage and its nutritional value for various species of wildlife. Today California Indians often refer to these practices as "caring about" the plant or animal. Traditionally, Indians did not consider their actions management per se; "management" is a Western term implying control. Rather, caring for plants and animals in the California Indian sense meant establishing a deeply experiential and reciprocal relationship with them.

For millennia native people used the vast diversity of California's flora and fauna as sources of food, medicine, basketry, weapons, tools, games, shelters, and ceremonial items. Numerous plant and animal species were integral to every facet of Indian culture—religious festivals and life events such as childbirth, puberty, and death. Plants and animals were talked to, prayed for, and thanked with offerings. When they gathered or hunted, Indians adhered to ancient rules and techniques that allowed for resource use while keeping the resource base intact. As a result, some traditional gathering and hunting sites are very, very old.

By virtue of their daily use of plants, California Indians acquired extensive and special knowledge of the life histories of plant species, and they understood how different harvesting strategies affected natural regeneration. Their concern about replacement and return means that their gathering may be called "judicious," because the act is conducted with calculated temperance and restraint. Many of the gathering and management strategies are potentially sustainable, allowing for repeat harvests. Potentially sustainable harvesting strategies included harvesting plants for their tubers after seeding, cutting shrubs during the dormant period, sparing individual plants for future regeneration, granting plant populations rest periods, and using appropriate, nondestructive technologies.

As horticulturists, California Indians tilled the soil, pruned shrubs, sowed the seeds of wildflowers and grasses, and, in some cases, set the fires that nourished their food and basketry crops. These techniques had subtle yet important ecological impacts at the species, population, community, and landscape levels within a multitude of habitats in different parts of California. In particular, the development of fire as a vegetation management tool enabled women and men to systematically alter the natural environment on a long-term basis and at a massive scale. What becomes clear is that California Indians had a profound knowledge of the plants, animals, and ecological processes around them. When historical indigenous interactions— both harvesting strategies and resource management practices—are investigated in depth, we find that by keeping ecosystems in a modest or intermediate level of disturbance, in many senses Indians lived in ecological harmony with nature.

CHAPTER 5

Landscapes of Stewardship

The more research that's done on reconstructing fire histories and stand structures of ancient forests, the more obvious it is that Indian burning crafted the structure and composition of pre-Eurasian [contact] vegetation in Mediterranean climates throughout California.

JAMES BARRY, Sierra Miwok, senior state park ecologist (2001)

Had California been devoid of human habitation during the Holocene, the vegetation in many places—its structure, composition, extent, and distribution—would appear quite different today. Desert fan palms would not be growing in many, perhaps all, of the Sonoran Desert canyons where they now form green oases. In some areas where valley grasslands exist today, there would be forests, woodland, or chaparral. Many of the rain-fed montane meadows would have long since vanished, engulfed by montane forest. In some oak savannas there would be fewer enormous valley oaks, and the lower montane forests of the Sierra would boast fewer towering sugar pines. The ranges of some animals would be different, too; for example, island foxes would be absent from the southern Channel Islands. In short, many of the natural features of the present-day California landscape are not, strictly speaking, "natural," but are rather in part the product of deliberate human action.

The vast treeless prairies of the northern and central coast are one of the most conspicuous examples of how California Indians shaped the landscape. Without the efforts of their Indian caretakers, these prairies would now be covered with coastal scrub, chaparral, or coastal coniferous forests. George C. Yount described how such prairies were maintained in and around present-day Benicia: "During 100s of years, fires have spread over all the country, in the dry season of the year, and by this means the timber has been destroyed, except on the tops of lofty mountains, and in the neighborhood of creeks and rivers, so that all is one widespread prairy, of richest soil, ready and waiting for the plough of the husbandman." "Natural" fires seldom occurred along California's coast, so these were set intentionally by the Patwin.[1]

The California landscapes that early explorers, settlers, and missionaries found so remarkably rich were in part shaped, and regularly renewed, by the land management practices employed by native peoples. Many of the biologically richest of California's habitats were not climax communities at the time Euro-Americans arrived but instead were mosaics of various stages of ecological succession, or fire subclimaxes, intensified and perpetuated by seasonally scheduled burning. In a very real sense, some of the most productive and carefully managed habitats were in fact Indian artifacts. In many cases these landscapes experienced far greater degrees of managerial care and ecologically sophisticated manipulation than are found today.[2]

Over time, indigenous peoples' investment of time and energy in tending many habitats produced real biological changes in those habitats. Important features of major ecosystems may have developed as a result of human intervention, and many plant communities (coastal prairies, valley oak savannas, and montane meadows) had essentially become dependent on ongoing human activities of various kinds.[3]

It is highly likely that over centuries or perhaps millennia of indigenous management, certain plant communities came to *require* human tending and use for their continued fertility and renewal and for the maintenance of the abundance and diversity patterns needed to support human populations. As a result, the removal of California Indians from their homelands to reservations and rancherias led to a gradual decline in the area, number, and diversity of managed landscapes.

Some native people, displaced during Euro-American settlement of their lands, returned to their homelands years after relocation only to find them overgrown and untended. Maria Lebrado Ydrte, granddaughter of Chief Tenaya of the Southern Sierra Miwok and part of the tribe that was driven out of Yosemite Valley by the Mariposa Battalion, returned to her beloved Yosemite after seventy-eight years. She shook her head and said, "Too dirty too much bushy." The open meadows she had known in her childhood were covered with trees and shrubs.[4] Her great-grandson, James Rust (Southern Sierra Miwok), echoed Maria's original disapproval of the way the land had been treated by non-Indians:

> In the old days there used to be lots more game—deer, quail, gray
> squirrels, rabbits. They burned to keep down the brush. The fires
> wouldn't get away from you. It wouldn't take all the timber like it
> would now. In those times the creeks ran all year round. You could fish

all season. Now you can't because there's no water. Timber and brush now take all the water. There never were the willows in the creeks like there are now. Water used to come right out of the ground. I remember Yosemite when I was a kid. You could see from one end of the Valley to the other. Now you can't even see off the road. There were big oaks and big pines and no brush. There were nice meadows in there. (Pers. comm. 1989)

Constance Gordon-Cumming, an artist who visited Yosemite Valley in the 1880s, shrewdly warned of the consequences of no human intervention:

> Indeed, there is a corner of danger, lest in the praiseworthy determination to preserve the valley from all ruthless 'improvers' and leave it wholly to nature, it may become an unmanageable wilderness. So long as the Indians had it to themselves, their frequent fires kept down the under-wood, which is now growing up everywhere in such dense thickets, that soon all the finest views will be altogether hidden, and a regiment of wood-cutters will be required to clear them.[5]

Galen Clark, who had lived among the Southern Sierra Miwok for many years, tried, in a letter dated 1894, to convince the commissioners of Yosemite Valley to reinstate Indian burning:

> My first visit to Yosemite was in the summer of 1855. At that time there was no undergrowth of young trees to obstruct clear open views in any part of the Valley from one side of the Merced River across to the base of the opposite wall. The area of clear open meadow ground, with abundance of luxuriant native grasses and flowering plants, was at least four times as large as at the present time. The Valley had then been exclusively under the care and management of the Indians, probably for many centuries. Their policy of management for their own protection and self-interests, as told by some of the survivors who were boys when the Valley was first visited by whites in 1851, was to annually start fires in the dry season of the year and let them spread over the whole valley to kill young trees just sprouted and keep the forest groves open and clear of all underbrush, so as to have no obscure thickets for a hiding place, or an ambush for any invading hostile foes, and to have clear grounds for hunting and gathering acorns. When the fires did not thoroughly burn over the moist meadows all the young willows and cottonwoods were pulled up by hand. . . . Since Yosemite has been under the care of the State of California it was for many years the policy of its manager to protect the Valley as much as possible from the ravages of fires and to preserve all the young trees from destruction. This constant vigilant care for the preservation of Yosemite has resulted in the whole Valley being overrun with dense thickets of young forest trees,

shrubbery, and underbrush, and an accumulation of a vast amount of
highly combustible material, which in the event of accidental fires, is
a fearful menace to the safety of property and the beauty of the landscape scenery. . . . Since then the forest growth has so far encroached
upon the borders of the meadow land that there is not one fourth of
that amount, and what there is left is becoming so thickly covered
with young willow and cottonwoods of four or five years' growth that
there are really not fifty acres of clear ground in Yosemite, except such
as has been under very recent cultivation. During the seasons of 1891
and 1892 men were employed to thin out some of the thickets of young
pines and cedars and clear up the old logs and combustible matter, in
order to be able to more readily control destructive fires when accidentally started. One hundred and fifty acres were in this way partially
reclaimed at an average expense of $20 an acre. It is of the utmost im-
portance that this kind of work should be continued from year to year
until the whole Valley is reclaimed from a threatened wilderness and
restored in some degree to its original condition.[6]

Much of the landscape in California that so impressed early writers, photographers, and landscape painters was in fact a cultural landscape, not the
wilderness they imagined. While they extolled the "natural" qualities of the
California landscape, they were really responding to its human influence.
The chalk drawings and paintings of Thomas Ayres, Albert Bierstadt, and
Thomas Hill, among others, reveal centuries of fire management. The
wildflower displays they depicted were edible plant gardens. The parklike
appearance of Yosemite Valley and the incredible fecundity of parts of the
great Central Valley were marks of centuries of human management.

Ironically, many of the first non-Indian visitors and settlers described the
countryside in ways that hinted of human intervention. They regularly compared the landscapes to gardens, parks, or orchards. The Belgian gold miner
Jean-Nicolas Perlot wrote that the oak woodlands near Stevinson (the acorn
collection grounds of the Yokuts) "presented an appearance of a prairie
planted with fruit trees."[7] Dr. Lafayette Bunnell, in an unpublished article
earmarked for *Century Magazine*, wrote of "a great variety of evergreen
and deciduous trees, planted by Nature's landscape gardeners" and observed
that because "the undergrowth was kept down by annual fires while the
ground was yet moist, to facilitate the search for game, the Valley at the time
of discovery presented the appearance of a well kept park." But according
to the European rationale, only raw, unspoiled, unused nature was capable
of emanating this kind of beauty. It did not occur to them that the size of
the trees and the open pattern in which they grew were products of human
intervention.

New Evidence for Shaping California

There are three broad realms in which California Indians acted as agents of environmental change in many plant communities in California:

Genetic modification. Indians modified the gene pools and genotypes of plants through selective harvesting, burning, and transplanting. Over hundreds to thousands of years, various characteristics of many useful plant species were undoubtedly selected for by California Indian groups, thus modifying their genotypes. These species probably still exhibit suites of character traits that are adaptations to small-scale human disturbance regimes.

Dispersal. Indians were conscious and sometimes inadvertent agents of plant dispersal, and their practices have rearranged the distribution of some species and created unusual polymorphisms.

Habitat modification. With various resource management techniques, Indians expanded and maintained habitat suitable for desired species, changing ecosystems in both time and space without necessarily altering the character traits of component species.

There are a number of ways in which we now recognize the profound influences of previous human modification of species and habitats in California. One can discern how different the vegetation was when Indians were the sole inhabitants by walking the land after a catastrophic fire. Old village sites reappear where impenetrable chaparral lay before the fire. Sandstone cave areas with indigenous pictographs are exposed after coastal scrub is burned back by wildfires.

More detailed reconstructions of interactions between indigenous people and the natural environment in North America have recently materialized through research in the field of historical ecology, which combines knowledge in the realms of the social, physical, and biological sciences and the humanities. This field relies on the compilation, analysis, and interpretation of findings from plant ecology, paleoecology, archaeology, pyrodendrochronology, and other disciplines to identify specific biotic resources, ecosystem types, and whole biomes that were likely to be significantly influenced by historic and prehistoric indigenous management practices.[8]

Invaluable information also comes from those California Indian elders who remember details of burning and other horticultural practices of their ancestors. These findings are incorporated into a comprehensive framework for interpretation, giving us the most complete window to the past. The examples of changes wrought by California Indians highlighted in this chapter demonstrate the potential far-reaching ecological consequences of human selection for specific traits, dispersal of plants and animals, and land management practices designed to modify habitats.

DOMESTICATING DEVIL'S CLAW

It is now recognized that genetic changes in plants can occur in relatively short durations of time through human selection. Indians in the Southwest, Great Basin, and California acted as potent selective forces on many of the plant species they found useful. An example is devil's claw *(Proboscidea parviflora)*. The two claw-shaped appendages on the podlike fruit provide fibers that are stripped for use as black-colored splints in the closed coil basketry of several tribes. The work of the ethnobiologists Gary Nabhan and Amadeo Rea has shown that in the *P. parviflora* gene pool there is a true domesticate—distinct from the wild genotype—that becomes robust when cultivated, watered, and fertilized. Domesticated varieties have longer claws and bear more fruit, and their seeds have a white rather than black seed coat. Emergence of domesticated devil's claw is a recent historical phenomenon that resulted from changes in the plant's availability in the wild and an increase in the marketability of baskets, which encouraged a shift to dependence on the plant's cultivation.[9]

ALTERING THE DISTRIBUTION OF PLANTS AND ANIMALS

California Indians widened the ecological amplitude of native species by introducing them to new areas. Some plants were actually transplanted, others were sown as seed in new locations, and some were dispersed as seed to new areas accidentally but then cared for after sprouting. Dispersal through Indian agency may have taken place over relatively short distances and may thereby have created the appearance that the normal distribution was somewhat more extensive than was actually the case. The inability of plant and animal geographers and authors of floras and faunas to distinguish natural and human-extended distributions has probably resulted in many mistaken assumptions about the "natural ranges" of plants and animals.[10]

As early as 1914 the ethnobotanist Melvin Gilmore claimed that the flora of a specific region was undoubtedly influenced by Indian populations and that plants viewed as native sometimes had been introduced and disseminated through human agency. Another ethnobotanist, Volney Jones, pointed out that the location and outlines of archaeological sites can often be ascertained by differences in the nature, quantity, or vigor of the plants growing on the site, as compared to those surrounding it. In addition, some phytoarchaeological studies have revealed the alteration of plant distributions by prehistoric cultural practices.[11]

There are many accounts around the world of plants established in association with historical sites of human occupation. *Agave parryi* exhibits an interesting association with archaeological sites in the western portion of the Apache-Sitgreaves National Forest in Arizona. Three species of culturally important plants, *Cleome serrulata, Lithospermum caroliniense,* and *Salvia subincisa,* are largely confined to Pueblo ruins in New Mexico. Nettles *(Urtica dioica)* are well established in enriched soils around camps of the Salish in Puget Sound and in village sites northward all along the coast.[12]

In California similar observations have been made. For example, jimson weed, or toloache *(Datura wrightii,* formerly *Datura meteloides),* a significant ceremonial and medicinal plant native to northern Mexico, is often associated with Indian village sites in San Diego County and other archaeological sites in southern and central California. The eminent botanist Willis Jepson says of jimson weed: "Since its occurrence in the Coast Range Valleys and Great Valley fifty to seventy years ago was characteristically associated with rich deep soils about old Spanish settlements and Mexican ranches, its introduction doubtless antedates the American occupation of California. The plant was sacred in certain religious rites of the native tribes. . . . It is therefore possible, by reason of tribal usage, that introduction may have been much earlier, that is, prior to the Spanish settlement of California."[13]

Charles Lummis told Jepson that the word *toloache* is "derived from the Aztec name of the Datura, and in its Spanish form is a familiar provincialism of the Mexican states of Chihuahua and Coahuila and is common among the Indians and Mexicans of California." The plant loves disturbance and appears in freshly plowed fields. Datura was used during boys' puberty ceremonies and other religious ceremonies and was administered as a drug for alleviating pain, particularly in the treatment of burns and broken bones.[14]

Individual plants or small colonies of the California walnut *(Juglans californica* var. *hindsii)* are often found growing in association with archaeological sites in Walnut Creek, Walnut Grove, and the Napa Range above Wooden Valley.[15] These walnuts, Jepson wrote, "are found in all cases about old Indian village sites and in the case of the Wooden Valley station the colony is gradually being exterminated by the advance of the primitive forest which is no longer held in check by the one-time occupants of the tribal settlement."[16]

The native tobacco *(Nicotiana clevelandii)* may have been planted in the enriched soils of the Indian middens around the camps of the Chumash in Santa Barbara. This plant also occurs on Santa Cruz Island, and the ethnobotanist Jan Timbrook points out the likelihood that the Chumash brought tobacco to the island. Paul Schumacher, an archaeologist, noted that he always found elderberry *(Sambucus mexicana)* shrubs in the neighborhood of ancient settlements or near graves.[17]

Traditional gathering sites for the edible bulb camas *(Camassia quamash)* are ancient in California. The low genetic variability in populations of the species in the Sierra Nevada is probably a result of the trading and selective harvesting practices of various tribes over long periods—a demonstration of both dispersal and selection (Susan D'Alcamo, botanist, pers. comm. 1993).

Paul Collins, a vertebrate zoologist, has concluded through examination of the archaeological record and morphometric analysis that Indians were responsible for establishing island foxes *(Urocyon littoralis)* on San Clemente, Santa Catalina, and San Nicolas Islands from northern Channel Island populations. Daniel Guthrie's faunal discoveries at archaeological sites on San Miguel Island indicate that the earliest record of the deer mouse *(Peromyscus maniculatus)* occurs at the same level as the first evidence of human occupation, suggesting accidental human transport of this species to the island.[18]

EXPANDING THE POPULATION SIZE AND DISTRIBUTION OF DESERT FAN PALMS

Groves of native desert fan palms *(Washingtonia filifera)* are widely scattered throughout the Sonoran Desert of southeastern California, within the territories of four tribes: the Cahuilla, Serrano, Tipai, and Chemehuevi. The desert fan palm was a welcome sight to people making the long trek across harsh desert environments. Palm oases often indicate the presence of water and frequently harbor a diverse array of plant and animal species. Archaeological sites, containing bedrock mortars, pottery sherds, and rock carvings, are associated with most oases. The desert fan palm has been very important to Indian economies in the form of fruit for meals, fronds for thatched roofs of round houses and other structures, and fibers for sandals and basketry. (See Figure 17.)[19]

As a group, palms have existed for at least forty million years in western North America. Fossilized imprints of palm fronds have been found embedded in rocks dating to millions of years before the present. Because of

Figure 17. Three women from the Palm Canyon Agua Caliente Reservation
in southern California carrying bundles of palm fiber *(Washingtonia filifera)*.
The photograph was taken before 1924. Desert fan palm oases in California may
be an artifact of planting and fire management by indigenous people. Courtesy
of the National Anthropological Archives, Smithsonian Institution, #57078.

the tropical affinities of its family and its presence in scattered, disjunct populations, the desert fan palm was labeled by the paleoecologist Daniel Axelrod a relict species. He hypothesized that it was much more widely distributed in the past when the climate was tropical and more favorable. Other ecologists have accepted and restated Axelrod's thesis without additional evidence.[20]

James Cornett, a biologist at the Palm Springs Desert Museum, has studied the desert fan palm for twenty years and has reached very different conclusions about its evolutionary history and biogeography. Cornett relied on a variety of methodologies in his studies of the fan palm, including ecological field studies, ethnographic interviews, ethnohistoric literature reviews, paleontology, and plant genetic studies. He found evidence that the species is not a relict at all. Reviewing the fossil record of palms in California, Cornett found no specimens that displayed spines on the petioles, often cited as an important characteristic of the genus *Washingtonia*. Without petiole spines, none could be classified in this genus. A relict species also would be expected to have a stable or shrinking geographic range—yet the range of *Washingtonia* is expanding.[21] Finally, a relict species with disjunct populations should display genetic divergence between populations due to isolation, but this does not appear to be the case. Electrophoretic studies conducted by Leroy McClenaghan and Arthur Beauchamp in the 1980s, in Anza Borrego State Park, reveal low genetic diversity, both within and among populations, pointing to recent colonization by fan palms.[22] Cornett's working hypothesis is that the fan palm is of recent occurrence and probably appeared within the present boundaries of the United States no earlier than the end of the Illinoian glacial episode. Further, his work suggests that the palm was introduced to the United States, specifically California, by indigenous people. Cornett states in a 1987 article: "It is not inconceivable that the entire status of palms in the Sonoran Desert might have been vastly different were it not for the Indians' use and management of this spectacular desert plant."[23]

The ethnographic literature and interviews with contemporary Cahuillas confirm that historically and prehistorically they enhanced palm populations through planting seeds and firing palm stands. Native elders assert that burning the palm oases increased fruit yields, decreased insect pests, cleared debris, and promoted seedling survival. These fires likely maintained the palms in far greater numbers than would have occurred under natural conditions, and influenced the age structure of the palm populations and the species composition of the oases. Jepson, in his 1910 *Silva of California*, spoke of the burning of the palm stands by the Indians in the vicinity of Palm Springs: "In order to get up the tree the more readily they were accustomed to burn off the dead . . . leaves. All the old trees in Palm Cañon show the signs of such burnings."[24]

The practice of planting seeds suggests that Indians probably acted as dispersal agents for the palm, carrying seed from one oasis to the next and expanding the palm's geographic range. Coyotes also disperse palm seeds: they eat the palm fruit, digest the fleshy part, and eliminate the intact seeds in their scat. But coyote dispersal cannot explain isolated groves seventy miles from the nearest palm oasis, because palm seeds are not likely to remain in a coyote's digestive tract for the six days it would require a coyote to travel this distance. Human dispersal is the only satisfactory explanation for the existence of these remote palm oases.[25]

In both an academic and a practical sense the knowledge of Indians' introduction of the fan palm into California has profound implications for various disciplines. For example, in plant biogeography, the assumption that the range and distribution of the desert fan palm is dictated by climate and geology alone is probably false. Further investigation of humans as conscious agents of plant dispersal in California floristic studies may help to explain additional distributions of plants to plant biogeographers.

MAINTAINING AND EXPANDING THE COASTAL PRAIRIES

Coastal prairies extend from north of Morro Bay to the Oregon border, cutting across the territories of many California Indian tribes, from the Salinan in the south up through Tolowa territory in the north. Native people have inhabited the northern coastal area for more than eight thousand years; the intersection of water and land and the proximity of food-rich marine communities made life inviting here.[26]

This biologically diverse community of the coastal prairie contains many plant species important to Indian economies for construction, cordage materials, fish poisons, foods, household items, and medicines. At what is now Point Reyes National Seashore, the tidytips, California buttercups, and gold fields—all annuals gathered for their edible seed by the Coast Miwok—still bloom every spring.[27]

In 1847 Charles Nordhoff, who spent eleven months on the California coast, noted a universal belief that the lack of trees on the coastal prairies was evidence of the soil's infertility. Years later, when he wrote a book on California for the would-be traveler, he described how this earlier assumption had been contradicted by actual experience: "But many of those treeless plains have since yielded from fifty to eighty bushels of wheat per acre, and there is no year in which some adventurous farmer does not discover some new product for which the climate and soil are specially adapted, and which pays better than gold-mining."[28] What people had missed was the Indians' role in preventing trees from growing in the prairies.

Originally the fertile prairies supported herds of tule elk, pronghorn antelope, and mule deer. More than one thousand tule elk composed a single herd. Scientists concur that this plant community very likely developed under relatively mild grazing pressure but with a short fire return interval. Fires from lightning storms might occur as rarely as once every eighty to one hundred years, not frequent enough to maintain the openness of the land. There is ample evidence from historical documents that local tribes maintained the coastal prairie through burning, usually in late summer or fall.[29]

The Russian naturalist Adelbert von Chamisso described such an indigenous-set fire in his personal diary on October 22, 1816: "At night, on the land behind the harbor [San Francisco], great fires burned. The natives [Ohlone] are in the habit of burning the grass in order to further its growth."[30] In 1818, near Fort Ross in Coastal Pomo territory, a Russian officer, F. P. Lutke, observed the following:

> When it was completely dark we had a very interesting spectacle: a certain extent of land near the settlement was all afire. The Indians who live in this area eat a wild plant which resembles rye, for which reason our settlers call it *rozhnitsa* (rozh, rye). When the kernels of the rozhnitsa have been harvested, the straw which remains is generally burned. This procedure makes the next year's crop bigger and more flavorful. The fires continued throughout the night.[31]

Archibald Menzies, the naturalist who accompanied George Vancouver on the *Discovery*, in his voyage around the world, landed on the west side of Tomales Bay on October 20, 1793. He jotted in his journal: "We landed on the west side & ascended the high ground which formd [sic] the bluff headland. . . . [T]he grass & brush wood on this headland had been lately burnd [sic] down so that I had little opportunity here to augment my botanical collection, the few plants I saw were not different from those I had before met with at San Francisco & Monterey excepting a few species of *Sisyrinchium* with yellow flowers of which I brought on board live plants for the garden."[32]

George Gibbs, acting as interpreter on Redick McKee's expedition through northwestern California in 1851 to make treaties with various tribes, commented in his journal on the lushness of the grasses: "Prairies of rich grass lie on their southern slopes, and especially on their tops, from whence their name of *Bald Hills* is derived. . . . Late in the season, however, the grass is often burned, and dependence cannot always be placed upon the usual grounds. . . . Elk are very abundant in these mountains, and the ground was marked everywhere with their footprints."[33]

The coastal prairies were burned to produce more food, reduce brush or trees, produce new grass for thatch, drive grasshoppers, enhance cordage materials, and increase forage for ungulates. Indian-set fires modified the grassland to fire-resistant species and expanded the grassland vegetation type. These fires also may have increased the productivity of the soil.[34]

Many of the records in the anthropological literature of Indian burning of coastal prairies connect the practice with a specific purpose. The Pomo, for example, burned areas of California fescue *(Festuca californica)* to create more new growth for future harvesting of thatch for covering winter houses. Cordage materials, including the stiff, heavy fiber from iris *(Iris macrosiphon)* leaves, were gathered from coastal prairies. Members of northwestern California tribes, including the Karuk, used a single fiber found in the center of iris leaves to make cordage used to create nets for fish, birds, and small mammals. Camping bags were also made from the fibrous leaves. All these products required the gathering of large numbers of plants, and burning the areas in which they grew was a way of ensuring an adequate supply.[35]

The Sinkyone, Cahto, Pomo, and Yuki burned grassy areas to drive and roast grasshoppers. Areas that harbored soaproot *(Chlorogalum pomeridianum)*, a plant of great importance to all coastal tribes for food, shampoo, fish poison, adhesives, and brushes, were burned by the Pomo. Grassy areas along the coast were burned annually by the Coast Yuki to facilitate gathering of tarweed seed.[36] Near Santa Cruz, Fray Juan Crespi noted in 1769: "The mesas . . . must be about one league wide, extending to some hills at the foot of the mountains. . . . Only in the watercourses are any trees to be seen; elsewhere we saw nothing but grass, and that was burned."[37] Alfred Kroeber recorded in his 1939 field notes: "The Yurok used to burn regularly prairies where there were grass seeds; also places where the white squaw-grass grows, and also hazel, both used for baskets. They would take turns, burning one piece of ground over and the next year another one and only in the second year would they come back to where they had burned first."[38]

Lucy Thompson, a prominent Yurok basket weaver, wrote and published a book in 1916 in which she described the usefulness of the coastal prairie flora and the necessity of burning along the Northwest coast: "All open prairies [were] for gathering grass seeds, such as Indian wheat, which looks similar to rye, besides other kinds of seed. . . . The Douglas fir timber they say has always encroached on the open prairies and crowded out the other timber; therefore they have continuously burned it and have done all they could to keep it from covering all the open lands."[39]

The coastal prairies included not only large areas of continuous prairie along the coast, such as the Bald Hills in Redwood National Park, but also many smaller prairie patches in the middle of coastal redwood forests. Fires that were set to maintain and expand these patches escaped into the surrounding coast redwoods, leaving a fire scar record in the trees' rings. Ernest de Massey, a French gold miner, recorded his trip through "a magnificent forest" in Trinity County in 1850: "Here were pines [redwoods] growing one hundred meters high that measured twelve to fifteen meters in circumference and were capable of making planks six meters or more wide. . . . Others were half-burned, maybe by lightning, by the Indians, or by some careless trapper. No doubt all these three causes contributed. The Indians, particularly in the spring and autumn, set the stubble in the pastures on fire to destroy the insects and reptiles, and to make hunting easier."[40]

The forest ecologist Susan Bicknell has accumulated evidence that many areas of the California coast now covered by forest and scrubland were once coastal prairie. She studied eight sites, including Salt Point, Sinkyone Wilderness, and Fort Ross, to determine presettlement vegetation patterns. In many instances she found that the grasslands were formerly considerably more extensive than they are today. She used a variety of methodologies, including vegetation and soil sampling, pollen analysis, charcoal analysis, ethnohistoric literature reviews, and opal phytolith analysis. Opal phytoliths are microscopic mineral remains of plant cells that remain unchanged in soils for a long time. They can be extracted and identified taxonomically by their shape and can tell us about the vegetational history of a site. For example, opal phytoliths displaying dumbbell shapes are characteristic of several native California grasses such as oatgrass (*Danthonia californica*). Soils from areas currently occupied by coastal scrub, bishop pine forest, mixed evergreen forest, redwood forest, and riparian woodland contain high concentrations of grass phytoliths, indicating previous grassland vegetation.[41]

Coastal prairies must have significantly dwindled in size since the arrival of Euro-Americans. Bicknell and other researchers have concluded that the reduction in these vegetation mosaics result in part from the absence of frequent burning by California Indians. The lightning fire regimes and ungulate grazing alone would not have been sufficient to keep small or large areas in coastal prairie in precontact times because the climate has been changing, becoming more favorable to the growth of forests.[42] According to paleoecologists, in cismontane California, oaks rapidly replaced conifers, primarily pines, during the transition from the late Pleistocene to the warmer Holocene. Oaks appear to have expanded their range significantly by the mid-Holocene, becoming the dominant arboreal form in many areas. Pollen records from the North Coast Ranges indicate that there was an expansion in the range and density of conifers, particularly Douglas fir, at the expense of oaks in most places, by about 2,500 to 2,800 years before the present. This expansion in conifers is interpreted as being associated with the transition to the present-day cooler, moister climate. The soil scientist James Popenoe supports Bicknell's conclusions: "Regular burning by resident Native Americans would have prevented Douglas-fir from becoming established in prairies and oak woodlands, even as the climate became more conducive to conifer forest."[43]

PRESERVING MONTANE MEADOWS

Strings of meadows in the Sierra Nevada, the Cascades, and other California mountain ranges served as pantries, trade grounds, and resting places for numerous tribes. Archaeological sites occur frequently at the periphery of large meadows. Old Indian trails lead into and out of meadows, and these networks of clearings dictated movement among villages and between tribes.

Montane meadows play important ecological roles. They break up the mixed evergreen and lower and upper montane forests, giving the landscape a quiltwork pattern and creating ecotones, areas of rich biodiversity where forest and meadow merge. The rare great gray owl forages primarily in montane meadows, and a host of other wildlife species, such as a grouse that feeds on knotweed (*Polygonum* spp.), regularly use and visit them. Ungulates, such as mule deer, gravitate to meadows, making these clearings opportune places for hunting.[44]

Many mountain meadows below an elevation of 7,500 feet, like the coastal prairies, are not truly natural but were influenced for millennia by California Indian burning practices and, in more recent times, by livestock and sheep grazing. Native people altered the size and composition of certain meadows through periodic burning. Setting fires in the ecotone areas surrounding the meadows discouraged the encroachment of more wet-tolerant lodgepole pine and other conifers, thus maintaining—and perhaps in some cases enlarging—meadow areas. Periodic burning also occurred within the meadow boundaries and influenced the species composition, density, and frequency of native plant populations. This contests the generally held view that Indians did not much influence meadow vegetation. Certain meadow plants such as deergrass, yampahs, and clovers were favored and maintained through burning. In addition to burning, Indians hand weeded young conifers and hardwoods to keep the forest from encroaching on meadows or prairies.[45] California Indians vividly recall these former practices:

Fires were set in the higher elevations sometimes to decrease the buck brush, whitethorn and manzanita. They burned in and around the higher mountain meadows to burn out the trees. (Ron Goode, North Fork Mono, pers. comm. 1989)

My grandfather, Jack Lundy, who was born in Yosemite, talked about burning. They would burn in the late fall, about October or November, because the onset of rains and snows would help extinguish the fires. They always burned from the bottom of the slope, never from the top of the slope, and always with the wind. They burned in the meadows— Paiute Meadow, Buckeye Meadow, and Sweetwater Canyon, the eastern part with the little meadows. Fires also would burn back the sagebrush and the tamarack [lodgepole], providing better pasture in the meadows for the animals. (Marshall Jack, Owens Valley Paiute/Central Miwok/ Washoe, pers. comm. 1990)

My father would cut the brush out of the meadows and on the edge of the meadows—Peckinpaw, Brown's, and Camp 10—and burn the brush. My dad always kept brush out of the meadows. (Sylvena Mayer, North Fork Mono, pers. comm. 1991)

Cattle ranchers and sheepherders continued the practice of burning meadows as early as the 1860s. They learned directly from the Indians.

The cattlemen and the sheepmen burned every year in the fall in October or the last of September. They got the idea from the Indians. It kept everything cleared out for the cattle, kept the brush down and green. The buck brush gets a big thorn when it's old and cattle won't eat it then. The fires would just smolder along. Fires were set in lower country, Arnold and on up through the Big Trees. Cattle were run towards Blue Mountain, South Fork, Black Springs, Big Meadow, Dorrington, Airola Meadow and Creek, and fires were set in these areas. Fires were set all around the meadows to keep them large and the trees from coming in. (Buster Riedel, pers. comm. 1990)

The ranchers would leave their campfires burning at 6,000 feet, and it burned a major part of different meadows like Muggler Meadow in Chiquita Basin. These burned in dry years. (Leroy Brown, pers. comm. 1990)

Pinky Bethel [North Fork Mono], Jim Tex [North Fork Mono], Ernie Goode [North Fork Mono], and Sam Pomona [North Fork Mono] told me that the Indians used to burn. Indians taught the ranchers how to burn. They burned to keep everything clean. The Indians kept the tamaracks [lodgepoles] burned out around the meadows. The Indians say there were none of those trees when we first came here. Fires were started in late fall about the last of September, October, or November. (Jim McDougald, pers. comm. 1990)

As the U.S. Forest Service and other government agencies began to suppress fires at the turn of the twentieth century, meadows that had been burned by California Indians began to fill in. Cattle ranchers speak of meadow encroachment with sadness.[46]

> My aunt and uncle up on the Sonora Pass, they used to burn. They always burned the skunk cabbage in the meadows in the fall and it also kept the willows down and next year they would come up from the roots again. They would burn every year, weather permitting, at the end of October or beginning of November. In the last twenty years I've seen such a change. There aren't any meadows anymore. Now there is nothing but willows ten feet tall. I don't even think there's a foot trail or horse trail anymore in the Kennedy Meadows area. (Rose Mitchell, pers. comm. 1990)

Meadows are an intermediate step in the chain of plant succession. Most meadows begin as lakes or ponds, gradually fill up with sediment to become meadows, and then slowly transition to forest as trees encroach from the edge. The soil moisture regime during the meadow stage is the single most significant factor that determines the rate of succession to forest.[47] In the absence of human influence, a meadow persists only as long as the ground is too wet for trees to encroach. Some meadows have had thousands of years to convert to forest, yet only in the past two hundred years have they been encroached by trees. With the absence of indigenous burning and weeding and the suppression of lightning fires, shrinkage of meadows has been occurring throughout the mountains in recent years. Large meadow areas bordered with aspen that used to have watercourses running through them are now stocked with thickets of conifers.

Should montane meadows be left alone to allow the "natural" process of succession to transform them into forest, or is there greater value in maintaining them as meadows, as indigenous Californians did for perhaps thousands of years? Yosemite National Park seems to have chosen the hands-off approach. Big Meadow lies at 4,300 feet in the mixed conifer vegetation zone in Yosemite National Park. It has resources that were important to the Southern Sierra Miwok, including deergrass for basketry. Big Meadow historically was an important Southern Sierra Miwok temporary summer and fall camp, and there are more than six hundred mortar holes in granite rock outcrops for processing plants within a quarter mile of the meadow (Bob Fry, pers. comm. 1986). A Yosemite National Park interpretive sign at a turnout off Highway 120 describes the succession of Big Meadow and ignores the cultural practices of the Southern Sierra Miwok that kept Big Meadow open: "The open area below is a former lake. Meadow land is only transitional. Even now trees are invading, slowly transforming the meadow to forest. From here, over the years, Big Meadow will appear as a pale green lens that shrinks and eventually disappears. National parks preserve not only panoramic vistas, but also dynamic natural processes that continue to shape the scene."

ENHANCING HAZELNUT FLATS

People of many tribes in California picked hazelnuts *(Corylus cornuta* var. *californica)* by the basketful and ate them raw or roasted. They were often hulled by hand at leisure in the village. The Wintu also hulled hazelnuts by beating them with a willow switch. Often a supply was dried, stored in the shell, and kept on hand through the winter and spring. The Yurok pounded hazelnut kernels into a flour, added warm water, and used the mixture to nurse sickly children or ill persons with weak stomachs.[48]

In addition to being a food source, hazelnut shrubs provided Indians with materials for many cultural uses—but only after the shrubs had been burned, because fire induced the sprouting of straight young shoots. The young hazelnut shoots that came up after fires were used by a number of northwestern tribes to make various kinds of baskets, including cradleboards and carrying baskets. The new growth was also twisted to make rope, and employed in the construction of fish traps for salmon and surf fish baskets. The Pomo and Ohlone used the straight new hazel branches for arrows. The Miwok and Yokuts selected sprouts that had grown for two years since the last fire and bent them into looped sticks for stirring acorn mush. The Karuk used older hazelnut wood (one-half inch in diameter), perhaps three years after the last fire, to fashion circular frames for snowshoes.[49]

Harold Driver recorded that the Karuk and Wiyot burned "hazel bushes" for "better basketry warp." Yurok weaver Lucy Thompson reported that the Yurok deliberately set fires to enhance nut production: "In taking care of the hazel flats, they go out in the dry summer or early in the fall months and burn the hazel brush."[50] Entire hillsides were also burned by the Karuk. Georgia Orcutt, a Karuk elder, told anthropologists in the early 1950s that there "used to be more hazel nuts than now" and that the "hazel shoots were better when the brush was burned down each year." "Now the brush has grown up all around," she continued, "and nothing is any good any more."[51]

The Thompson Indians of southern British Columbia also burned hazelnut bushes to enhance nut production, and this may have been a common practice in much of the Pacific Northwest.[52] Burning probably enhanced the food supply for myriad wildlife. Squirrels, chipmunks, woodrats, and other small mammals as well as birds feed on the nuts. Grouse commonly eat the catkins. Rabbits and deer browse on the whole plants. Today weavers in California still gather young hazelnut shoots, in many cases substituting pruning for former burning practices, to encourage production of desirable material.[53]

CULTIVATING TOBACCO PATCHES

California's native tobaccos, both *Nicotiana attenuata* and *Nicotiana quadrivalvis*, were widely used by tribes in rituals, as offerings, and medicinally to heal cuts and as an emetic.[54] Burning, pruning, and sowing of areas of native tobacco were common practices throughout California, and there is evidence that in some areas the care of tobacco patches approached a level resembling that of agriculture.

The Western Mono loosened the earth around favored tobacco plants with digging sticks when it became dry and carried water to soften it. The tips of the plants were pinched off to encourage the growth of big leaves.[55] The ethnobotanist Maurice Zigmond described the leaf pruning process among the Kawaiisu: "It was relatively late in the summer when the women embarked upon a series of prunings of the individual tobacco plants [*Nicotiana quadrivalvis*] and may also have done some weeding at the same time. There were three prunings a week apart as leaves were approaching maturity. On each occasion the small weak leaves, the new growth at the junctures of the large leaves and stalks, and the flowering tops were broken off. After the third pruning there were left only the large healthy leaves on the stems. About five days after the last pruning, when these leaves were picked off, only the bare stems remained." Zigmond noted further that "sometimes the ground about the plants was burned to make them grow better."[56] (See Figure 15 in Chapter 4.)

Enhancing tobacco growth was one of the most consistently recorded reasons for indigenous burning in California. For example, Driver recorded that the Western Mono, Foothill Yokuts, Panamint, Kawaiisu, Tubatulabal, and Owens Valley Paiute all pruned tobacco to increase leaf size and burned over the fields where the tobacco grew.[57] Omer Stewart recorded burning by the Pit River (Achumawi): "When the grasslands, with their weeds and herbs, dried in the late fall they were set on fire nearly every year, because the Achumawi recognized that burned-over plots produced tobacco and wild seeds more abundantly than the areas not burned."[58]

Not only were tobacco plants pruned and the areas in which they grew burned, but seeds of tobacco were sown. The Sierra Miwok understood the environmental conditions required by native tobacco, so they sowed seeds on north-facing slopes.[59] The Yurok cultivated tobacco in the following manner: "[After] selecting a proper place, pile brush over the ground and then burn it, which would leave the ground with a loose layer of wood ashes. Over this, while the ashes were yet dry and loose, they would sow the seed and protect the crop by putting around it a brush fence. From year to year they would select from the best stalks, seed for the next year, and at times to hold the seed for a number of years if necessary, for if kept properly it will grow after being kept for a long time."[60]

Burning off shrublands to plant tobacco seeds was common among various tribes:

> Tobacco grown [by the Tolowa]; burned off clump of brush, planted seed, covered with aromatic leaves, fir boughs, etc., to impart good flavor; patch sheltered by brush windbreak, to prevent wind from blowing away strength of leaves.[61]

> Tobacco was cultivated [by the Shasta]; every spring after burning logs and brush, wild tobacco was planted. There was a tobacco garden at Butler Flat and others elsewhere.[62]

> Where logs have been burned the best ones grow. They [the Karuk] never sow it [tobacco] in an open place. Upslope under the trees is where they sow it. . . . And where they are going to sow tobacco, too, they burn it, too. . . . It is in summer when they set fire to the brush, at the time when everything is dry, that is the time that is good to set fire, in the fall before it starts in to rain.[63]

Patches of tobacco growing on open shrublands and in forests were unique habitats ecologically. Today they are no longer present; instead, only a few scattered plants of native tobacco can be found.

SHAPING VALLEY OAK SAVANNAS

Valley oaks *(Quercus lobata)* reach the largest size of all California's oaks, with trunks up to nine feet in diameter. These majestic giants dot many stream courses in valleys and the surrounding foothills and can live for six hundred years. Their acorns were valued by many tribes for food and were often mixed with acorns from other species. It was in reference to their watery acorns (which made a watery mush) and proximity to streams that various tribes called them "water oaks." The bark was used by the Concow Maidu as a dye to blacken redbud strands for basketry, and medicinally by the Yuki to alleviate diarrhea. The Kawaiisu used valley oak logs for house construction, and the acorns were made into tops for children.[64] To native people, these regal trees marked the seeming timelessness of the earth and also the continuity of life. They were massive fixtures on the landscape bearing witness to many generations of humans. They served as landmarks and meeting grounds, provided shade for work stations, and were the source of plant materials to satisfy human needs.

Riparian forests once covered nine hundred thousand acres on the alluvial fans and floodplains of the Central Valley, and valley oaks were a significant part of that vegetation. Historically valley oaks occurred not just on the upper sandy benches along rivers but also in miles-wide swaths radiating out from the river corridors. Away from the immediate river channels, valley oaks formed savannas with luxuriant grass cover underneath. Lieutenant George Derby of the U.S. Topographical Engineers surveyed the Sacramento Valley from the American River northward to Butte Creek in 1849, recording a band of valley oaks two to six miles wide along all the major watercourses.[65]

There are numerous other early accounts describing this parklike landscape. During an exploring expedition in May 1817, Fray Narciso Durán described the oaks along the Sacramento River: "All along this river it is like a park, because of the verdure and luxuriance of its groves of trees."[66] Traversing the east side of the San Joaquin Valley along the Merced River in 1833, Zenas Leonard provided this account:

> The land is generally smooth and level, and the plains or prairies are very extensive, stretching towards the setting sun as far as the eye can reach; whilst a number of beautiful rivers, all heading in this rugged mountain, running parallel with each other thro' the plain, also to the west, with their banks handsomely adorned with flourishing timber of different kinds. . . . This grove of timber may be found along the river at any point, and generally extends about four miles into the plain.[67]

Edwin Bryant commented on the valley oaks of the Sacramento Valley near Sutter's Fort in 1846, mislabeling them "evergreen":

> This plain exhibits every evidence of a most fertile soil. The grasses, although they are now brown and crisp from the periodical drought, still stand with their ripened seeds upon them, showing their natural luxuriance. Groves or parks of the evergreen oak relieve the monotony of the landscape, and dot the level plain as far as the eye can reach.[68]

After European settlement, this plant community formed what L. T. Burcham called "some of the finest grazing land available in the State." The soil scientist Eugene Hilgard and colleagues noted in 1882 that the grand trees seen in almost all the fertile valleys of California "interfere very little with the growth of crops under their widespreading branches."[69]

Like many of California's other plant communities, the vast, open, and highly productive oak savannas encountered by early Euro-American explorers, then grazed and farmed by the first settlers, were largely the products of conscious management by Indians. Jepson was one of the first non-Indians to tie the structure of these oak woodlands and the size of their magnificent trees to California Indian practices:

> The long inhabitation of the country by the Indians and the peculiar local distribution of the Valley Oak in the rich valleys are in some way connected. These oak orchards, of great food importance to the native tribes, indicate plainly the influence on the trees of Indian occupancy of the country. The extent and nature of the relation of Indian tribal culture and the habit of the oaks cannot yet, if ever, be completely defined, although it is clear that the singular spacing of the trees is a result of the periodic firing of the country—an aboriginal practice of which there is ample historical evidence.[70]

Jepson's conjecture that native peoples' use of fire was responsible for the structure of the oak savannas is backed by several lines of evidence. Observances from early visitors and settlers confirm indigenous burning:

> There are at this spot sixty oak trees and a few willows in the bed of the stream. The forage was extremely scanty, and that the country appeared to have been burned by the Indians [travel diary dated September 27, 1806].[71]

Jacques-Antoine Moerenhout, the French consul in California between 1843 and 1856, witnessed a fire on the San Joaquin plains on July 15, 1848:

> While this fire was very lively, it was only consuming the grass and wild oats which, being very thick and dry, threw up a bright flame and were burning so fast that it would have been impossible to stop or to extinguish the fire. Moreover, several trees were afire and some, partly burned, had fallen across our way. This was but a slight obstacle to our progress, however, and the only one, for since the road was quite wide and free of grass and trees, the fire had stopped at its edges. We noticed even several places where this slight break had cut the line of the fire and had sufficed to save some considerable areas from destruction. In less than half an hour we had passed the fire. That is to say, that as we approached the Moquelames [Mokelumne] River the fire was burned out, but everything was black and dismal. No more pasturage, no more grass, no more verdure—even the leaves of the trees had either been burned or withered by the fire. It was no longer the same country. This desolation extended for about a quarter of a league from the river—as far as the lowlands which are overflowed in the season of high water, for there the grass and all the plants were still green, and the fire could not touch them.[72]

He had this to say about the enormous oaks: "One remarkable thing that I have observed in all parts of Upper California in which I have traveled is that one never finds any of the large oaks broken down, fallen into decay or partly consumed by age and weather. All are sound and vigorous."[73]

California Indians themselves recount stories about how fire was used in the oak savannas. Omer Stewart's unpublished anthropological field notes from 1935 describe an interview with Lucy Lewis, a Pomo woman:

> The valleys were filled more or less with large oaks. The grass burned each year [by the Pomo]. The brush would burn. Trees were just scorched. Fires were started, and they were just allowed to burn every place. Special spots were prepared to serve as safe spots. Game were caught this way. The big trees were not killed, smaller ones which were covered with moss hurt. She [Lucy Lewis] has witnessed two fires. Burning was to make the grass grow better. Brush was kept down. All forests were burnt out. Burning was special to get grasshoppers. No method of control of fire except to get to a safe place.[74]

Thirty-nine years later David Peri and Scott Patterson would again interview Pomo people in the region of Dry Creek and Cloverdale, recording that they burned under not only valley oaks, but also tan oak *(Lithocarpus densiflorus)*, Oregon oak *(Quercus garryana)*, and scrub oak *(Quercus dumosa)*.[75]

Even today, native people remember what their elders told them about burning in the oak savannas during the 1800s:

> Burning was in the fall of the year when the plants were all dried up when it was going to rain. They'd burn areas when they would see it's in need. If the brush was too high and too brushy it gets out of control. If the shrubs got two to four feet in height it would be time to burn. They'd burn every two years. Both men and women would set the fires. The flames wouldn't get very high. It wouldn't burn the trees, only the shrubs. They burned around the camping grounds where they lived and around where they gathered. They also cleared pathways between camps. Burning brush helped to save water. They burned in the valleys and foothills. I never heard of the Indians setting fires in the higher mountains, but don't take my word for it. (Rosalie Bethel, North Fork Mono, pers. comm. 1991)

It is highly likely that another primary reason for indigenous burning in valley oak savannas was to promote the growth, density, and abundance of the grasses and wildflowers that grew under the oaks and provided an important food source in the form of small grains and edible seeds. Valley oak acorn production varies considerably from year to year. So-called mast years, when the trees produce heavily, are followed by two or more years of relatively low production. Wildflowers and grasses, on the other hand, will produce abundant seeds and grains almost annually—providing a consistent, dependable food supply.[76]

In 1775 Captain Pedro Fages wrote about the different kinds of grass grains eaten by the Valley Yokuts:

> Mention should first be made of rice, which occurs in three or four different species; distinguishing them by color, they are yellow, whitish, blue, and black. The last named variety has a pasty color beneath its bark or pellicle. There is another kind of rice similar to turnip seed, the plant of which is like the wild amaranth, which is found commonly in the canyons of the mountains. There is also a grass seed having a stalk like wheat, which when sufficiently compressed, yields a rich flour, being of the oleaginous variety.[77]

Edwin Bryant met with Captain Sutter at Sutter's Fort in 1846 and related in his journal: "He [Sutter] told me, that several times, being hemmed in by his assailants, he had subsisted for many days upon grass alone. There is a grass in this valley which the Indians eat, that is pleasant to the taste and nutritious."[78]

In 1843 Thomas Farnham hypothesized that Indians fired the oak savannas not to promote the valley oaks but rather to keep them from engulfing the grassland and shading out the understory:

> These grains [wild oats and rye], resowing themselves from year to year, produce perpetual food for the wild animals and Indians. The plains are burned over every year by the Indians; and the consequence is, that the young trees, which would otherwise have grown into forests, are destroyed, and the large trees often killed. Nevertheless, the oak, the plane tree, of immense size, [and] the ash . . . fringe the stream everywhere, and divide the country into beautiful glades and savannas.[79]

Today members of a number of California tribes recall that burning was practiced to enhance production of various small grains and seeds, some of which formerly grew in the Central Valley. For example, Lalo Franco, Wukchumni Yokuts, recalls, "I know that down in the low country in Yokohl Valley [southern San Joaquin Valley] the Indian people would burn that whole valley. It produced grass seeds—a type of wild rye. That grass seed you can eat and it was gathered and then burned" (pers. comm. 1991).

Recent studies indicate that in the absence of fire, grassland ecosystems become choked with detritus, and productivity and reproduction fall drastically. Other studies show that grain production in most native perennial grasses dwindles in the absence of some kind of intermediate disturbance, such as herbivory, fire, or flooding. Furthermore, many of the herbaceous plants with edible seeds have high light requirements and grow only in open grasslands or light gaps in forests and shrublands.[80] Studies have shown that the increasing density of forest species reduces forage production in the understory. This reduction is due in part to shading by the overstory but also to the increased deposition of forest litter on the ground.[81]

This management practice had other positive effects. To be excellent acorn producers, valley oaks must not be crowded. Fire promotes a stand structure of trees with broad, rounded canopies that bear many more acorns. And older trees are more productive: a fifty-year-old valley oak may produce only five pounds of acorns, compared to five hundred pounds for a mature tree.

Today 90 percent of the former valley oak savanna habitat is gone—converted to farms and orchards and, more recently, housing developments. According to the authors of *Oaks of California*, "It is now rare to see extensive groves of this monarch species, and a regal California heritage may soon be lost."[82] The problem is that seedlings are not living long enough to become trees. While there is acorn germination, the life history is cut short, as the trees often are killed at the sapling stage. Small mammals gnaw on the roots, undermining the saplings. Before the construction of levees and dams, when streams naturally overflowed their banks, floodwaters may have been essential for depositing rich silt and flushing out the small mammals. Acorns are also consistently attacked by insect pests before they are able to germinate.

PROMOTING OPEN MONTANE FORESTS STUDDED WITH LARGE SUGAR PINES

Dubbed "the most princely of the genus" by the botanist David Douglas, sugar pine *(Pinus lambertiana)* was valued by various tribes for the construction of earth lodges, sweat houses, and war bows. The pitch was used in Kashaya Pomo whistles and was an adhesive in Karuk households. Nuts were fashioned into jewelry by northwestern tribes. The sweet resin exuding from wounds and fire scars, which contains the sugar pinitol, was relished by different tribes; it was eaten on the spot or dried for later use. It also had medicinal properties: dissolved in water, it was used as a wash for sore or blind eyes by the Sierra Miwok. The Mono made a tea of pine sugar for relaxing muscles during tedious labor. The Yokuts concocted a medicine that was heated and put on the abdomen of the parturient by midwives. The Kawaiisu ate the dried powder as a laxative.[83]

The nuts of the sugar pine provided a concentrated source of fats and protein. The nuts can be pounded into a flour and made into a soup. Many tribes gathered large quantities of nuts. One Miwok family might collect as much as half a ton of cones each year. They were stacked upside down by the women and covered with dry pine needles, then ignited to burn away the pitch.[84]

In August 1903 Merriam watched an old Northern Miwok woman at West Point preparing sugar pine nuts. The cones, still green, contained nuts that were as yet hardly ripe. They were roasted for a short time in a fire, after which they were removed and split lengthwise with a knife, making it easy to get at the nuts between the scales. The nuts were then shucked and the meats removed and pounded in a small portable mortar. The nut flour was made into a soup.[85]

Men in many tribes skillfully climbed the trees just before the cones opened and swayed the limbs, which caused the pendent cones to snap off. Some of the grandest trees can rise more than two hundred feet, as this pine is the tallest and largest of the world's one hundred species of pines; further, horizontal limbs often do not occur until one-third of the way up the trunk of mature trees. Thus climbing the largest of these trees was a heroic feat. Native people adapted their climbing approach and apparatus to the stature of the tree. The Sierra Miwok, for example, propped a dead tree against the trunk to serve as a ladder, or used a special climbing pole. To retrieve the cones on young trees, the Pomo lashed a deer antler to the end of a straight pole.[86] These harvesting practices left the resource base intact. Mature trees that reached an age of more than five hundred years must have had grandfathers, fathers, sons, and grandsons for many generations climb their trunks.

Sugar pine, along with giant sequoia *(Sequoiadendron giganteum)*, ponderosa pine *(Pinus ponderosa)*, and white fir *(Abies concolor)*, dominate the montane forests in certain areas of the Sierra Nevada.[87] Many of the Sierran sugar pines undoubtedly would not have grown so large had it not been for indigenous burning practices, which kept the brush down and minimized the possibility of catastrophic crown fires. According to the ecologists Neil Sugihara and Joe McBride, "recurrent non-stand-replacing fires are a key factor favoring sugar pine in white fir-dominated mixed-conifer forests. Frequent fires during the pre-settlement period favored sugar pine regeneration. These fires also enabled sugar pine to become an important component of the large trees, typically becoming the largest trees in the stand."[88] (See Figure 18.) Virginia Jeff (Central Sierra Miwok, pers. comm. Aug. 1989) recalls that in the 1920s her father took two precautions to protect large trees before setting fires in the fall: he put sand or dirt on the pitch of pines so that they would not burn and raked the debris from around the oaks. Why set fires? "They burned so they would not have big fires," says Virginia. "It looks terrible today—mountain misery [an understory plant] is high and dead limbs are everywhere."

Figure 18. A timber "cruiser" checking the height and condition of a sugar pine *(Pinus lambertiana)* near Susanville, Lassen County, circa 1915. Large-diameter sugar pines and ponderosa pines *(Pinus ponderosa)* were desirable to Indians and non-Indians alike. California Indians harvested the nuts for food and jewelry, the resin for medicines, the roots for baskets, and the wood for structures and war bows, often leaving the trees intact. Non-Indians felled sugar pines and ponderosa pines for lumber. Courtesy of Special Collections, Meriam Library, California State University, Chico, #SC15570.

Scientific studies have shown that competition from brush severely retards sugar pine seedling establishment and growth. Overtopped trees decline, being intolerant of shade. Forest ecologists classify sugar pine as a fire-dominant species, adapted to grow in the forest gaps caused by light- to medium-intensity fires. Without fire, it succeeds to white fir (Wayne Harrison, pers. comm. 2004).

In 1955 Herbert Mason pointed out that "the practices of civilized man in his association with the sugar pine differ from the practice of aboriginal man. This difference tends to create conditions favoring incense cedar and white fir at the expense of sugar pine." He went on to say:

> There is evidence that fire, whatever its cause, has occurred frequently enough in forests to have resulted in a relatively stable "fire type" vegetation. This was pointed out long ago by Jepson. In other words, certain types of vegetation cover are able to maintain themselves in the face of recurrent fire so long as the fire is not of too great intensity at any one time. Most low-and middle-altitude vegetation of California has this "fire type" character. The recurrence of minor fires prevents the building up of conditions conducive to disastrous high-intensity fires. Thus recurrent fire, regardless of cause, is and has been a part of the normal disharmonic fluctuation of environmental conditions.[89]

Many California Indians talk about the way the lower and mid-elevation montane forests used to look in the old days and how the Indians used to burn:[90]

> They always started the fires at the bottom of the slope—never at the top of the slope. The fires rarely got into the crowns of the trees—it mainly stayed in the brush—burning only the undergrowth. Fire sometimes would scorch the trees. Fires were set in the same areas on a five to ten year period, closer to five years. The fires were set for many reasons: to increase visibility, to give nutrients, to decrease brush, for the regrowth of certain plants such as sourberry and manzanita, to thin vegetation, to attract animals to new feed, and to increase food production. Fires were set between 1,500 and 6,000 feet elevation from the San Joaquin to the Granite Creek area. . . . They'd also burned gathering sites, along the edges of creeks, and up the mountain from the trail, and at the south end of canyons. Whatever was going to burn the easiest was set on fire, the duff or grass. Duff or downed wood was never raked. (Ron Goode, North Fork Mono, pers. comm. 1989)[91]

Ironically, several of these individuals have lost their homes—including baskets, mortars and pestles, and other valuable cultural items—to fires that burned out of control because fire suppression policies made possible the buildup of fuel.

Fire scar studies conducted by Tom Swetnam and Anthony Caprio show that before the era of fire suppression, the fire return intervals on southwest-facing slopes where sugar pines occur were only five to ten years. The ecologist Norm Christensen concludes that lightning-ignited fires occurred only once in fifty or more years on most sites, leaving native peoples as the only possible cause of the more frequent fires.[92] Richard Reynolds, Bruce Kilgore, and others have argued that California Indians augmented the frequency of fires and that this human influence was great enough to alter the fuel and vegetational structures characteristically found in prehuman montane forests.[93]

Imagine a carpet of wildflowers and grasses surrounding noble, widely spaced sugar pines and other conifers. This open forest structure would have been palatable to native ungulates and would have provided a source of seeds, bulbs, and greens for human foods and medicines. According to Melba Beecher and her sister, Lydia Beecher (Mono), wildflower seeds and grass grains were as important as acorns years ago. Some of these, such as the farewell-to-spring flowers *(Clarkia williamsonii)*, were burned to enhance abundance: "The Indians used to burn to increase the seed crops such as tarweeds [*Madia* sp.], *Clarkia* spp., snowdrops [*Plagiobothrys nothofulvus*], chia [*Salvia columbariae*], wild sunflower [Hall's wyethia, or *Wyethia elata*], and red maids [*Calandrinia ciliata*]. A burn makes these plants come out more" (anonymous North Fork Mono elder, pers. comm. 1991).

Early written accounts reinforce the idea that small seeds were significant in the diets of tribes of the Sierra Nevada and elsewhere where sugar pine grows. An understory of grasses would expand the aerial extent of a fire while not burning too hot around large sugar pines and California black oaks, both of which were extremely important food resources.

Melba Beecher remembers that her mother and other elders gathered the grains of a wild grass at higher elevations in Jose Basin. On one field trip with her sisters; Iliene and Steven Cape; and a Forest Service botanist, Joanna Clines, in the Sierra National Forest (at about 4,500 feet elevation), she pointed out this edible grass, which turned out to be California brome *(Bromus carinatus)*. She also mentioned that the flower stalks were broken off at the joint and used in basketry. She said, "There used to be a lot of it in the hills by Pine Ridge and where that bridge is washed out by Mersick Peak and below Stevenson Mountain. It *[Bromus carinatus]* was all through the Jose Basin" (pers. comm. 1997).

Native ryes, needlegrass, and bromes (*Elymus, Leymus, Nassella,* and

Bromus spp.), as well as other native grasses and wildflowers, probably formed a forest understory before the arrival of Euro-Americans. These plants have excellent forage value and would have attracted the sheepherders and cattlemen who came with large herds beginning in the early 1860s.[94] There are, in fact, numerous early accounts of the rich pasturage of the montane forests in the Sierra Nevada and the burning of areas by sheepherders and cattlemen to improve the annual browse. John Muir noted, "Incredible numbers of sheep are driven to the mountain pastures every summer. Running fires are set everywhere, with a view to clearing the ground of prostrate trunks, to facilitate the movements of the flocks and improve the pastures. The entire forest belt is thus swept and devastated from one extremity of the range to the other."[95]

Betty Jamison asserts that ranchers learned from the Indians about the necessity of burning in the forests:

> My dad [John O'Neal] learned some of the burning techniques from the Indians. The Indians herded sheep for my father. The Mono Indians at North Fork would go over to Mammoth to collect pinyon nuts. Any dried limbs or pine needles in the forests they would set fire to on their way back about October or November. Back then you had a forest you could ride through. They burned to clear the masses of little trees and duff which increased the grasses for forage for deer and other animals. It also cleaned up the area. They burned from the bottom of the slope. They'd light various spots and it would creep, not blaze up. The flame lengths were six inches to two or three feet. With gusts once in awhile it would go up a pine tree and burn the lower dead branches. They lit the fires every year. Now there are fewer and fewer gooseberries and wild strawberries. We used to collect them up in the high country. I've never heard of their fires getting out of control. Areas surrounding Soquel, Basaw, Muggler, Jackass, Clover, Soldier, and Reds meadows were burned by the Indians and later years by ranchers. (Pers. comm. 1990)[96]

Although they continued the practice of burning, ranchers and sheepherders could not duplicate the Indian fire regime.[97] Areas with grass and forb cover that provided the fine fuels to carry light surface fires under indigenous burning were likely severely diminished under heavy livestock grazing. Thus non-Indian fires burned very differently. One of the most telling of accounts is from George Sudworth's comments on his reconnaissance work on the Stanislaus and Lake Tahoe Forest Reserves in 1899:

> Excepting in high mountain meadows, all of which are fenced and which are grazed by cattle, the principal forage for sheep and cattle on the open forest range consists of a few very hardy shrubs and low

broad-leaf trees. There are practically no grasses or other herbaceous plants. The forest floor is clean. The writer can attest the inconvenience of this total lack of grass forage, for in traveling over nearly 3,000,000 acres not a single day's feed for saddle and pack animals was secured on the open range. This is in striking contrast to the rich forage range in the timber forests of the Rocky Mountains. Barrenness is, however, not an original sin. From a study of long-protected forest land in the same region, and from the statements of old settlers, it is evident that formerly there was an abundance of perennial forage grasses through-out the forests, whether grazed or not by cattle and horses. It is also true that the severest annual surface fires kill these plants only to the ground; unless uprooted they sprout up the following season. It would seem that this bare condition of the surface in the open range has been produced only through years of excessive grazing by millions of sheep— a constant overstocking of the range.[98]

Ronald Lanner writes that large, old sugar pines are a thing of the past: "White pine blister rust in combination with timber management practices in much of California, will have the long-term effect of preventing sugar pines from becoming old and large."[99]

An Engendered Land

In granite outcrops in the highest windy passes of the Sierra Nevada one can still find mortar cup depressions formerly used by California Indians for processing foods. Projectile points can be found along ancient lakebeds at prehistoric camp sites in the Great Basin. Mescal pits lined with stones lie undisturbed in the Sonoran Desert. Rock paintings decorate the walls of basalt outcrops on the Channel Islands. These records provide evidence of former human occupation. More difficult to discern, yet more wide-ranging and significant, are the signatures that California Indian occupation left on the vegetation of California.

In prehistoric and early historic times, fire management practices and a host of other techniques were successful in promoting certain plant communities, modifying the structure of others, and enhancing habitat heterogeneity in many areas of California. Subtle enough to be thoroughly missed by modern ecologists, the changes wrought by native practices nonetheless constitute a pathway of vegetation change different from what would have occurred in wild nature. The evidence strongly suggests that the prehistoric Indians' effect on the environment can no longer be ignored by scientists and government agencies charged with stewardship of our natural resources. To better understand how Indians influenced the landscape, and how this knowledge can be used for management today, we need to encourage more of the interdisciplinary studies described in this chapter that collect data from diverse sources, such as pollen-charcoal analyses, phytolith studies, fire scar research, photographic studies, ethnographic information, archaeological findings, and plant ecological studies.

What we have already learned about the ways in which native people shaped the landscape has clear implications for modern wildland management. Removing California Indians from traditional economic and land management roles in California has not led to a prehuman state of nature in our wildland areas. Instead, the hands-off approach to management of wilderness preserves is jeopardizing the long-term stability of many plant communities. For example, the coastal prairies of the northwestern coast and the montane meadows of the Sierra Nevada are being encroached on by woody vegetation, homogenizing the landscapes and lowering biodiversity. This is due not only to fire suppression and habitat degradation and loss from modern land uses but also to an absence of the former indigenous influences.

CHAPTER 6

Basketry

Cultivating Forbs, Sedges, Grasses, and Tules

My mother started the fires. I watched her but I got scared because we weren't supposed to set fires and I thought the fire would spread. But she said don't set the fires if it's too brushy, but if you have a clear place around it—it's O.K. to set it. She burned one plant [of deergrass, *Muhlenbergia rigens*] at a time. In the olden days they'd let the fire go. She burned in January or February. Burning makes finer straws. It also makes them bud out more—and you get more straws. It also makes more plants too.

HAZEL HUTCHINS, Mono (1992)

Basketry captures the apotheosis of California Indian cultures. In the homes of a bustling Indian village, there were dozens of baskets of multifarious shapes that were put to hundreds of uses. They ranged in size from tiny one-inch gift baskets to five-foot salmon storage vessels. Baskets stored bulbs, knocked seeds from their pods, hauled dirt, ladled acorn mush, snared woodpeckers, and carried grasshoppers. Cone-shaped burden baskets held oak firewood; fan-shaped baskets fanned embers into flames. Large boat baskets ferried humans across rivers. Open-twined baskets stored trash. Twined quivers carried arrows. Baskets caged birds and captured fish, and funerary urn baskets kept human remains.[1]

A well-equipped Indian kitchen contained basketry pots, pans, and serving dishes: some winnowed and parched seeds, others held stone-boiled soup and strained manzanita cider. There were baskets for washing eels, for winnowing acorn meal, sifting flour, and cleaning seaweed, even baskets with rough interiors to aid in removing the brown skin on the yampah, or *ipo* (also spelled *epo*) (*Perideridia* spp.), an edible potato-like tuber. Basket hoppers kept seed meal from scattering during pounding. Boat-shaped bowls and oval trays were used for serving food, and basketry dishes were used for eating and drinking.[2] Miguel Costansó, an engineer and cosmographer who was a member of an expedition for the Spanish crown in 1770, described Chumash women and their basketry: "These are the women who make the trays and receptacles of rushes, to which they give a thousand different forms and graceful patterns, according to the uses for which they are meant, whether it be for eating, drinking, storing their seeds, or other purposes."[3]

A vast array of plant parts—roots, rhizomes, stems, branches, leaves, seed pods, culms, bark—were woven into baskets. Useful material grew from tidal salt marshes at sea level (spiny rush, *Juncus acutus* subsp. *leopoldii*) up to subalpine meadows just below the tree line (shrubby Geyer's willows, *Salix geyeriana*). Each type of habitat produced unique materials: from the desert came plant parts from the Joshua tree, desert fan palm, and devil's claw; coastal coniferous forests produced spruce roots, bear-grass leaves, Oregon ash branches, maidenhair and woodwardia fern stems, and hazelnut sticks; chaparral provided materials from redbud, buck brush, deer brush, oak, and sourberry. Many of the most useful species were riparian or wetland plants: willows, cottonwoods, sedges, tules, cattails, maple, maidenhair fern, dogwood, and rushes. Ruth Merrill recorded that seventy-eight species of plants, representing thirty-six families, provided materials for California Indian basketry.[4]

"The greatest mechanical ingenuity displayed by the Indians is in the construction of their baskets and bows and arrows," wrote Father Palóu, in reference to the Ohlone.[5] But the basketry craft, all over California, was more than a mechanical exercise, and baskets were more than utilitarian objects. They were absolutely central in the worldviews, cultures, and everyday lives of native people in California.

Basketry was (and still is) an art form, a means of self-expression. A basket was so much a part of the person who made it that a weaver's peers could often recognize the weaver by a basket's subtle style, its grade of weave, and its designs. Basket makers say that a decorative pattern is not preplanned; "it just happens." The ethnobotanist David Prescott Barrows commented on this phenomenon among the Cahuilla: "No model or pattern is ever used; the basket takes ready shape under her skillful fingers, and is always symmetrical and shapely, and the intricate regularity of pattern carefully preserved."[6]

Baskets had great symbolic meaning in native cultures. Given as gifts, they strengthened kinship ties. A woman's place in society might be determined by how well she made baskets, as in Karuk society, where a good basket maker commanded a high bride-price. Baskets were laid to rest with the dead, or burned at special mourning ceremonies, as a way of paying respect. Baskets full of seeds, bulbs, or other foods were peace offerings to early missionaries and explorers.[7] According to Modoc legend, the world started as a flat disk, then the Creator, Kumookumts, working as if weaving a basket, enlarged the world until the final disk was made.[8] "A basket untouched and unused will die," says the Rumsen Ohlone weaver Linda Yamane, who worries about baskets sitting alone and unused in museums.[9]

The basketry craft was a concrete expression of native peoples' seamless connection to the natural world. "The basket begins in the roots," write David Peri and Scott Patterson, referring to the cultivation and gathering of roots for basket making and also expressing their belief that the activities of gathering and tending fostered a kind of cultural rootedness to a place. Weavers tended and gathered from the same places generation after generation, applying ancient cultural knowledge in the process of directly converting plant materials into cultural objects. The Cahuilla say that baskets are *temalpakh*, meaning "from the earth."[10]

Alfred Kroeber called basket making "the most developed art" of native people in California.[11] Basketry that produced pieces that are widely considered aesthetic marvels required consummate skill. If you watch an experienced weaver's hands today they seem as sure of their course as the hands of a concert pianist—confident, independent, and rhythmic. Hands that tugged, twisted, turned, and tightened plant lacings became very strong and dexterous. "When I used to make baskets quite a bit," Norma Turner Behill, a Mono/ Dumna woman, told me, "I would shake someone's hand, they would say, 'Owwww you almost broke my hand'"(pers. comm. 1992). Over time weaving became as natural as eating, walking, or cooking. Veteran basket makers could split, clean, trim, and soak material and weave almost blindfolded— confirmed by the many near-blind women who continued to weave into old age. Karuk women with dim eyesight could still weave expertly with the aid of special markers for shifts in rows or colors of material.[12]

A basket's fineness of weave and perfection of design were not achieved by skill alone. Its beauty began with tending the native plants that were to *become* the basket. Whether it was iris in coastal prairies, bracken fern in red fir forest, sourberry in coastal scrub, deergrass in chaparral, or oak in blue oak woodlands, almost every type of sedge, wildflower, fern, bush, tree, or grass used in a basket was fussed over and meticulously groomed by weavers. It is ironic that the many late-nineteenth- and early-twentieth-century non-Indian collectors who wrote of the beauty and grandeur of California baskets were unaware that their splendor had as much to do with the painstaking grooming of plants in the wild as with the way the materials were split, trimmed, sized, and woven. (See Figure 19.)

This chapter discusses the indigenous management of grasses, sedges, rushes, and forbs for basketry material. Although woody materials from

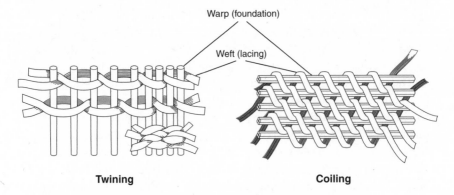

Figure 19. The two basic basketry weaving techniques: twining and coiling. Achieving a tight, neat, fine weave using either of these techniques was linked to harvesting plant material that was uniform, straight, and flexible.

shrubs and trees were also used extensively in baskets, the management of these plants is discussed in the next chapter.

Inducing Plants to Make Good Basketry Materials

Branches, stems, leaves, roots, and rhizomes, if left to grow willy-nilly, end up crooked, brittle, and short and are rejected by weavers. It is no coincidence that in a basket weaver's home, the whorls of basketry material on the shelves are as orderly and regular as neatly coiled rolls of synthetic string, or that long, uniform branch rods are as neatly bundled as stacks of wooden dowels. Nature just doesn't grow that way. On close examination, these materials exhibit features usually not present in the wild: they have no blemishes, no insects, no side branching. To gather plant parts like these by the ton, weavers had to cultivate.

Rigorously weeded and tilled gardens of sedge and bracken fern were tended for "white root" and "black root"—long, straight rhizomes prized in basketry. Many kinds of shrubs were carefully burned or pruned to stimulate the production of long, supple sprouts, cherished for lacings and foundations of baskets. Native bunchgrasses were burned to encourage abundant flowering culms, prized for the stuffing of coiled baskets. It is no wonder that Ruth Merrill wrote that "the materials in old, worn specimens, or even in unused bundles prepared for weaving [in the Phoebe Hearst Museum], bear little resemblance to either the living plant or the carefully pressed herbarium specimen."[13] (See Table 4.)

TABLE 4

Basketry Plant Parts and Their Management

Plant Part	Description	Management
Dry fruits	seed pods of devil's claw	cultivated, irrigated, fertilized (see Chapter 5)
Flower stalks	culms of native grasses, including deergrass, alkali sacaton, and California brome	burned or pruned periodically
Rhizomes	underground swollen stems of bracken fern, sawgrass, sedges, tule, and horsetail	cultivated with tilling, thinning, weeding, and soil management
Roots	underground plant parts of willows, Joshua trees, white and red alder, cottonwood, wild grape, and various conifers	selectively harvested (only a portion taken) with rest periods between harvests
Sprouts	young shoots of oak, sourberry, willow, maple, hazelnut, dogwood, redbud, and many other shrubs	stimulated with burning, pruning, and coppicing (see Chapter 7)
Stems	stipes of tule, cattail, juncus, maidenhair fern, woodwardia fern, honeysuckle, silver wormwood, nettle, wild grape	encouraged with cutting and pruning
Tillers	young leaves of bear-grass, desert fan palm, desert agave, Mojave yucca, and iris	encouraged with cutting and pruning and/or burned periodically

Pruning basketry shrubs can be likened to cutting back a garden of roses, and the impact of thinning rhizomes from bracken fern and sedges is not unlike that of periodic flooding. However, often the weavers' sphere of influence reached beyond the individual plants to the surrounding ecosystems. Burning in particular affected the abundance and densities of many other animal and plant species. Thus this craft left its mark on myriad ecosystems: coastal prairies, oak savannas, coniferous forests, chaparral, wetlands, and streamsides.

California Indians' cultivation of "wild" plants for basketry blurred the line between gathering and agriculture. When one considers the spectacular numbers of plant parts produced for basketry (a small Western Mono gift basket required more than 1,000 deergrass flower stalks; one sedge basket required 1,000 rhizomes),[14] the cultivated plants come close to fitting definitions of a "crop." Large numbers of sprouts, flower stalks, rhizomes, and tillers appeared at one time after tending and were harvested and stockpiled in the village. In addition, the continuous intergenerational visits of native people to the same gathering sites point to semisedentary lifestyles, similar to those of New World farmers.

Much of what has been written about the history of basketry is deficient because it describes and evaluates the finished product but does not take into account the biological context that supported the enterprise. We will now follow weavers into their heartlands to get a more detailed glimpse of the gathering and tending of four important basketry materials: bear-grass, sedges, deergrass, and tules.

Bear-grass

Bear-grass (*Xerophyllum tenax*), although grasslike, is not a grass at all but a member of the lily family. The thick, tuberous rhizomes of bear-grass, like those of many lilaceous plants, are edible; they are best boiled or roasted in a pit for two days.[15] They pleased the palate of the Hupa and Maidu alike and were eaten by black bears, who also like to wallow among their dense tufts. Roosevelt elk graze on the new, young growth.

This herbaceous perennial was harvested by many tribes in northern California and on up the Pacific Coast as far as the Olympic Peninsula and southeastern British Columbia (the bear-grass used in southwestern basketry is different, as it belongs to the genus *Nolina*). Tribes as diverse as the Modoc on the Modoc Plateau, the Yurok in the northern Coast Ranges, the Maidu in the northern Sierra Nevada, and the Shasta in the Cascades gathered the young, fresh tillers to provide a soft background or decorative overlay material in twined baskets. Bear-grass is still harvested today by weavers.

After it is harvested, bear-grass turns from pale green to sand colored. It is used to make the plaited tassels on Hupa dance aprons and the shiny decorative patterns in Karuk and Chimariko hats. Braids of bear-grass, pine nuts, and abalone dangled from buckskin bands tied to Karuk women's hair. Some northwestern California women's dance skirts had front aprons with multiple fringes of a hide, each entirely covered with a braidlike plaiting of bear-grass interspersed with blackened pine nut beads.[16]

Used for ornamentation, the tawny bear-grass strands stand out against black maidenhair fern stems and red alder–dyed backgrounds, and they add texture. Quivers, trinket baskets, flour trays, hoppers, sifters, and dippers all carried its leaves. Today, in addition to baskets, bear-grass is used in necklaces, earrings, and other pieces of jewelry, providing a glossy, golden brown patina. Florists and nurserymen also admire its sleek, delicate lines, and thus the leaves have become a popular part of prepackaged bouquets and customized floral arrangements. Commercial harvesters compete with native gatherers, cutting thousands of pounds of the grass in the Cascade Mountains to sell for floral arrangements in the United States and Asia.[17]

Bear-grass colonies of dozens or sometimes hundreds of clumps are found in the foothill belt from 2,300 to 4,500 feet in chaparral, red fir forest, and mixed evergreen forest in the northern Coast Ranges, northern Sierra Nevada, and Cascade Mountains. Some of the best bear-grass grows in the shadow of large, old sugar pines, ponderosa pines, or Douglas-firs. The tough, wiry leaves grow thickly at the base of the plant and are long-lived unless burned off. Leaves of bear-grass that grow in shade are reputed to be less brittle and not so bleached. After five to seven years of not flowering, a bear-grass plant sends up conspicuous white or cream colored feathery, fragrant plumes that resemble suspended flames on long, stout stalks.[18]

HARVESTING AND PROCESSING

Bear-grass is harvested, processed, and used today much as it was in aboriginal times. To harvest bear-grass, a weaver grasps the longer center leaves near their upper ends and pulls them from the plant. Hupa weavers seek out the leaves that have turned white at the tip. Basket makers assert that gathering some of the center leaves does not harm the plants. The harvested leaves are then dried for two or three days in sun or semishade. The barbed midrib running down the center of the underside of the leaf is removed. Strands of dried bear-grass are sorted by length and width, the thick bundles tied at the base and hung or stored on shelves inside dwellings. The edges of dried leaves can be razor-sharp and cut careless fingers. Before being used in weaving, the leaves are soaked in water, but if soaked too long they turn green. The Hupa occasionally dyed (and still dye) the bear-grass leaves a rich yellow, using either boiled wolf lichen *(Letharia vulpina)* or the root of Oregon-grape *(Berberis aquifolium)*.[19]

TRADITIONAL BURNING

Bear-grass is fire adapted; it vigorously sprouts back from its woody rhizomes after a fire. This characteristic is reflected in its nickname, "fire lily." The Yurok, Karuk, Hupa, Chilula, and other tribes burned the bear-grass clumps periodically and then harvested leaves from the burned clumps one to three years later in July or late summer, when the tips began to show white. The new green leaves that sprout after a burn are more easily picked and worked, as they are stronger, thinner, and more pliable.

According to early accounts, burning took place from July to early fall, before the rains began. Pliny Goddard, in his fieldwork among the Hupa in the early 1900s, recorded that "the ground is frequently burned over and the spot visited on the second or third year after." An inspector with the Bureau of Indian Affairs on the Hoopa Reservation, circa 1918, was well aware of the Hupa's burning practices and commented in his annual report: "There is a feeling among Indians and whites that forest fires are necessary to good grass. The Missionary gave the only really adverse comment upon this theory that I heard while at Hoopa. He stated that he had discouraged the making of Indian basketry because the Indians had maintained that the old grass must be burned off in order that new grass may grow, such as is fit for basket material." Interestingly, the inspector also reported that the general political climate was favorable to Indian burning, perhaps in recognition of its benefits: "On the other hand, among Indians, white neighbors, and apparently among employees, there seems to be no vigorous sentiment adverse to the annual burning over the forests. It was stated that last summer the Valley was smoky throughout almost the entire season as a result of adjacent fires, and it was apparent at night that these had begun again this year."[20]

During his explorations of the Chilula territory in 1851, Redick McKee recorded the burning of bear-grass in his journal:

> Leaving the river, we ascended a long spur of mountain to the top of the dividing ridge between it and Redwood creek, through alternate forest and prairie land. . . . Prairies of rich grass [*Xerophyllum tenax*] lie on their southern slopes. . . . The Indians used the stalks in their finer basketwork. . . . Late in the season, however, the grass is often burned, and dependence cannot always be placed upon the usual grounds [for animal feed].[21]

In 1932 Harrington recorded burning among the Karuk:

> And the bear lilies also they [Karuk] burn off, they pick them the next summer, in July; that is the time that they pick the bear lily. . . . It is in summer when they set fire to the brush, at the time when everything is

dry, that is the time that is good to set fire, in the fall before it starts in to rain. At different places up back of the people's rancherias they set the fires.[22]

Seven years later, in 1939, Edward Gifford jotted in his field notebook: "The Karok habitually burned the brush with the idea that better growth resulted the following year. Hazel, iris and *Xerophyllum* were burned off regularly to produce better growth. July and August were [the] best months for burning off country."[23] Also in 1939, Alfred Kroeber was recording similar practices among the Yurok, another northwestern California tribe: "The *Xerophyllum* was also burned in fall, and was ready for picking again the next season. . . . This grows in all kinds of places, but they only use what grows in certain spots which they burn over."[24]

Sedges

The rhizomes of various types of sedges (*Carex* spp.) were among the most widely gathered basket materials by native people. It was used as a lacing material for myriad baskets by such diverse groups as the Pomo, Yuki, Patwin, Cahto, Valley Maidu, Tubatulabal, Mono, Sierra Miwok, and Ohlone. Many weavers still gather it today.

Sedges, which resemble tufts of grass, grow along rivers and creeks, at the edges of marshes, and in wet meadows, where they control erosion, serve as nesting cover for wood ducks, and provide a food source for rails, grouse, seed-eating songbirds, and other birds. Sedges have flat, grasslike leaves, stems with a triangular cross section, and flowers arranged in spikes. Although there are at least 131 native species of sedge in California, only the ones with creeping rhizomes—especially the widespread Santa Barbara sedge (*Carex barbarae*)—were used in basketry.[25] Rhizomes are actually stems that travel horizontally under the ground and pop up in new places to grow "spur" plants. They are also called runners, reflecting their rapid growth and wandering habit. As they wind their way through the earth, creating a vast, underground network, they help to anchor the soil. In recent years, restorationists have recognized the tremendous benefits of native sedges and have planted them to help stabilize riverbanks and reduce erosion.

Sedges are well adapted to alternate periods of flooding and desiccation and will flourish even when their roots are submerged. Flooding is necessary for revitalizing the sedge habitat: although it can sometimes tear out the beds, it also rejuvenates the habitat by depositing sand and silt rich in organic matter. Weavers prefer to dig rhizomes in the sandy loam and fine silt soils that are deposited at the outer edges of floodplains. Sedges that grow in heavy clay soils are less desirable to basket weavers because they produce short, kinky rhizomes.

Sedge rhizome, usually called white root, is a weaver's delight: it is as soft, fluid, and yielding as the river that nourishes it, yet it is also extremely strong. V. K. Chesnut commented that the rhizomes dug by Pomo weavers were "nearly as flexible as string and can scarcely be broken by hand." Like water, the rhizomes adapt to any container, even baskets as small as one inch across. Using white root, a weaver is able to achieve a fineness of stitch, fifty-eight stitches to the inch, that is unsurpassed with any other material. In the construction of a basket, split rhizomes as fine as threads are pulled tightly around deergrass culms, branches of shrubs, or other inner foundation material, and the same strength that holds the soil in place holds the basket together. One cooking basket can use as many as 300 rhizomes; a burden basket can require more than 1,000 rhizomes.[26]

HISTORICAL HARVESTING AND USE IN BASKETS

Worked and carefully managed sedge beds used to occur along miles and miles of streams, often growing under the luscious shade of enormous valley oaks. In precontact times, large numbers of women went on root-digging expeditions along many major watercourses, including the American, Cosumnes, Feather, Kaweah, Kern, Kings, Merced, Mokelumne, Navarro, Russian, Sacramento, San Joaquin, Stanislaus, and Tuolumne Rivers. The smaller creeks and tributaries also were visited.

Rhizomes were dug in spring, summer, or fall. Using a digging stick, a weaver exposed the rhizomes so that she could select the straightest and longest ones, usually one or two years old. Once she chose a rhizome, she dug to reach both ends and cut each end from the plants to which it was attached. She was careful to leave behind the plants and to replace the disturbed soil.

Sedge rhizomes had almost sacred status for different divisions of the Pomo and other tribes. They were used in burden, seed beater, gift, cooking, storage, and other types of baskets. In the 1800s sedge baskets caught the eyes of wealthy collectors and basket dealers, and they bought large numbers of them. For many Indian weavers, basket weaving became the primary means of family support. The serenity, perfection, and profound beauty emanating from these objects continued to captivate hundreds of non-Indian basket collectors and dealers through the 1900s. In 1900 a basket commanded a price as high as $125.00. Often the basket dealer—the go-between—was the recipient of most of the profit.[27] But gathering white root and other materials for baskets could be tough going. In the 1930s and 1940s some Anglo landowners charged Foothill Yokuts women $2.50 per day for the privilege of coming onto their land along the Kings River to gather sedges and deergrass.[28] As early as 1899 John Hudson recorded that many ranchers forbade Pomo weavers to dig sedges on their lands.

CULTIVATION OF SEDGE BEDS AND ITS EFFECTS

Behind the pliant, symmetrical coils of sedge in a weaver's storehouse lay many hours of tending and soil management. While harvesting itself is an important component of the cultivation of sedge beds, weavers also take time to remove stones, sticks, roots, whole plants, and debris from the beds, resulting in loose, uniform soil with no impediments to the fanning out of new runners. The Pomo also rake around the plants to remove dead branches, leaf litter, and other debris. After harvesting rhizomes in a bed, weavers allow the bed to rest for two to three years so that the rhizomes can replenish themselves.[29]

As a result of this tending, cultivated sedge beds become very different from wild, untended ones. Although tended sedges look like wild plants aboveground, belowground there are marked differences. The rhizomes of wild sedges are gnarled and short, with many kinks and branches, whereas the tended ones are long, straight, uniform, and unbranched. Mature, cultivated beds show a density of about one plant per square foot, with no or very few other species present, and have rhizomes that are longer and easier to remove.[30]

According to David Peri, the cultivation of sedge beds for production of basketry materials results in "increased aeration, water condensation, the stimulation of new root growth, and increased plant vigor." He goes on to say:

> While cultivation loosens the soil, at the same time it mixes surface nutrients into the ground, improves drainage during winter months and allows better absorption of moisture during the growing season. Summer cultivation interrupts the capillary action of water to the soil surface, allowing water to remain in deeper, cooler soil layers where roots thrive. Loose soil at the surface also insulates because of its air content. When the roots are not impeded by compact soil, crops increase in size and quality. The removal of mature or older roots in a tract also stimulates the growth of new roots, and increases the size of the tract.[31]

The prominent Pomo weaver Elsie Allen claims that pruning the white root in the ground prevents the roots from "growing so thick they over use the nutrients in the soil" and also prevents the roots from "growing into tangled masses that are so hard on basketweaving diggers."[32]

Weavers will tell you that while wild sedges are important to the ecology of California's streams, cultivated sedges are even better. Elsie Allen, for example, says, "[M]an needs to learn that the sedge root is a vital part of the harmony of nature and preserving of the soil. It is especially useful in preventing creek banks from washing away, and can be encouraged to grow by all property owners. Digging of the roots, when correctly done and leaving behind about half of those found, actually strengthens the growth and soil-holding properties of the roots."[33]

The combination of weeding, sparing rhizomes for future production, thinning the competition, tilling the soil, and allowing for rest periods reflects a complex and labor-intensive wild plant management system. Frank Latta, a schoolteacher who learned about white root gathering and tending on expeditions with the Wukchumni Yokuts elder Wahnomkot, dubbed root digging a "scientific farming operation."[34] Thus the native sedge cultivation system, little changed since aboriginal times, is another native practice that blurs the distinction anthropologists make between agriculture and hunting and gathering. Although sedge is not domesticated, its cultivation is, and was, every bit as labor-intensive as the cultivation of a crop in an agricultural system.

Deergrass

One of California's largest and showiest native perennial bunchgrasses, deergrass has recently become popular in urban and rural gardens and public plantings. As one drives along the arboretum on the University of California, Davis, campus, for example, the long, golden flower stalks of deergrass wave in the breeze above the large tufts, magnetically drawing the eye.

Deergrass (*Muhlenbergia rigens*, formerly *Epicampes rigens*), was important in precontact times to deer and weavers alike. The deer used it for shelter and shade, especially during the fawning period, and Indian weavers used it to form the warp, or foundation—the "stuffing"—of coiled baskets. Since a deergrass plant spreads laterally over time, sprouting from vegetative buds at its base, it can be very long-lived.

Deergrass was once a major associate in the perennial grassland of the Central Valley and elsewhere.[35] The grasslands of aboriginal California were very different in terms of appearance and ecological functioning from the grasslands present today, which are dominated by non-native annual grasses. Countless interconnected tufts of deergrass, forming slight mounds, covered meadows, prairies, and valley grasslands, and large patches were interspersed in chaparral, oak woodlands, riparian areas, and lower montane forest. Grasslands made up of deergrass and other perennial grasses supported populations of insects, birds, mammals, reptiles, and amphibians, providing forage, cover, and homes. Two native butterflies, umber skippers *(Poanes melane)* and California ringlets *(Coenonympha tullia california)*, lay their eggs in deergrass, and the larvae feed on the blades. Elk and pronghorn antelope grazed its tufts when they were widespread residents.

GATHERING AND USE IN BASKETRY

Deergrass flower stalks were (and still are) valued by basket makers for their flexibility and length. (See Figure 20.) The material was also important, says the Mono/Dumna weaver Norma Turner Behill, "because it expands when water goes through it, along with the white root" (pers. comm. 1991). When immersed in water, baskets made with flowering culms of deergrass became watertight as the stalks expanded, making them desirable for water jugs and cooking baskets. Acorn mush cooked in baskets with a foundation of deergrass stalks took on the wonderful earthy taste of the grass. Today, some weavers still prefer to cook acorn mush in deergrass baskets, rather than modern pots, because, as Norma explains, the deergrass "has a good little scent to it and that adds to the flavor of the acorn" (pers. comm. 1991).

Tribes in more than half the state eagerly sought deergrass, another indication of just how widespread it once was. When Prince Paul of Wurtemberg visited Sutter's Fort on the Sacramento River in summer 1850, he noted that the Indians "make use of grass-stalks to make watertight baskets." These grass stalks were probably the flower stalks of deergrass. Barrows reported in the 1890s that deergrass was "found everywhere" in the homes of Cahuilla, "dried and tied into bundles."[36]

A Western Mono cooking basket required about 3,750 flower stalks, the yield of at least three dozen large bunchgrass plants, figuring about 100 stalks per healthy plant. Early photographs showing many orderly bundles of deergrass at the sides of weavers hint at the bunchgrass's ancient importance and indicate the vast numbers of stalks that were needed. (See Figure 21.)

Gathering the flower stalks of deergrass was a relatively simple if time consuming process, and it is carried out today much as it was two hundred years ago. In mid- to late summer, weavers seek out the partially green or light golden flower stalks from the current year. The stalks are ready to pick if they let go easily from their sheaths with a small squeak when gatherers tug on them. If harvesting in late summer or fall, gatherers cut the stalks or break them off at the lowermost node. The stalks are then sized and neatly bundled, often tied together with old cloth scraps. No storage

Figure 20. Mary Jack, a Yokuts weaver, making a basket with a deergrass foundation. Deergrass *(Muhlenbergia rigens)*, once a widespread native bunchgrass, was used for basketry by more than half the tribes in California. Note the diamondback rattlesnake design of red (redbud) and black (dyed bracken fern) on the basket. Courtesy of the Huntington Library, San Marino, California, Grace Nicholson Collection, #56-D-17–4.

period is required to cure the material, and it does not have to be soaked in water. After the Shoshone weaver Florence Brocchini gathers her deergrass, she lays the stalks in the sun to dry. They are ready for use after she pulls a piece of leather or deerskin along the upper stalk to remove the panicle of seeds.[37]

Mono weavers often incorporated a design reminiscent of a rattlesnake's diamondback patterning in their deergrass baskets. Norma Turner Behill says

Figure 21. Maggie Icho, a Wukchumni Yokuts weaver, with two large bundles of deergrass, each of which contains more than a thousand flower stalks. Over three thousand flower stalks were required for certain baskets. Courtesy of the Autry National Center/Southwest Museum, Los Angeles, #30530.

that Mono weavers have long associated deergrass and rattlesnakes in nature. The coiled basket shape nicely accommodates the coiled rattlesnake pose.

Some weavers have special areas that produce particularly good quantities of flower stalks. Nonetheless, the grass is gathered less and less because it is becoming more difficult to find. As early as the 1940s Francisco Patencio, a Cahuilla leader, warned of the dwindling populations of deergrass. "The places where the basket grasses grew are all ranches and the cattle stamp on the ground, and eat down the grass," he said. "There is not any more to be gathered."[38]

Justin Farmer, a contemporary Ipai weaver, laments: "With the encroachment of cowboys and the longhorn cow, the plant was grazed upon unmercifully and retreated to marshy areas along the side of the highway where stock couldn't feed on it."[39] Further alteration of the habitat occurred with the spread of agriculture and exotic grasses, the damming of streams and rivers, and then suburban housing development. As recently as the 1950s and 1960s, Mono, Foothill Yokuts, and Miwok women would make trips down to the Central Valley to harvest deergrass along the irrigation canals. "Farmers and ranchers burned every year along the canals in the Clovis and Jefferson school area. Indians would collect the straw grass for basketry after the fires. Now I don't see any of that grass along the canals," said one North Fork Mono weaver (pers. comm. 1991). Today small, remnant colonies of the grass border streams and creeks, wet rocky ledges, and some highways—a mere shadow of its former range and distribution.

TRADITIONAL BURNING AND MANAGEMENT

In addition to being pushed out of its former habitats by development and other alterations of land for human use, deergrass is no longer receiving the care once given it by native weavers and gatherers. Today thousands of acres of chaparral, tangles of oaks, and thickets of conifers are engulfing patches of deergrass that once thrived as active collection sites. Mere skeletal plants are, surprisingly, hanging on in the shade cast by chamise, manzanita, buck brush, and other shrubs. As the shrubs advance, however, the deergrass will die out, as it needs full sun or partial shade to survive. Poorly growing plants packed with thatch can be found at palm oases and along rivers such as the Kern. Old leaves keep piling up, shutting out sunlight to the new growth. The ratio of dead to live material is high, and only thin outer rings of active growth exist. Flower stalk production is often scanty, because the plants are smothered in so much dead material. "When they get this way they need to be burned," said Virgil Bishop, remembering how his mother, Ida Bishop, and grandmother, Annie Gibbons, made baskets with the *monop* (the Mono name for deergrass). "There used to be some back in there [along Cascadel Road in North Fork] that my mother and grandmother gathered. But it's being covered over with trees [ponderosa pine], so it's disappearing," said Virgil.[40]

In aboriginal times, native weavers in different parts of California burned stands of deergrass, usually in fall or early winter, both to prevent the encroachment of other vegetation and to improve the quality of the deergrass. Burning was conducted every two to five years. Traditionally, the Mono, Foothill Yokuts, Luiseño, and Kumeyaay burned areas of deergrass in the

Sierra Nevada and southern California to maintain openings, promoting a vegetation structure that was conducive to the sunlight requirements of deergrass.[41] (See Figure 22.) "When I was a kid," said Virgil Bishop, "my mother and grandmother used to burn the bunchgrass in the fall, about September or October. Then in the spring the new growth began. The ashes go into the ground and help it to grow. They burned every two years or so. [My mother] would touch the edge of one of the plants and it would burn the whole area. On Cascadel Road and Kingsman Flat, she used to burn there around springs after World War II. Fifty-by-hundred-foot areas were set on fire" (pers. comm. 1991). Burning modified the morphological growth of the stalks, enhancing the qualities preferred by weavers. For certain baskets, tribes preferred the long, finer culms that would grow after burning. Fires also made the stalks stronger and more flexible. Clara Charlie, a Chukchansi/ Choynumni Yokuts elder, said that the flower stalks "are less brittle when the ricegrass [deergrass] is burned" (pers. comm. 1989).

Western Mono elders say that burning increases the quantity of deergrass bunches. According to Hazel Hutchins (Mono), burning "makes them bud out more—and you get more straws. It also makes more plants too" (pers. comm. 1992). The same idea was echoed by Grace Tex (North Fork Mono), who told me that you burn deergrass back "the same time of year [fall] as the redbud, to have more shoots come up. It makes more of the plant" (pers. comm. 1990).

Periodic burning may also have prepared the seedbed, lowering plant competition and increasing seed germination rates. Melba Beecher (Mono) informed me that "the seeds grow better after a fire and it also makes more plants after a fire" (pers. comm. 1991). Results of a burn study in Cuyamaca State Park substantiate the spread of deergrass to new areas following the application of low-intensity fires.[42] Light burning will sweep over a deergrass patch and burn very unevenly. Portions of the base of large deergrass plants remain undamaged and alive. This is where buds will sprout, forming new stems and leaves the following spring. Fire has other benefits, too: it destroys insect pests, recycles plant nutrients, and eliminates competitive grasses, shrubs, and trees.

In addition to burning, gathering itself may have benefited populations of deergrass. When the weavers harvested basketry material they lifted away the dead material on and around the plant, allowing more sunlight to reach the plant and stimulating it to renew itself: "If you don't pick it, it won't grow," says Norma Turner Behill. "You can't use the old ones, and they get in the way and push the young ones out. It grows better when you pick it. When they're green, about June or July, you can pull them out and they

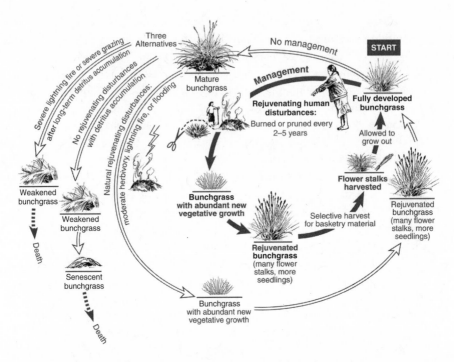

Figure 22. Coppice grass system: a conceptual model for indigenous management of deergrass. Management techniques mimic natural disturbances such as flooding and lightning fires and can be applied over long periods to manage deergrass in situ. In the absence of disturbance or with a disturbance that is too severe, the life span of the bunchgrass may be shortened and the number and quality of flower stalks may quickly decline, making it difficult to find enough good material for basketry.

squeak. I never cut them. The stalks are more flexible when they're pulled rather than cut" (pers. comm. 2004). That gathering benefits the plant is supported by the fact that deergrass responds positively to disturbance. Deergrass is healthiest in torrent-swept gorges in the mountains above the desert, on scoured banks of major rivers, or in patches of newly burned chaparral.

Tules

Freshwater marshes and vegetation along streams provided many plants useful to California Indian tribes. Some of the most valuable were the tules or bulrushes belonging to the genus *Schoenoplectus* (formerly *Scirpus*). These were multipurpose plants: the Kumeyaay, Pomo, and Yokuts harvested the long, spongy stems of three tule species (*Schoenoplectus acutus, S. cali-*

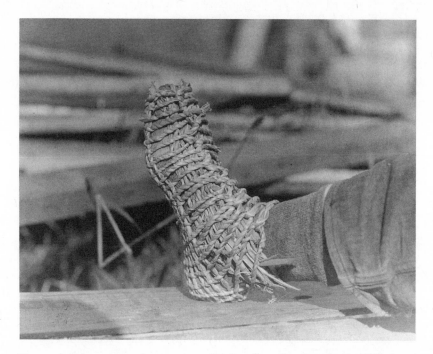

Figure 23. Tule moccasin for winter use by the Klamath Lake Indians, Klamath Reservation, California. Tule marshes were burned periodically to clean out dead plant material, recycle nutrients, and stimulate new tule growth. Courtesy of the Phoebe Apperson Hearst Museum of Anthropology and the Regents of the University of California, #15–4069.

fornicus, and *S. tabernaemontani*) for house construction, many tribes used them for reed boats, clothing, footwear, and duck decoys, and the Paiute and other tribes boiled down the roots to use them for sweetening. (See Figure 23.) One of the most widespread uses for tules, however, was in basketry. Many groups harvested, in addition to the stems, the long, brown horizontal rhizomes—called "black root"—of some tule species (*Schoenoplectus fluviatilis, S. maritimus,* and *S. robustus*) for making baskets. These rhizomes were dyed black and used as a design material.[43]

California Indians managed tules by cutting the stems, digging the rhizomes, and burning. Each of these practices had specific ecological effects. Because tules have succulent shoots that arise from rhizomes or perennial rootstocks each year, the same sites can be harvested year after year. Cutting tules at their base on a routine basis, California Indians assert, cleared out the older, dying growth, helping to revitalize the habitat, both for wildlife and human long-term use. Digging and cutting tule rhizomes probably was

Figure 24. Upper Lake Pomo woman gathering tules. New tules *(Schoenoplectus acutus, S. tabernaemontani,* and *S. californicus)* were used by many tribes for products as diverse as baskets, boats, clothing, furniture, and mats. Cutting tule stems, roots, and rhizomes removed dying and dead material and may have activated new stems and rhizomes. Courtesy of the National Anthropological Archives, Smithsonian Institution. Photograph by Edward Curtis, #76–4130.

also beneficial to maintaining and enhancing black root areas, because it stimulated new rhizome growth. (See Figure 24.)

The stems of some tule species are deciduous and die back to their bases each winter. These dead tules do not decompose very quickly, and within a short time a thick mat of dead tules will accumulate, shutting out sunlight to the new shoots that will arise in spring. Thus some tribes burned freshwater marsh and areas along streams to enhance the growth of tules for basketry.[44]

Fire assisted in the destruction of disease spores of rust fungus, which if allowed to grow would form black spots on the stems of tules, rendering them useless for basketry. Rust fungi have resting spores that overwinter on the dead material or live plants and then germinate to produce tiny spores that are windborne. If they land on the plant in the presence of moisture they will germinate and infect the plant (Bob Raabe, pers. comm. 2002). Rust-infected material is sometimes invaded by secondary organisms. Slugs feed on the tissues that are rusted because these contain a chemical that is attractive to them.[45]

There are some historical accounts of tules on fire. John Sutter and his clerks kept a record of daily events at Sutter's Fort between 1845 and 1848. In three separate entries the clerk noted that the tules were burning in the Sacramento Valley. These entries also noted the arrival or the departure of Indians. For example, on December 1847 the ledger reads: "Maximo and a few other Moquelemnes arrived. The Tular on the left bank of the Sacramento in fire." In January 1848 a clerk recorded: "A pleasant day, great fire in the Tule on the other side [of the Sacramento River]. Olimpio & Jeronimo returned from Nicholas'es." In 1849 Bayard Taylor, having just passed the valley where Livermore now stands, wrote of the "hazy air" made more dense by the smoke arising from "the burning tule marshes" as he stood at the divide overlooking the San Joaquin plains.[46]

By managing freshwater marshes and riparian areas for the harvest of tules, California Indians maintained the productivity of the marshes and enhanced their value as wildlife habitat. Bulrushes are extremely important for wildlife. Their hard-coated seeds are among the more important and most commonly used foods of ducks and of certain marsh birds and shorebirds. The stems and underground parts are eaten by muskrats and geese. Further, bulrushes furnish important nesting cover for waterfowl such as the least bittern. By clearing out reed-choked marshlands, burning reduced plant density, created edges useful for waterfowl movement and nesting activities, and enabled greater plant diversity because it allowed more light to reach the soil surface. In the absence of periodic burning, many wetland areas fill with organic matter, modifying the site conditions so that they become less suitable for the current plant assemblage.[47]

Some tribes burned tule areas as a means of hunting. The Northern Sierra and Plains Miwok hunted golden beaver (Castor canadensis) by burning off the tule around its pond. According to a Tachi Yokuts informant, "No particular species of small game [was] hunted separately. Rabbits, squirrels, rats, etc., lived among the tules, which were set afire and all species present burned to death, clubbed, or shot."[48]

Cultivating Beauty

California Indian women have been described as the "world's best basket makers."[49] The exquisite beauty, delicate lines, and virtuosity of the baskets achieved by weavers in California bespeak an intimate association with the landscapes from which the material was, and is, gathered. At the root of the beauty of California Indian baskets is the tending of the basketry plants and the care of the places that harbored them. Basket makers weeded, pruned, coppiced, burned, and cultivated the soil around perennial herbs, ferns, shrubs, trees, grasses, tules, and sedges to produce superior material for weaving. By exploring the intricacies of this craft—from cultivating and gathering material to weaving and using finished baskets—we gain insight into how the demands for high-quality material for baskets shaped the way that Indians managed the land. Restorationists, ecologists, and land managers interested in re-creating some of the landscapes and habitats that existed two hundred years ago must not ignore the profound influence the basketry craft had on those landscapes and the plants that made them up.

CHAPTER 7

From Arrows to Weirs

Cultivating Shrubs and Trees

My grandmother Susy Hammond [Chukchansi Yokuts] would burn in the fall for sourberry. She'd burn to make the straight sticks used for cradles and winnowing baskets. The gooseberries with a white flower and stickers were also burned in the fall so they would grow straight for arrow shafts.

LUCILLE HIBPSHMAN, Southern Miwok/Chukchansi Yokuts (1989)

In southwestern England in 1970 a six-thousand-year-old wooden walkway was uncovered by peat cutters clearing weeds from drainage ditches in the boggy flatlands. Well preserved, it consisted of thousands of pegs of hazel and ash. Archaeologists meticulously studied the pegs and at first were puzzled by their uniform characteristics: their lengths, ages, and diameters were similar, and they were all very straight and lacked the side branching typical of hazel and ash branches.[1] What the peat cutters had uncovered was more than just an old track. The peculiar architecture of the pegs revealed coppicing, an early form of woodland management that was undertaken by Neolithic farmers to encourage the rapid growth of many rodlike shoots from ash and hazel rootstocks.[2] Branches from unmanaged ash and hazel would have been unsuitable for building a walkway, because they would exhibit tapering, crookedness, side branching, and weaknesses caused by the exit holes of insects.

At the same time that the Neolithic people in Britain were harvesting their coppiced shoots to build their walkway, native people in California were managing redbud, elderberry, willow, gooseberry, and other shrubs and trees in a similar way to create long, slender shoots for use in baskets, traps, weirs, arrows, and other objects. These human-produced shoots poke out of half-finished baskets in historic photographs, lie buried as plant remains in caves and rock shelters, and sit in cultural museums as tied bundles of whole shoots or coils of split branches. (See Figure 25.)

Everywhere in California individuals from various tribes gathered these branches for the construction of myriad items. Yuki women harvested the young burnt sienna sprouts of redbud *(Cercis orbiculata)* for basketry in the oak woodlands of the northern Coast Ranges; Kumeyaay men gathered young shoots of elderberry *(Sambucus mexicana)* for flutes and clapper sticks in the chaparral of southern California; Paiute fishermen harvested the slender new growth of gray willow *(Salix exigua)* near streams for the latticework in fish weirs; Foothill Yokuts hunters gathered the first- or second-year growth of gooseberry *(Ribes* sp.) and California button willow *(Cephalanthus occidentalis* var. *californicus)* for arrows. The young growth of many other kinds of shrubs and trees—hazelnut, mock orange, wild rose, dogwood, oak, bitter cherry, chokecherry, maple, honeysuckle, buck brush, deer brush, sourberry—was gathered and turned into cultural products as well.

To induce the native shrubs and trees of California to produce young shoots with properties desirable for construction, native peoples took advantage of these plants' natural ability to sprout from trunks, roots, and burls after disturbance. They mimicked disturbance by burning the plants or pruning them mechanically, and the shrubs and trees responded by producing long, straight shoots, called epicormic shoots or adventitious branches by plant morphologists, sprouts by horticulturists, or simply "sticks" by California Indians.

Desirable Properties of Woody Basketry Materials

California Indians were not trained in wood product science or plant morphology. Nevertheless, they had an intimate knowledge of the properties of wood. For the construction of baskets, clothing, weirs, traps, and many other objects, they valued flexibility, straightness, presence of anthocyanins (reddish pigments), absence of bark blemishes, long length, uniform diameter, and absence of lateral branching.[3] Using young shrub and tree growth that fit these characteristics vastly improved an object's function, helping arrows to fly straight, cooking baskets to hold water, sifting baskets to sift, and cordage to maintain its strength. It also enabled weavers and other craftspeople to create objects of great beauty, intricate design, and remarkably fine quality.

In selecting buck brush *(Ceanothus cuneatus)* branches for the hoods of cradleboards, rims and warps of winnowers, or the warp of seed beaters, Norma Turner Behill looks for "brownness, roundness, length and no lateral or side branches. . . . [T]he old sticks are white and sometimes they have little branches on them and they're dry—not flexible."[4] Other California Indian weavers stress the importance of selecting the "right" material. For example: "Weavers like their sticks without any side branches. They're easier

Figure 25. Plant architectures of managed and wild redbud *(Cercis orbiculata)* shrub branches. *(Bottom)* Branches that are used for basketry have a specific morphology. They must be flexible, straight, and long, with no lateral branching, lichens, pathogens, or insects, and sometimes with a special pigment. These requirements are found in young shrub and tree growth managed with burning or pruning. *(Top)* Old growth, in contrast, is brittle, crooked, and short, has lateral branching, and may harbor insects and/or diseases.

to split, and with sourberry, they're easier to scrape the skin off" (Grace Tex, North Fork Mono).[5]

The desirable properties of the woody materials from shrubs and trees are described more fully below.

Flexibility. Many products demanded pliant material for manufacture because they were made from shoots that had to be bent or twined. For example, in acorn granary frameworks, or the frameworks of certain hemispherical house types, branches had to be woven in and out through the framework. Fish weirs, under the continuous pressure of a stream current, required branches that were resilient under tension. The split branches used in twined basketry, woven like sewing thread, had to exhibit extreme pliancy. In coiled baskets, whole shoots or branches were bent in an arc around and around a central core.

Straightness. The value of a straight-growing branch is fairly obvious. Straight branches will make longer rods in foundations, split more easily and evenly, and make stronger, more uniform baskets.

Presence of anthocyanins. Materials with bright or contrasting colors were important for making designs in baskets. The pigment present in the epidermis of very young growth in certain shrubs and trees was valued for this purpose.[6] A good example is the wine red color that occurs in the young branches of redbud, a widespread shrub in California utilized by numerous tribes.

Absence of bark blemishes. Whole shoots of willow, redbud, sumac, and other woody plants were sometimes selected for basketry material with the "bark" (actually epidermis) left intact. The epidermis could exhibit no insect or disease damage or other blemishes, and it had to be smooth and homogeneous.

Length. The length of a branch was extremely important because longer branches meant less work for the weaver. Splitting a few long shoots into sewing strands (weft) for baskets was easier than splitting many short shoots. In addition, longer shoots meant fewer new sewing strands, which saved time and made a basket with a neater appearance and fewer breaks in the continuity of stitches.

Uniform diameter. Woody materials forming the foundation of a one-, two-, or three-rod basket had to be uniform in diameter, with no tapering, or the basket would bulge where the foundation material was too thick and sag where the foundation material was too thin. Twined baskets, such as Western Mono sifting baskets, needed shoots of uniform diameter in order to exhibit their fine latticelike appearance and serve their functions. Split branches that become the sewing strands or weft of a basket are trimmed to get many even and similar-width sewing strands, but there is less trimming if all of the branches that will become sewing strands for a basket are of a similar size to begin with. The Mono/Dumna basket weaver Norma Turner Behill says, "You've got to get all the same size. Same roundness, same length. It takes a long time to gather enough fine ones for a sifting basket or burden basket."[7]

Absence of lateral branching. The juncture of a lateral branch to a main stem is the region that harbors the densest wood. The cells in this region are short and small in dimension.[8] (See Figure 26.) Even if a lateral branch is cut off and the stub sanded down flush with the main stem, the extremely dense wood at the joint will have different properties from the rest of the branch and make the branch asymmetrical in weight (Richard Dodd, pers. comm. 1992). The asymmetry in weight would drastically affect the aerodynamics of arrow shafts, lances, harpoons, atlatls, and spears, and the dense wood of the joint

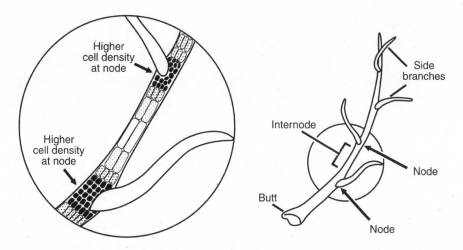

Figure 26. Side branches are undesirable to weavers because the juncture of a lateral branch to a main stem (node) is the region that harbors the densest wood. The cells in this region are relatively short and small, while the cells in the internodes are long and large. Thus, when the basketry material is seasoned by storage, the plant material will dry unevenly and undergo differential shrinkage. This will cause the material to warp and crack and render it unsuitable for weaving. The juncture area is also characterized by varying grain angles, which may have an important effect on properties.

would make it unsuitable for splitting to make basketry weft. In addition, the dense wood at the joint would shrink differently when drying and be less elastic than the rest of the branch. The juncture region is also characterized by varying grain angles, and this may also have an important effect on properties. For all these reasons, shoots with no side branches were the only ones that could be used for many cultural products.[9]

Some of these desirable qualities were inherent in certain species of shrubs and trees and not others. Certain plants were known to be excellent or poor for making particular objects. For example, a young shoot of desert-willow (*Chilopsis linearis* subsp. *arcuata*) was reputed to be so flexible it could be bent nearly in half without breaking or cracking, making it suitable for use in domed structures and in the making of storage granaries.[10] The durability of a material—how long it could be used before wearing out—was generally known for each plant species. For the most part, however, woody plant materials exhibiting more than one of these desirable qualities could come only from shrubs and trees that had been managed for that purpose. Older

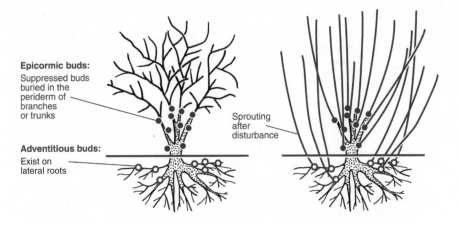

Epicormic buds:
Suppressed buds buried in the periderm of branches or trunks

Adventitious buds:
Exist on lateral roots

Sprouting after disturbance

Figure 27. Epicormic and adventitious shoots. Epicormic branches develop from suppressed axillary buds buried in the outer bark of woody stems. The buds activate if the stem is injured, cut, or burned. Adventitious shoots arise from buds that form through differentiation of parenchyma tissues not associated with axillary buds. Such shoots arise de novo, most commonly from roots and wound tissue.

branches, and those starting as epicormic or adventitious shoots but allowed to grow for several years, were shorter, lacked anthocyanins, had more side branches and bark blemishes, varied more in diameter, and were more brittle and crooked. And they were often less accessible than the young managed shoots arising from the base of a plant.

Vegetative Sprouting as an Adaptation to Frequent Disturbance

California Indians depended greatly on gathering young, straight, flexible shoots that arose as epicormic branches or adventitious shoots from shrubs and trees. Epicormic branches develop from suppressed axillary buds buried in the outer bark of woody stems. The buds activate if the branch or stem is injured, cut, or burned. Adventitious shoots arise from buds that form through differentiation of parenchyma tissues not directly associated with axillary buds. (See Figure 27.) Such shoots arise de novo—most commonly from roots and wound tissue.[11]

Both epicormic branches and adventitious shoots are adaptations that may have arisen in response to natural disturbances such as fires, floods, herbivory, windfalls, and landslides. Growing in environments in which they were commonly burned, uprooted, or eaten, shrubs and trees evolved these adaptations as defense mechanisms for survival.

Today, when one visits the aftermath of a wildfire of light or moderate intensity, appearances may be deceiving. Burned shrubs—blackened, lifeless, and ugly—appear destroyed. Yet given time, the shrubs rise anew from their ashes. Hidden inside the charred branches, both below- and aboveground, are tiny, unharmed suppressed buds that can miraculously transform into new shoots. These sprouts spurt from the bases of their trunks, roots, or sometimes from woody stools or burls.[12]

After its apical meristem is killed by fire, changes in internal physiological controls signal the release of suppressed buds. Rapid elongation of new shoots occurs. These shoots are characterized by an upsurge of vertical growth and a retardation of lateral or side branches. A shrub's branch tips might grow a mere five inches a year under normal circumstances. Yet after a fire it can grow branches that tower five feet from its base in less than six months—a 1,200 percent increase in growth.

The adaptive significance of vegetative reproduction in shrubs has long been a major topic of inquiry by ecologists and evolutionary biologists, but many unanswered questions remain.[13] It is a characteristic that is widely displayed in the native flora of California, suggesting that many different habitats have been exposed to recurrent disturbance for a very long time.

Scientific studies show that only when shrubs are exposed to flooding, browsing by deer, or burning to the base by fire do elastic new branches arise vigorously from old stock. Without disturbance, the buds remain suppressed on individual shrubs and trees and can lie dormant for many years. If left undisturbed, wild shrubs exhibit spreading growth forms and have branches that twist and turn like a braiding stream.

Spurring Sprouts

California Indians could not count on random natural disturbances to consistently provide woody material that was suitable for cultural items. Nor could they rely on natural disturbances to produce the enormous quantities of young growth they needed. One Western Mono cradleboard needed between 500 and 675 straight sticks of sourberry. A larger Western Mono cone-shaped burden basket might require 1,200 sprouts of sourberry. One twined seed beater needed 188 sticks of buck brush.[14] (See Table 5.) A diligent basket weaver had to harvest as many as 10,000 sticks during the winter months to make half a dozen or more baskets in one year (Craig Bates, pers. comm. 2001). The approximately twenty-five weavers in a prehistoric village of one hundred persons might harvest on the order of 250,000 shoots in a single season.

The amount of material needed for basketry in a prehistoric village was

TABLE 5

Western Mono Basket Types, Plant Species Used, and Numbers of Useful Shoots from Managed versus Unmanaged Shrubs

Basket Type	Plant Species Used	Number and Length of Sticks per Basket	Number of Unmanaged Plants Required per Basket	Number of Managed Plants Required per Basket
Burden	Ceanothus cuneatus (buck brush)	2 (3 ft.)	10 shrubs	1 shrub
	Rhus trilobata (sourberry)	1,200 (4 ft.)	400 patches	12 patches
	Cercis orbiculata (redbud)	25 (6 ft.)	50 shrubs	1 shrub
Coiled cooking	Cercis orbiculata	75 (2.5 ft.)	150 shrubs	6 shrubs
Full-size cradleboard	Rhus trilobata	675 (3.5–4 ft.)	102 patches	6 patches
	Cercis orbiculata	75 (6 ft.)	150 shrubs	10 shrubs
	Ceanothus cuneatus	15 (4 ft.)	65 shrubs	10 shrubs
Twined seed beater	Ceanothus cuneatus	2 (for rim; 3 ft.)	10 shrubs	1 shrub
	C. cuneatus	188 (2.5 ft.)	376 shrubs	15 shrubs
Twined sifter	Rhus trilobata	1,000 (3.5 ft.)	333 patches	10 patches
	Cercis orbiculata	25 (6 ft.)	50 shrubs	2 shrubs
Winnowing	Ceanothus cuneatus	2 (for rim; 4.5 ft.)	10 shrubs	1 shrub
	C. cuneatus	376 (3 ft.)	752 shrubs	31 shrubs
	Cercis orbiculata	30 (5 ft.)	20 shrubs	4 shrubs

SOURCE: Based on discussions with Norma Turner Behill (Mono/Dumna).

so vast as to make gathering wherever one might find the right material prohibitive. Moreover, gathering basketry material was a collective enterprise. Great efficiency was needed to produce and gather enough materials yearly to comply with the strict standards for the manufacture of many cultural items. Most of the basketry materials cannot be used immediately but must be stored for a period to season them. This period varies from one to four years depending on the plant species (Margaret Mathewson, pers. comm. 1992).[15] Women had to gather one year's new growth for a basket they might make two or three years hence. To rely on natural fires from lightning to induce production of large numbers of desirable shoots would be chancy, as lightning could strike in the wrong plant community type, not strike in a location with suitable kinds of plant species, or strike many miles from the village. Thus California Indians had to manage the vegetation to obtain sufficient, regular supplies of materials.

It probably did not take indigenous cultures long to figure out that the new growth that appeared after a lightning fire provided superior material for making vessels, arrows, and other items. Similar to the Neolithic farmers in England, California Indian gatherers ingeniously took advantage of plant adaptations by applying three horticultural techniques that mimicked lightning fires and other natural disturbances: burning, coppicing, and selective pruning. Burning was by far the most popular and effective practice, and light surface fires were set every year in different plant communities to ensure a reservoir of pliant, straight new suckers suitable for baskets, game sticks, granaries, headdresses, household utensils, and many other items.

The Evidence That California Indians Managed for Young Growth

Many lines of evidence point to the widespread use of burning, pruning, and coppicing to encourage young growth in aboriginal California. Old photographs, for example, can tell us a great deal about these practices. (See Figure 28.) Indigenous women in the process of making baskets were a favorite subject for professional photographers in the late nineteenth and early twentieth century. Hundreds of old photographs housed in library and museum archives across the United States depict tall, straight shoots, tied in bundles in the background, surrounding the weaver, or protruding out of half-made baskets. Today, it would be difficult to find in nature shoots that are as long, straight, and uniform as those in the photographs.

Finished artifacts made by native people and now shelved in museum collections also hold hints about the horticultural practices of the people who made them. Careful inspection of these objects, from old fish traps to musical instruments, reveals that young, specialized growth was often used. (See Tables 6 and 7.) By counting the growth rings clearly exposed on the ends of flutes and clapper sticks of elderberry, for example, one can see that the materials were from one to four years old when they were harvested. Examination of the bundles of rods in baskets reveals that the rods are unusually long and straight and lack lateral branch scars. Those rods that have retained their bark have no blemishes, lichen or moss growth, insect exit holes, or bark fissuring. The bark of the neatly wrapped coils of split rods

Figure 28. Sarah Knight, a Pomo weaver, before 1903. Observe the long, straight branches protruding from the basket, which is evidence of management—either burning or pruning at least one year before harvest. Courtesy of the National Anthropological Archives, Smithsonian Institution. Photograph by G. W. Tilton, #T-1762.

created for use as basketry sewing strands may still contain anthocyanins, and the rods have no lateral branch scars.

Historical accounts—ethnographic reports, diaries, the writings of native weavers—are another source of evidence for horticultural management of trees and shrubs. Ethnographic accounts typically labeled the woody materials that native people considered suitable for baskets, fish weirs, arrows, headdresses, and other objects "withes," "sprouts," "young growth," "suckers," "wickerwork," "year-old," or "second growth," suggesting that new growth specifically was required. Galen Clark, for example, reported that the Yosemite Miwok wicker basket traps for catching fish were constructed of "long willow sprouts" loosely woven together and closed at the pointed lower end. C. Hart Merriam noted that the split strands used by the Southern Sierra Miwok "for twining the rods together" are "black oak *Quercus californica [Q. kelloggii] (teleeli),* mostly young shoots which have great strength."[16]

In a 1923 letter to Walter Peirson, Grace Nicholson, a basket dealer, makes

TABLE 6

Ages of Adventitious Shoots and Epicormic Branches Used in Baskets

Tribes	Basket Element	Plant Species	Age of Growth
Western Mono, Foothill Yokuts, Southern and Central Miwok	warp and weft	*Cercis orbiculata* (redbud)	1 year
Southern and Central Miwok	warp rim stick	*Ceanothus integerrimus* (deer brush)	1 year 1–3 years
Western Mono, Foothill Yokuts, Southern and Central Miwok	warp rim stick	*Ceanothus cuneatus* (buck brush)	1 year 1–3 years
Southern and Central Miwok	weft	*Acer macrophyllum* (big-leaf maple)	1 year
Westem Mono, Foothill Yokuts, Southern and Central Miwok	warp weft	*Rhus trilobata* (sourberry)	1–2 years 1–2 years
Foothill Yokuts, Southern and Central Miwok	warp	*Cornus* spp. (dogwood)	1 year
Western Mono, Foothill Yokuts, Southern and Central Miwok	warp weft	*Salix* spp. (willow)	1–2 years 1–2 years
Southern and Central Miwok	warp weft	*Corylus cornuta* var. *californica* (hazelnut)	1 year 1 year
Western Mono, Foothill Yokuts, Southern and Central Miwok	warp and weft rim stick	*Quercus* spp. (oak)	1 year 1 year
Southern and Central Miwok	warp	*Prunus virginia* var. *demissa* (western chokecherry)	1 year

SOURCE: Based on discussions with Norma Turner Behill (Mono/Dumna), Margaret Mathewson, Ruby Cordero (Chukchansi Yokuts), Margaret Baty (Western Mono), Gladys McKinney (Mono), Grace Tex (North Fork Mono), Craig Bates, and Clara Jones Sargosa (Chukchansi).

TABLE 7

Ages of Adventitious Shoots and Epicormic Branches Used in Cultural Products, Excluding Baskets

Tribes	Cultural Item	Plant Species	Age of Growth
Western Mono, Foothill Yokuts, Southern Miwok	looped stirring stick	*Quercus* spp. (oak)	1–3 years
Foothill Yokuts, Southern and Central Miwok	arrow	*Calycanthus occidentalis* (spicebush) *Ribes* spp. (gooseberry)	2–3 years
Washoe, Paiute, Foothill Yokuts, and Western Mono	harpoon and spear	*Salix* spp. (willow) *Fremontodendron californicum* (flannel bush) *Umbellularia californica* (California bay) *Cercocarpus betuloides* (mountain mahogany)	2–4 years
Western Mono, Foothill Yokuts, Tubatulabal, Southern Miwok	digging stick	*Quercus* spp. (oak) *Cercocarpus betuloides* (mountain mahogany)	2–4 years 4–6 years
Paiute, Foothill Yokuts, Southern Miwok	fish trap	*Salix* spp. (willow)	1 year
Western Mono, Paiute, Tubatulabal, Foothill Yokuts	fish weir	*Salix* spp. (willow)	1 year
Southern and Central Miwok	fire drill	*Aesculus californica* (buckeye)	1–3 years
Western Mono and Chukchansi Yokuts	cordage	*Fremontodendron californicum* (flannel bush)	1–3 years
Foothill Yokuts, Southern and Central Miwok	flute	*Sambucus mexicana* (elderberry)	1–3 years

Table 7 *(continued)*

Tribes	Cultural Item	Plant Species	Age of Growth
Western Mono, Foothill Yokuts, Southern and Central Miwok	clapper stick	*Sambucus mexicana* (elderberry)	1–4 years
Western Mono	cooking tongs	*Fremontodendron californicum* (flannel bush)	3–4 years
Foothill Yokuts	pigeon decoy cage	*Cornus* spp. (dogwood)	1 year

SOURCE: Based on discussions with Bill Horst, Norma Turner Behill (Mono/Dumna), Margaret Baty (Western Mono), Craig Bates, and Margaret Mathewson.

the connection between the prohibition on burning and the lack of good basketry material:

> This weaver, Mary [Pomo], lives in Mendocino county, California. . . . This basket has been four years in the making, which accounts for the perfection of the weave and the design and the shape. It is most difficult to secure any good basket materials, as the Indians are not allowed to burn off underbrush. Mary and her husband made trips to the coast and into Yolo county to find enough material of the desired quality to finish this basket.[17]

While living among the Luiseño people in southern California in the 1930s, Eleanor Beemer noticed that a weaver, Juana Rodriguez, burned off the sumac *(Rhus trilobata)* growing by the river "to force them to produce more canes."[18] We would probably have much more historical evidence of Indian management of shrubs and trees if it were not for the fact that early explorers, missionaries, settlers, and anthropologists generally did not connect the practice of Indian burning, which they observed frequently enough, with the goal of stimulating plants to make better material for utensils, cordage, or basketry. When indigenous burning is mentioned, it often appears as an isolated observation of one or two lines in a book or report.[19]

When one talks with many native weavers, it becomes clear that native people know how to manage the burning–harvesting cycle and the pruning–harvesting cycle to achieve the best results without harming the plant, pointing to a long history of managing growth in this way. "Never cut the shrubs in the late spring or summertime when the sap is rising," one Miwok elder advised. "You want to prune back redbud when the plant's energy is in its roots." For the most part, native gatherers cut back willow, dogwood, maple, sumac, redbud and other woods during the dormant period, knowing that cutting at this season causes the least harm to the plant.

Elders of the Western Mono and Foothill Yokuts tribes remember their relatives applying fire to shrubs to induce long, straight sprouts for basketry material:

> They burned the redbud, sourberry, ferns at lower elevations, and the red willow. They burned because they had no way of pruning. It was to improve the new growth, and the areas were not so overgrown. (Clara Charlie, Chukchansi/Choynumni Yokuts, pers. comm. 1989)

> My mother or father would cut the whole redbud bush down in the fall, about October or November, and the new ones would grow. They would pile leaves on top of sourberry bushes and burn them in the fall. Next year the shoots would come up nice and smooth with no knots. (Margaret Baty, Western Mono, pers. comm. 1992)

> Our sourberry patch is up on the hill and we used to burn. But the laws say we can't burn anymore. A long time ago my uncle and my father used to burn it for my mother, grandmother and great aunt, in October or November, after they'd get their sticks. They burned sourberry in the 1930s before World War II. They burned the grass all around the bluebrush [Ceanothus cuneatus] and white oaks [Quercus douglasii]. This was to keep it clean under the oaks. My uncle and father cut down the bluebrush and the next year, the shoots would come up straight and long and were used in baskets. They'd pile brush on the sourberry and light it on fire with a match. Mother, grandmother and my great-aunt would harvest the little ones [sourberry] the following August to January for the tops of the baby baskets. They like them after a burn because these are the ones that grow right from the ground and they're nice and slender (Norma Turner Behill, Mono/Dumna, pers. comm. 2004)

To summarize, these and other elders say that the habitats of buck brush, deer brush, sourberry, redbud, button willow, black oak, chokecherry, blue oak, interior live oak (Quercus wislizenii), willow, and leatherwood or flannel bush were set afire in the past.[20]

Craig Bates (pers. comm. 1992) was given a description of how two Maidu women, Mary Jones and Marie Potts, burned maple (Acer macrophyllum) for basketry: "They said that they went with their mothers to cut the maple shoots, and after they cut the shoots they spread dry leaves on the stumps. Then they set the leaves on fire. One of them said that they sat down and ate a sandwich, and about the time they were done eating, the fire was out and they left. I believe that this was done in the winter or early spring."

The practice of inducing shrubs and trees to produce long, straight shoots is ancient. We have no direct evidence of how far back in time this type of management was practiced in the area that is now California, but there is evidence from surrounding areas that suggests indigenous peoples were pruning, coppicing, and burning for young growth for at least several millennia before the arrival of Euro-Americans.

Based on the study of a split-twig animal figurine retrieved from an old village site, the southwestern archaeologist Vorsila Bohrer is convinced that ancient people from the Grand Canyon area of Arizona and southern Utah burned bushes to stimulate production of abnormally long and straight sucker shoots that could be used for making handcrafted items such as the fetish, which was made with a 6-foot-long sprout of *Rhus trilobata*. Radio-carbon dating shows that the shoot was harvested as early as 3500 B.P.[21]

The antiquity of the basketry craft is another reason to suspect the early presence of shrub and tree management in California. Anthropologists have retrieved basketry fragments dating to 10,900 B.P. from Fishbone Cave in Pershing County, Nevada, on the edge of Lake Winnemucca, and we have archaeological evidence that baskets were used in California four millennia ago. Even older basketry fragments, preserved in areas rich in bird guano and salt, were recently discovered in a shell midden at Daisy Cave on San Miguel Island, off the southern California coast. These fragments carry ra-diocarbon dates of between 8600 and 9900 B.P.[22] Although the very oldest basketry fragments are not of woody materials, it is likely that the crafting of baskets and the use of fire as a management tool went hand in hand. All the oldest woody-material basketry fragments found in Oregon, Nevada, and California show workmanship as fine as that found in historical-period baskets in museum collections. The exacting qualities, the fineness and even-ness of stitch and the smooth texture, are undoubtedly a product of the selec-tion of shrub material that had uniform qualities and was straight-growing without lateral branches—the type of material produced by a shrub after it has been pruned or burned.

Cultural Uses of New Woody Growth

Among tribes in California, a majority of material culture objects required the special types of young growth produced from burned or pruned shrubs and trees. These objects encompassed ten cultural use categories: basketry, ceremonial items, clothing, cordage, games, musical instruments, cages and traps, structures, tools and utensils, and weapons. (See Table 8.) The hard and inelastic wood derived from older woody growth was required for some ma-terial culture objects, but the list of these is rather short. Most of the items—

TABLE 8

Cultural Items Made with Adventitious Shoots and Epicormic Branches Harvested after Burning or Pruning by the Sierra Miwok, Foothill Yokuts, Paiute, Maidu, Western Mono, and Tubatulabal

Basketry	burden baskets, cooking baskets, cradleboards, dippers, drying baskets, feast baskets, gift baskets, parching baskets, plate-form baskets, seed beaters, sifters, storage baskets, winnowing baskets
Cages and traps	duck traps, fish basketry traps, pigeon and flicker traps, quail basketry traps, cages for decoy pigeons, deadfall or trigger traps for small game
Ceremonial items	feather plumes, feather sticks, healing sticks, whipping sticks
Clothing	rod armor, snowshoes
Cordage	withes for lashings
Games	basket rackets, goal posts, game counters, hoops, lances, poles, shinny sticks, looped sticks for ball games, blunt arrows, toy arrows, ground sticks, whipping sticks, popguns
Musical instruments	clapper sticks, flutes
Structures	acorn granaries (lashings, in and out weave on sides), assembly houses (lashings), fish weirs, houses (lashings), sweat houses (lashings)
Tools and utensils	cooking tongs, digging sticks, fire drills, hooked sticks for catching rodents, looped stirring sticks, netting shuttles
Weapons	arrows, harpoons, spears

SOURCE: Based on discussions with Margaret Mathewson, Norma Turner Behill (Mono/Dumna), and Craig Bates.

TABLE 9

Cultural Items Made with Branches of Older Growth by the Sierra Miwok, Foothill Yokuts, Paiute, Maidu, Western Mono, and Tubatulabal

Ceremonial items	pipes, rattles, wands
Clothing	ear plugs, hairpins, nose plugs
Firewood	firewood
Musical instruments	bull roarers, foot drums, musical bows
Snares and traps	deadfall traps, deer traps, quail brush fence snares
Structures	house frameworks, support beams, thatch; acorn granary frameworks, support beams, thatch; fish weir vertical sticks, hunting blinds
Tools	arrow straighteners, fire hearths, knocking poles, ladders
Toys	dolls
Weapons	bows, spear handles, wooden-handled knives

pipes, rattles, ear and nose plugs, hairpins, bull roarers, foot drums, deadfall traps, ladders, and bows—required only small quantities of older wood. (See Table 9.) Only three use categories—transportation, firewood, and building structures—required large amounts of old growth or dead wood.[23]

BASKETRY

Since basketry represented a full 50 percent of the plant material culture of most native tribes before the arrival of Euro-Americans (Margaret Mathewson, pers. comm. 1992), vast quantities of shoots and sprouts were needed just for baskets. One finds occasional references to these quantities in the ethnohistoric and ethnographic literature:

> These twigs or shoots of willow or red-bud were gathered [by the Northeastern Maidu] in large quantities by the women.[24]

They [Yurok] gather these sticks [hazel for basketry] by the thousands and take them home, where the women, children and men all join in peeling the bark off the sticks.[25]

They [the Southern Sierra Miwok] opened several large sacks and threw down on the ground for me to see coils of split willow strands and bundles of rods for baskets. They had been most industrious and had a large stock on hand.[26]

Some baskets were subject to considerable wear and needed repair or replacement frequently (Craig Bates, pers. comm. 1993). Cooking baskets, dippers, and basket traps had constant contact with water or damp substances (such as acorn mush) and wore out quickly. Burden baskets for carrying firewood, seeds, berries, and bulbs also had heavy use. Parching baskets had live coals in them, blackening the interior and perhaps causing deterioration in a short time. Sifting and winnowing baskets also had heavy and constant use.

Certain baskets were purposely disposed of according to custom. For example, baskets used for cradles were generally used only for one child and were not passed down but rather abandoned or destroyed after being outgrown. Baskets of a deceased person were buried or burned with the individual upon death.[27]

Every family in a village had a large contingent of baskets, each serving a different purpose for the household. According to Larry Dawson, "The baskets of each tribe were made in accordance with a set of traditional standardized forms having designated uses, weaves, materials, and names. Every household had more or less a complete set of up to twenty-two forms with several new replacements in progress for those about to wear out. Old photographs of Indian camps clearly show the numerous baskets in, around and upon the shelters."[28]

CAGES AND TRAPS

Cages, constructed almost entirely of young shoots, held live birds and other animals as pets or for ceremonial purposes or food. The Tubatulabal made cages of flexible willow poles in which they could keep the young golden eaglets they took from nests in steep cliffs. The Michahai, Wukchumni, and Waksachi Yokuts made cages of open-twined basketry for snake doctors to house their snakes for the Snake Ritual. These cages had pointed bottoms, were covered with a lid, and hung from trees.

Basketry traps made of young shoots were used for capturing both birds

Figure 29. Maggie Icho, a Wukchumni Yokuts, with a cage for band-tail pigeons, 1945. Courtesy of the Autry National Center/Southwest Museum, Los Angeles, #22978.

and fish. The Foothill Yokuts captured bandtail pigeons using a pigeon decoy and a spindle-shaped cage made of young growth; they also made pigeon snares with sprouts six feet long. (See Figure 29.) The Pomo made a cylindrical cagelike trap for catching woodpeckers. It was twenty inches long by four inches wide and made from twined sprouts. The Pomo also used sprouts to make a quail trap that consisted of several plain twined woven tubes attached to one another to make a long tunnel about six inches in diameter and four feet long. It was broadly flared at one end, and the other end was connected with a purse net (Craig Bates, pers. comm. 2001). Weirs of brush extended from the flaring entrance to guide the driven birds into the trap.[29]

Rough-twined traps constructed of the shoots of willow and other species were employed by many tribes for catching fish. Typically made by male fishermen or hunters, they were placed in a series in tandem with fish weirs, often to capture anadromous fish in quantity. The Northern Paiute set conical basketry traps at the openings of fish weirs and placed bipointed basketry minnow traps in the shallows of streams. Fish were caught by the Yokuts using weirs of stones or willows set across a stream with a basket trap or sack in a small opening. The Kacha-Poma (Redwood Valley) made cone-shaped baskets of "hazelnut wands" for catching fish. The Washoe made basket traps of "willow wands" that were intended to completely block streams. Sometimes small fish were taken with the aid of baskets made of young shoots: Yaudanchi Yokuts women used scoop-shaped open-work baskets to capture small fish, and the Entimbich and Wobonuch (Western Mono) used special "sieve" baskets. Northern Paiute women fished the shallows and behind weirs and dams in low water, retrieving small fish with shallow twined parching trays made of young shoots.[30]

CEREMONIAL DRESS, CLOTHING, ARMOR, AND FOOTGEAR

New, flexible stems were used to make various types of clothing and regalia. The inner sapwood of maple, cottonwood, and willow shoots, which split easily into thin, uniform strips, was used to make paper-thin skirts, usually tied around the waist with cordage and worn by Quechan, Yokuts, Maidu, Mojave, and Luiseño women and girls. Wintu men wore dance regalia made of the inner bark of maple. Feathered headdresses were attached to a foundation hoop made of young shoots of hazelnut, willow, or dogwood. They adorned the heads of men in dance ceremonies. One specific type of headdress used by the Valley Maidu was constructed with fifty to one hundred fifty young willow shoots, each two to three feet long.[31] Young uniform sticks or split shoots of willow, hazel, oak, or ocean spray, harvested after a fire, were twined together to make handcrafted rod armor that was strong enough to withstand the blow of enemies' arrows. Walter Hough described the rod armor of the Hupa: "[It is made of] wattles and twine, woven and bound with buckskin. This is worn in battle to protect the body; it is tied across the breast from left to right. The red lines denote the number of enemies slain or captives taken; also the rank of the wearer."[32] Snowshoes were made from the leader (central topmost) branches of oak or serviceberry saplings or shoots that started as epicormic or adventitious branches and had grown several seasons.[33] The selected branch had to exhibit extreme elasticity, as it is bent to a roughly circular shape and secured with buckskin or some other type of lashing. This footgear transported people across the snow-covered mountains in winter.

CORDAGE

Cordage is made by twisting together "separate fiber strands into a single, long twined string or rope." The making of string or cordage may be the oldest fiber art in North America. Native peoples likely brought cordage technology with them when they first entered California. Since cordage had to be both flexible and strong, adventitious or epicormic branches served well as a cordage material. The young growth of maple, cottonwood, and flannelbush (Fremontodendron californicum) was split into thin strips and used as lashing to hold structure frameworks solidly in place. Cordage made from young shoots held together ground mats, roofing mats, and wall mats that covered building frameworks; it was used for wrapping looped stirring sticks, attaching prongs to fish spears, and lashing ladders. Willow withes held together the bundles that made up tule boats.[34]

The North Fork Mono elder Rosalie Bethel described the use of young flannelbush (leatherwood) for cordage: "We used to cut the branch and split it in half and remove the inner bark and scrape the outer bark. We used it for tying. After a fire, it grew nice and tall and straight and we used it for nets and for holding together the stirring sticks. We tied different types of cone-shaped storage bins for acorns and manzanita berries with this plant" (pers. comm. 1989).

Norma Turner Behill also recalls the use of the flannelbush material that would sprout after burning: "Slippery elm [Fremontodendron californicum] for the ties for the granaries, soaproot brushes, and house structures would grow real straight and tall after a fire. The best time to gather it is in the spring, when the bark comes off easily" (pers. comm. 2004).

Although it is not a woody plant, dogbane, or Indian hemp (Apocynum cannabinum) must be mentioned in this context. Indian hemp is a perennial herb found in seasonal wetlands below 7,000 feet in many plant community types in California. It had widespread use throughout the state as a cordage fiber.[35] The hemp stems contain soft and sturdy "bast" fibers. These bast fibers were collected, extracted, and made into cordage that was used to construct many items, including nets and lines for fishing, deer nets, rabbit

TABLE 10

Quantities of Cordage Material (*Apocynum* and *Asclepias* spp.)
Gathered for Cultural Items by Tribes of the Sierra Nevada

Tribe	Cultural Item	Use	Dimensions	Total Cordage Length Gathered	Number of Stalks Gathered
Washoe and Northern Paiute	gill net	fishing	1/16 in. 2-ply 100 ft. × 4.5 ft. × 1.5 in. mesh	12,022 ft.	60,110
Washoe and Northern Paiute	bag net	fishing	1/16 in. 2-ply 2.5 ft. × 2.5 ft. × 2.5 ft. × 1 in. mesh	885 ft.	4,425
Washoe and Northern Paiute	A-frame dip/lift net	fishing	1/16 in. 2-ply 7 ft. sq. × 4 ft. [× 4 panels] × 1 in. mesh	7,890 ft.	39,450
Sierra Miwok	feather cape	ceremonial uses	1/16 in. 2-ply 1.75 in. mesh	100 ft.	500
Sierra Miwok	deer net	hunting	1/8 in. 2-ply 40 ft. × 6 ft. × 4 in. mesh	7,000 ft.	35,000

SOURCE: Adapted from Lindstrom 1992; Anderson 1993b.

nets, netting bags, tumplines, slings, flicker-feather headbands, hairnets, feather capes, feather skirts, belts, cord belts for women's aprons, bow strings, and shredded fibers for women's skirts. Both Indian hemp and milkweed are still gathered today.[36]

Large quantities of Indian hemp (as well as plants in the genus *Asclepias*) were needed to make these different items. (See Table 10.) Passages in journals and reports comment on the vast fields of "flax" or "hemp" in California. According to Craig Bates (pers. comm. 1992), five plant stalks were needed to make one foot of string. A feather cape used for ceremonies by the Sierra Miwok required 500 stalks; a forty-foot deer net would require 7,000 feet of string or 35,000 plant stalks.[37] For the most part Indian hemp is found today only in small scattered colonies; in former times, it must have existed in dense stands to accommodate these indigenous industries.

Indian hemp plants were gathered mostly in late fall or winter after the stalks had died. As long as the living underground organs that would produce the new growth in spring were not disturbed, the old aerial parts could be harvested with no harm to the plant. Hemp patches were managed to maintain and enhance their density, size, and frequency. The primary management tool was fire: the patches were periodically burned by the North Fork Mono, Pomo, Wukchumni Yokuts, and probably other tribes to decrease accumulated dead material, provide increased access for harvesting, allow more sunlight to reach the new growth, and recycle nutrients. Fires also prevented trees from encroaching on the stands and blocking the sunlight.[38]

Long, straight stems of Indian hemp with no laterals, growing in direct sun, made the best material for cordage (Craig Bates, pers. comm. 1992). Plants produced more of these quality stems after burns. Lalo Franco, Wukchumni Yokuts, told me, "There's another plant that benefited a lot by fire, my dad said. It was the Indian hemp. Because they used that for rope and it could get really gnarled up. The hemp plant would get all congested. The whole plant would burn all the way down to the roots, but then it would come up the next year straight. And then you need those straight long stems so they could crack them open and get their fiber out of there" (pers. comm. 1991).

According to Rosalie Bethel of the North Fork Mono, "The Indian hemp comes up better after they set the fires in the fall" (pers. comm. 1989), after stems had been collected and the seeds dispersed. Indian-set fires probably activated rhizome production, increasing the size of the tracts. Because large quantities of Indian hemp were harvested for the making of various cultural items, the cumulative acreage burned to maintain productive collection sites was probably substantial.

GAMES

Some games were played with blunt arrows, darts, or lances that were thrown at or through a target. These arrowlike objects required a strong, light, easily straightened cylindrical piece of wood that carried well, frequently a young shoot. Other games involved contests of distance throwing and required aerodynamic young shoots called ground lances. These lances were made of young willow, wild cherry, or spicebush. In many games, each side kept track of its winnings with counter sticks made of the first-year growth of various shrubs such as willow, maple, and wild cherry.[39]

A hoop-and-pole game was popular among many tribes; the object was to throw a pole or lance through a rolling hoop or to toss a hoop and hang it on a lance point. The lance was a withe of willow, wild cherry, or maple, and the hoop was made of flexible shoots of willow, chaparral, wild cherry, flannel bush, or oak wrapped with split shoots, bast fiber, or buckskin. A type of basketball was played by Sierra Miwok women. The goal posts were made of flexible willow wands, and a single buckskin ball was manipulated and thrown with the aid of handled basket rackets made of first-year growth. In one type of ball game the Yokuts made a "spoon stick" of two willow saplings with an oak loop lashed on the lower end with sinew. Players forwarded mistletoe-root balls with the spoon sticks to their mates, who sent them on to the nearest pair of opponents. The Yokuts made a fifteen-inch popgun for shooting wads, using elderberry for the cylinder and a young shoot of maple for the piston.[40]

MUSICAL INSTRUMENTS

Many tribes used young shoots of elderberry to craft delicate, thin-walled flutes capable of playing soft, melodic tones.[41] Clapper sticks, also made of young elderberry, consisted of long, split sticks that were hollowed out for three-fourths of their lengths and were struck against the hand or trembled in the air with the wrist to make clapping sounds during dances. The shoots selected for flutes were one to three years old; those selected for clapper sticks were one to four years old (Craig Bates, pers. comm. 1992).

Ben Cunningham-Summerfield (Mountain Maidu), a user of traditional plants, including elderberry, has observed in Bidwell Park and Yosemite National Park: "After burns elderberry vigorously resprouts and you get longer growth between the leaf nodes, which makes better flutes and clappersticks. The larger pith, straighter stems, and thinner walls all contribute to better material for traditional musical instruments and other uses. Elderberry also responds well to coppicing, producing vastly improved material for musical instruments" (pers. comm. 2004).

STRUCTURES

Dwellings, acorn granaries, fish weirs, hunting blinds, and dance houses were fashioned from timber, saplings, and the young growth of shrubs and trees. Hemispherically shaped structures, which were commonly built by many tribes, required access to large numbers of saplings for their construction. These saplings were bent into a dome shape and were strong enough to support bark, mats of willow sprouts or tule, brush thatch, or earth. The upright framework of the hemispherical house was secured by horizontal withes, forming tiers. These tiers strengthened the frame and supported the outer covering. Frank Latta described the hemispherical house of the Foothill Yokuts as a "wicker-like dome." Wheat described a Northern Paiute house as follows: "To build a house, Indian women required only a dozen long, strong willows to make the frame, large bundles of cattail leaves and willows for making mats, and willow withes, strings of sagebrush bark, or strips of old cloth to tie the building together."[42]

The conical house used at the higher elevations by the Sierra Miwok, Western Mono, and Yokuts tribes was a structure of straight poles tied together in a cone shape, spread like a tripod, and covered with mats of tule or willow, brush, bark, or earth. The poles were sometimes reinforced with smaller poles, and the whole upright structure was held in place by horizontal bands of withes made of young oak, willow, grapevines, or some other shrub material. For the Wukchumni Yokuts conical house, three oak poles were tied together near their tips and set up as a tripod. The poles were held in place by horizontal bands of pliable willow. The thatch covering was made by twining together small bunches of fine willow or long fine grass in long sections. The skirtlike sections were held in place by long willow strips running around the outside of and parallel to the inner horizontal frame bands. A panel door was made by tying a specially made mat of tule or willow to a framework of willow or young alder.[43]

Fish weirs, in their simplest form, were made by planting a row of vertical stakes firmly across a stream or creek and then weaving pairs of first- or second-year shrub shoots between them, sometimes with a half twist in each space. The result was a latticework of intertwined shoots, impenetrable to migrating fish. Northwestern California tribes, such as the Tolowa, used communal fish weirs made of hazel wickerwork. Many other tribes built these structures as well. Anna Gayton describes a Wukchumni Yokuts weir:

"Weirs of willow were set across rivers or streams at suitable places. The willows used were about finger thickness and 3 feet long. They were interwoven with willowbark rope. Into the weir and facing upstream were set cylindrical baskets 6 to 10 feet long and 12 to 18 inches in diameter; as many as three of these might be used."[44]

TOOLS AND UTENSILS

Different kinds of tools were made of woody shoots produced through burning or coppicing. The digging sticks used for prying bulbs, earthworms, and other foods from the earth were among the most important of these tools. According to Lalo Franco (pers. comm. 1991), "Fires stimulated strong shoots that would come up from the bottom rather than high in the shrub. Those were used for the digging sticks." Rodent sticks were also made from such shoots. These were flexible sticks with hooked ends used for retrieving rats, muskrats, gophers, ground squirrels, rabbits, and chipmunks from their burrows or hiding places. The looped stirring sticks used by many tribes were made of "tough sprouts" coming up from the base or along the boles of different shrubs or trees such as willow, chokecherry, or oak. One end was bent to form a round, open loop and then tied back on itself to the handle. This tool was used for stirring acorn mush and for fetching and retrieving the fire-heated stones that were dropped in the acorn mush to cook it. California Indian elders remember the association of fire with the young growth used for household utensils:[45]

> They used the new growth of the black oak for spoons. Where it's cut or where wind has broken it and there starts new growth, that's what they would harvest. These would be gathered in burnt over areas or cutover areas. (Lydia Beecher, Mono)

> The young shoots of the slippery elm [Fremontodendron californicum] were used to make tongs for cooking our acorn. These were harvested three or four years after a fire. (Norma Turner Behill, Mono/ Dumna)

WEAPONS

Straight branches of shrubs, one or two years old, made the best material for crafting arrows. Shrubs such as mock orange, button willow, snowberry, gooseberry, willow, alder, hazelnut, and spicebush were often selected, especially those that grew away from well-watered areas. According to Bill Horst, a non-Indian arrow maker, the drier an area, the slower the shrub

Figure 30. Fernando Librado, a Chumash, straightening arrow shafts. Arrows were often managed at the organism scale—pruning or burning individual shrubs to create young, straight growth for aerodynamic material. Courtesy of the Autry National Center/Southwest Museum, Los Angeles, #37698.

grows—reducing the inner, pithy core and heightening the strength of the perimeter wood.[46]

Arrow-making plants were pruned or burned one to two years prior to harvest, according to Bill Horst (pers. comm. 1992) and at least two years prior to harvest according to Craig Bates (pers. comm. 1992). Shrubs were pruned or burned between November and January and the shoots gathered during that same period one to two years later. Citations describing the necessity of pruning or burning before harvesting of branches for shafts are rare in the ethnohistoric and ethnographic literature. The branches of spicebush *(Calycanthus occidentalis)* and snowberry *(Symphoricarpos rivularis, S. mollis)* were pruned by the Pomo to induce straight sucker growth. Florence Shipek documented that the Luiseño people burned shrubs every third year to maintain the desired quality for bows and arrows. Edith Van Allen Murphey wrote of the tribes of the Great Basin: "Small bird arrows were made from the long shoots of the snowberry *(Symphoricarpos racemosus)*. . . . This was cut down in Fall, so that it would send up shoots the following spring and be straight and smooth by autumn."[47] (See Figure 30.)

Clara Charlie (Chukchansi/Choynumni Yokuts) recalls burning button willow to "make it nice for arrow shafts." Bill Franklin (Sierra Miwok) recalls that they "burnt shrubs in the fall to make straight shoots for arrows."[48]

Horst says that button willow sprouts harvested for arrows must be harvested within two years of pruning, otherwise they "get too long and heavy and you start getting branching. Branching starts to kink the shaft and causes problems with straightening and in trimming it. Where there is a lateral branch you have a weak spot there."[49]

To create the longer, wider-diameter shoots needed for fish harpoons or spears, some of the sprouts created by burning or pruning were purposely left to grow for several years.[50] According to Horst, Yokuts men would pick all the leaves off these shoots every year to retard the development of side branches.

Shaping Shrubs and Shaping Ecosystems

In carrying out the practices necessary to produce the large quantities of woody shoots needed for their material culture, indigenous people in California became agents of habitat modification. They expanded and maintained suitable habitat for the desired sprouting species in both time and space. Burning, coppicing, and pruning had many ecological consequences. To this day, the signatures of human management of sprouting shrubs and trees are evident in the individual plants and in the land and habitats surrounding them.

ECOLOGICAL EFFECTS AT THE PLANT LEVEL

For many native shrubs and trees, repeated pruning or burning was not only harmless but beneficial as well. Their repeated resurrection, while appearing to take tremendous energy, may have paradoxically kept them young and vigorous. Norma Turner Behill pointed out a redbud in the Sierra Nevada foothills that has been coppiced for more than forty years without harm to the plant. According to Sylvena Mayer (North Fork Mono), "You can't kill a redbud. Cut it to the ground and it just keeps getting thicker" (pers. comm. 1991). Elsie Allen, a consummate weaver, said that a white gentleman once came and told her she should not take up basketry because it would destroy a lot of plants. "He did not understand what I knew very well," Elsie said, "that the cutting out of roots and trimming of shrubs actually helped spread the growth and there was no danger as long as the digging and cutting was not overdone at any one place."[51]

Repeated pruning or burning keeps shrubs and trees in a physiologically "young" state, often prolonging their life spans considerably. Without some kind of disturbance, shrubs support an increasing number of dead or dying branches. As the proportion of dead to live aboveground biomass in-

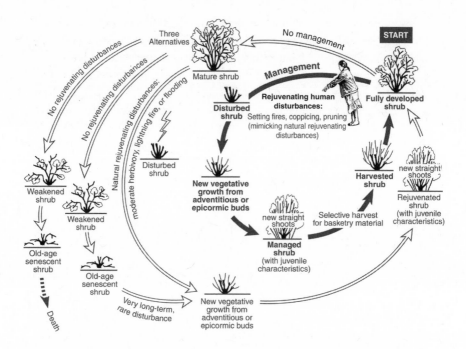

Figure 31. Coppice shrub system: a conceptual model for indigenous management of shrubs. Management techniques mimic natural disturbances such as flooding and lightning fires and can be applied over long periods to manage shrubs in situ. In the absence of disturbance the shrub's life span may be shortened, and it quickly becomes unsuitable for basketry material.

creases, the assimilation efficiency of the shrubs (i.e., their ability to convert absorbed radiation into photosynthate) is reduced. More and more photosynthate is utilized for maintenance respiration rather than converted into growth and reproductive biomass, thus reducing plant vigor.[52] Pruning, coppicing, and burning, as forms of disturbance, reverse this process and stimulate the growth of more live, healthy branches and increase the leaf area available to the plant. Coppiced or burned shrubs and trees often have extensive root systems that are capable of supplying large amounts of both nutrients and water to the new shoots, and the increased leaf area, in turn, may allow the enlargement of the root surface area, which accommodates the additional water loss from the leaves in transpiration.[53] (See Figure 31.)

When burning was used to stimulate resprouting, it had several other effects as well, among them the control of diseases and harmful insects. Willows and other species of shrubs are subject to attack by a wide range of leaf-eating, stem-sucking, and wood-boring insects. Female sawflies (*Euura* spp.),

238 · INDIGENOUS LAND MANAGEMENT

for example, lay eggs in elongated shoots of willow and one year later exit as wasps (Steve Haydon, pers. comm. 2002). Most of these pests, which typically rendered material from a shrub useless as basketry material, were controlled by a combination of regular burning and well-timed harvesting: "The men would burn sourberry, redbud and chaparral *[Ceanothus cuneatus]* in the fall after the women picked their sticks. The smoke would decrease the insects on the sourberry" (Ruby Cordero, Chukchansi Yokuts/Sierra Miwok, pers. comm. 1992). Scientific studies confirm that fire is the most effective agent for combating many kinds of diseases and insects.[54]

Regular burning of shrubs also helped to recycle nutrients in the system. It made available to the plant nutrients that would otherwise be held in dead and dying branches. And because burning typically was done after the active growing season, fire recycled nutrients in the leaves after they had contributed to root growth.

ECOLOGICAL EFFECTS AT THE COMMUNITY LEVEL

The effects of managing for young growth extended beyond individual shrubs and trees to the community and landscape levels. When Indian women or men set hillsides on fire, they not only spurred the growth of young sprouts from shrubs and trees but also opened up areas to increased sunlight, heightened the structural complexity of forest, woodland, and shrubland habitats, stimulated the seed germination rates of seral and serotinous species, recycled nutrients for the whole community, altered insect populations, and promoted increased biodiversity.

Periodic burning encouraged native annuals, grasses, and herbaceous perennials to grow under shrubs and trees, creating a healthy understory that enhanced the permeability of the soil surface, checked surface erosion, increased rates of nutrient cycling, enhanced soil fertility, and provided food and habitat for animal species, thus increasing biodiversity and the possibility of mutualistic community interactions.[55] After burning, nutrients such as phosphorus were freed and stimulated the growth of many plants and mushrooms, including those that have important mycorrhizal relationships with shrubs and trees. Although nitrogen was largely volatilized by the fire, nitrogen-fixing plants, such as wild peas, locoweeds, lupines, clovers, and deer brush, quickly moved into burned areas or sprouted from fire-activated seeds, replenishing this lost element. Burning opened up areas to increased sunlight, allowing shade-intolerant herbaceous plants—some of which fire ecologists dub "fire-followers"—to come to life from hidden seed banks and quiescent bulbs. Sun-loving plants such as lilies, brodiaeas, soaproot, and wild onions appeared and attracted numerous wildlife species such as deer, bears, and gophers.

The light fires characteristic of the indigenous style of burning increased the structural complexity of communities in two dimensions. Vertically, they increased the variety of plant physiognomies, helping to establish layers of herbs, shrubs, and trees at different, distinct heights. Horizontally, they increased the patchiness of the community, ensuring greater heterogeneity of leaf cover and species composition. For certain species, these fires also functioned to maintain a greater variety of age and size classes of individuals. For example, in chaparral and oak woodlands where fires have occurred periodically, redbuds of all ages are found. Where fire has been excluded, there are no redbud seedlings or saplings, because the seeds need fire to crack their seed coats and spur germination.

Plant Renewal and Cultural Renewal

The beauty and antiquity of basketry and other material arts in California is testimony not only to successful land management practices but also to a successful way of life. Like other great spiritual and philosophical traditions around the world, California Indians recognized that life is dynamic, relying on cycles of destruction and renewal. Plants and plant communities required periodic tending, which often involved burning, to retain their health and vigor over time and provide Indian peoples with the resilient and enduring materials on which their ways of life were based.

The burning, pruning, and coppicing that California Indians practiced extensively in at least ten plant communities across the state constitute an important part of the land's ecological history. Some of these management techniques were applied to the same areas for long periods and were probably sustainable. Hardwood shrubs and trees that vegetatively resprout cannot be dated by plant ecologists, but it is conceivable that some individual redbud trees or sourberry shrubs alive today could be hundreds or even thousands of years old, having been used and managed for basket sprouts by many generations of women.

Only recently have scientists begun to realize that managing for young growth may be a universal practice among indigenous cultures throughout the world. Today coppice woodland systems are being revived in England, Ireland, Sweden, Japan, and the United States. These indigenous management practices may help us to restore and maintain biodiversity on certain public and private lands.

CHAPTER 8

California's Cornucopia

A Calculated Abundance

Seed beating chia *[Salvia columbariae]* plants helped scatter the seeds around the area and helped the crop. Otherwise the seeds just stay in the vessel and with no collecting the insects get the seeds.

WILLIAM PINK, Luiseño/Cupeño (1997)

Aboriginal California was rich with wild produce. Buttercup seeds, manzanita berries, cactus pads, Indian rhubarb, wild onions, and hundreds of other plants nourished the native peoples. Millions of grains from native grasses became flour for Indian bread, and millions of plump tubers became raw or baked "potatoes" cooked in Indian earth ovens. If one could look inside a Chunut Yokuts tule house in 1700, one would see baskets of dried blackberries, mushrooms, and different kinds of seeds, as well as jackrabbits, elk, antelope, and fish hung on poles. "The house was full of things to eat," said Yoimut, a Chunut informant interviewed by Frank Latta.[1] Southern Paiute women beat the small edible grains of grasses into burden baskets in the Mojave Desert; Kitanemuk women gathered wild currants in chaparral areas on what is now Tejon Ranch; Yuki women pinched sweet clover leaves for greens in meadows in northern California; Pomo women plucked and dug tules in the marshlands of Clear Lake for their edible stems, roots and rhizomes; Chumash women squeezed the oozing liquid from milkweed stems for chewing gum.

By all accounts, the seed gathering grounds of groups across California were extremely productive, even in the desert regions. Barrows described the Cahuilla landscape in the 1840s: "[E]very bush or tree was dropping fatness. The desert seemed a land of plenty, here the manna fell at each man's door." Walton estimated that a single Cahuilla worker could gather about 175 pounds to the bushel or eight and one-half bushels of mesquite pods per day and that one acre well covered with trees could produce one hundred bushels per year. The famous horticulturist Luther Burbank spoke of the abundance of yampah (*Perideridia* spp.): "There are places where the plant grows almost like grass, so that hardly a shovelful of dirt can be turned over without exposing numerous roots."[2]

The land's abundance benefited the wildlife as well as the people. Songbirds, black bears, mountain beavers, California spotted owls, and beetles all relied in part on human stewardship of food plants to keep their bellies full and feed their young. Native elders point out that bears and other large mammals that now have to roam great distances to feed themselves could find plenty of food in a smaller area before the transformation of the landscape by Euro-Americans.

Early European and American explorers and settlers saw in the productive landscape an ever-full horn of plenty that gave the native people no need to be industrious. In their eyes, native people were merely the reapers of this abundance, not the sowers. Army sergeant James Carson, for example, suggested that the land was lying fallow when he wrote: "I have had intercourse and consultation of the value of California as an agricultural country, or to what purposes its rich lands could be converted from the stillness in which they have lain through ages past, and made to swell our comerce and trade, and enrich our people."[3] Years later, anthropologists would make the same mistake, arguing that the rich suite of plant and animal species and their high densities and numbers squelched any motivation to cultivate food plants and thus develop agriculture.

Previous chapters have described in detail how California Indians tended native plants so that they would produce the materials needed for basketry and other cultural items. The native approach to food was no different: indigenous people burned areas to stimulate food-producing plants and discourage their competitors, sowed seeds to enlarge populations of seed-bearing annuals, carefully tended and pruned seed-bearing shrubs and trees to increase their productivity, and dug fleshy tubers, bulbs, and corms in ways that ensured their existence and vitality over time. To a considerable extent, therefore, the abundance of the California landscape was an anthropogenic phenomenon, based on human labor and knowledge.

Many "wild" edible plants bore more profusely when they were helped along in some way. Berry bushes produced more fruit when they were burned or pruned. Bulbous and tuberous plants that reproduce through offsets thrived in the aerated, weeded soils softened by digging sticks. Plants producing edible leaves were rejuvenated with pinching and periodic fires. The edible seeds of annuals were plumper, more viable, and less affected by insect pests when collected from areas that had been treated with fire.

The sophisticated management of honey mesquite *(Prosopis glandulosa* var. *torreyana*) by the Timbisha Shoshone, documented by the ethnobiologist Catherine Fowler, is just one example of the tending of a food source. The Timbisha pruned the dead limbs and lower branches of their mesquite trees, kept areas around the trees clear of undergrowth, and kept the blowing desert sands from engulfing the trees. These practices made the mesquites much more productive than they would have been had they not been cared for.[4]

Ancient California Cuisine

A tremendous variety of edible plants, animals, algae, and fungi was available, and nearly all of it found its way into the diets of the indigenous people. Since the flora and fauna varied greatly from region to region, so too did the cuisine. Particular ways of preparing and combining foods were truly regional in that they stemmed from both the people of the place and the unique plants and animals of the locality. Within each group or village, Indian epicureans helped each generation of people to get acquainted with the local flora as it applied to food: this plant serves as a condiment; those berries make a scrumptious sauce; those leaves are poisonous and must be leached and boiled before eating.

DIET AND NUTRITION

Plants provided 60 to 70 percent of the primary nourishment for most tribes in California. Stephen Powers estimated that plants constituted as much as four-fifths of the native peoples' diet in some parts of the state. Alfred Kroeber noted that "the California Indians were perhaps the most omnivorous group of tribes on the continent."[5] The diversity of plant foods in indigenous diets included four plant parts: seeds and grains; bulbs, corms, rhizomes, taproots, and tubers; leaves and stems; and fleshy fruits. (See Figure 32.) These plant parts provided a rich and balanced diet. Table 11 shows just a small sampling of the hundreds of species used for food in California, divided by the type of plant part eaten.

Although animal foods were consumed in less volume than foods derived from plants, they were still a significant part of the diet, especially for some groups and at certain times of the year, and were important sources of protein and fat. Tribes along the coast indulged in at least a dozen varieties of shellfish and a diverse array of ocean fish. Animal meats included rabbit, elk, bighorn sheep, rattlesnake, bear, deer, gopher, squirrel, pronghorn, and many kinds of birds. Lizards, amphibians, and insects were also eaten. Native diets also included condiments and sugars. Salt came from the globules adhering

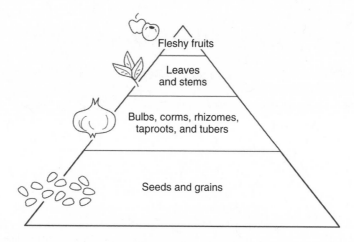

Figure 32. Plant food pyramid. Small and large seeds, such as acorns, pine nuts, and chia, and grass grains such as Indian ricegrass constituted the foundational food. These were followed in importance by bulbs, corms, and tubers such as wild onions, lily bulbs, and brodiaea corms. Leaves and stems of plants such as miner's lettuce and clovers were important greens. Last, fleshy fruits such as wild strawberries, wild grapes, and blackberries were also eaten.

to saltgrass *(Distichlis spicata)*, mineral deposits, or kelp, or was scraped from rocks on the beach. In northern California, the Yuki rolled the leaves and stems of coltsfoot *(Petasites frigidus* var. *palmatus)* into balls and placed them on a thick piece of redwood bark with slow coals around them to slowly change the leaf balls into ashes containing a high proportion of salt. Sugar came from pitch gathered from pine trees, or from sugar crystals that formed on common reed *(Phragmites australis)*. A sticky, sweet substance with a taste like that of honey was collected from willows by the Kawaiisu.[6]

RECIPES AND FLAVORS

A cookbook of ancient California cuisine would have hundreds of recipes, utilizing perhaps a thousand of California's native plant species (in contrast, modern Western diets rely on only about thirty plant species out of the many thousands with food potential).[7] The Yurok prepared a dish of smelt with a sauce of raw salal berries *(Gaultheria shallon)*. The Concow Maidu sprinkled salmon fillets with a flour of deer brush seeds and pulverized laurel leaves (similar to store-bought bay leaves) before baking them in an earth oven. Manzanita cider was sometimes employed as an appetizer to stimulate ap-

TABLE 11

Partial List of California Indian Plant Foods

Plant Parts	Examples of Species
Dry fruits, seeds and grains	walnut, hazelnut, California bay, buckeye, chinquapin, acorn, pinyon pine, sugar pine, gray pine, blue wild rye, California brome, wild oats, Indian ricegrass, coast tarweed, common madia, lupine, cocklebur, California lomatium, woolly sunflower, peppergrass, chia, mule ears, witchgrass, cattail, valley tassel, desert ironwood, yellow pond lily, evening primrose, desert nama, white navarretia, arrow-grass, meadow foam, melic grass
Bulbs, corms, rhizomes, taproots, and tubers	bracken fern, wild onion, camas, soaproot, purple amole, balsam root, yampah, pussyears, mariposa lily, sanicle, common goldenstar, harvest brodiaea, Kaweah brodiaea, blue dicks, ookow, snake lily, wild hyacinth, leopard lily, redwood lily, Washington lily, bear-grass, white brodiaea, golden brodiaea, ituriel's spear, fire cracker flower, small ground-cone, Hartweg's orchid
Leaves and stems	agave, yucca, mountain dandelion, wild onion, pigweed, angelica, wild celery, clover, cattail, tule, nettle, violet, American vetch, mule ears, narrow-leaved milkweed, red maids, jewelflower, tansy mustard, shooting star, dudleya, horsetail, soaproot, miner's lettuce, bractscale, goosefoot, Indian rhubarb, California ground-cone, thistle, candy flower, nude buckwheat, cow parsnip, prince's plume, fiddleneck, sweet cicely, watercress, water parsley
Fleshy fruits	manzanita, madrone, Oregon grape, wild strawberry, wild plum, islay, sourberry, lemonadeberry, wild currant, serviceberry, gooseberry, wild rose, elderberry, buffalo-berry, California bay, desert fan palm, desert apricot, thimbleberry, blackcap raspberry, California blackberry, desert peach, salmonberry, huckleberry, purple night-shade, western chokecherry, wild grape, redberry, twin-berry, bridal honeysuckle

petite; the Sierra Miwok sucked their cider through straws made of hawk tail feathers.[8]

The flora provided foods with all five flavors, sour, bitter, sweet, pungent, and salty. Human tastebuds were tantalized with the sweet, thirst-quenching taste of manzanita cider, the puckery sour taste of sourberries, the pungent taste of saltgrass, the salmony taste of bracket fungi, the spiciness of tarweed seeds, and the slightly bitter taste of Mormon tea bread. The succulent stalks of century plant *(Agave deserti)* and the sweet fruits of beavertail cactus *(Opuntia basilaris)* were savored by desert tribes; Wintu and Valley Yokuts ate the salty brine shrimp and salamander larvae from vernal pools. Dozens of kinds of wildflower and grass seeds, each having a distinct taste, were lightly roasted and pounded and eaten separately.

Balsam-root *(Balsamorhiza sagittata)* seeds made into a mush tasted like popcorn. The seeds of a kind of tarweed were as "rich as butter," and the parched flour of buttercup seeds *(Ranunculus californicus)* had the taste of parched corn. Purple nutsedge *(Cyperus rotundus)* was reputed to have a taste reminiscent of both fresh coconut and raisins. The taste of the cooked flower buds of barrel cactus and the Mojave yucca (William Pink, pers. comm. 1997) was not unlike that of artichokes. Today, some of these foods are still relished and are the touchstones of cultural identity in many tribes. Mushrooms are added to stews and made into thick gravies. Acorn mush sits in refrigerators in fall, and pinyon nuts are offered as afternoon snacks.[9]

FOOD KNOWLEDGE AND SURVIVAL

California Indians had an intimate knowledge of how food could be procured from the landscape. John Muir, who spent long periods alone in the wilderness, recognized this as a great advantage over the cultural knowledge of Westerners. "Strange that in so fertile a wilderness we should suffer distress for the want of a cracker or a slice of bread," he wrote in his diary, "while the Indians of the neighborhood sustained their merry, free life on clover, pine bark, lupines, fern roots, etcetera." Muir showed similar admiration for the native peoples' knowledge of ethnobotany: "Like the Indians, we ought to know how to get the starch out of fern and saxifrage stalks, lily bulbs, pine bark, etc. Our education has been sadly neglected for many generations."[10]

Every Indian adult had an in-depth knowledge of what in nature was good to eat and was skilled in procuring and preparing food wherever he or she went. Native people were able to walk almost anywhere without fear of starvation or dehydration. The Kumeyaay kept wads of lemonadeberry *(Rhus integrifolia*) leaves in their mouths to assuage thirst on long journeys. A thirsty Cahuilla traveler would cut the top off a barrel cactus *(Ferocactus cylindraceus)*, remove a portion of the pulp to create a depression, and then squeeze the pulp in the depression until water was released from the spongy mass. A small handful of chia seeds taken from belt pouches and kept in the mouth and chewed would sustain an Indian runner for hours.[11]

In spring, if food was scarce, California Indians would eat the buds on the ends of the limbs of pine trees, the sweet-tasting inner bark, or the core of the green cones. Remarkably, a Tongva woman survived on her own on San Nicolas Island between 1835 and 1853—a span of eighteen years—after her relatives were removed by force. One of her "rescuers," Captain George Nidever, recorded one of the foods she ate as a root called "corcomites"—probably blue dicks *(Dichelostemma capitatum)*, an abundant wildflower with an edible corm.[12]

Food and Native Culture

The traditional themes associated with food among California tribes were positive and life-affirming: mutual care, generosity, sharing, neighborliness, festivity, gratitude, abundance, and religious ceremony. Food was often used as a peace offering and gift. The cultural significance of food is apparent in the many examples of food being given as a gift to early explorers and settlers. In 1602 Vizcaíno received from the Indians of Santa Catalina Island "prickly pears and a grain like the *gofio* of the Canary Islands, in some willow baskets very well made." The Mono Lake Paiute gave William Brewer of the first Geological Survey of California edible larvae of *koochahbie*, a brine fly *(Cirrula hians)*. John C. Frémont and his men ate acorn bread. James Lynch and the Stevenson Regiment camped on the banks of the Stanislaus River in 1849 and the Indians gave them "all the pinenuts" they cared to take. The Donner party and gold miners in the Sierra Nevada were offered foods to revitalize their health.[13]

SHARING OF FOOD AND OWNERSHIP OF FOOD RESOURCES

In most California Indian societies, at least some of the harvest was shared communally. This served as a wealth-spreading device, a way to both dissolve competition and strengthen social bonds. The chiefs divided important harvests such as yampah tubers among families so that none went hungry. In many tribes, when a hunter hunted alone or with others and brought home game, it was distributed among all of the village members. This suggests that various mechanisms existed to diminish any inequities that might arise among families.

Among the Wiyot, if an animal was killed in the territory of another local group, the meat was given to the owners of the tract, who gave part of it—perhaps half—back to the hunter. Among the Desert Cahuilla, the first deer a young man killed was given to the clan of his mother. On communal hunts for rabbits and small game, a boy gave his kill to some other family, usually his mother's. The husband alternated in providing game for his own and his wife's family. When a whale or sea lion drifted ashore, a Mattole chief was summoned to take the lion's share, including the head and jaw, which contained a lot of fat, but the rest was divided up among others in the local group. Among the Tolowa, all seafoods—fish, shellfish, mammals—obtained in canoes were distributed communally to all who stood by when the canoes landed.[14]

The anthropologist Thomas Garth described food sharing among the Atsugewi: "Certain men, because of extraordinary wealth and hunting and fishing prowess, became the leaders (Bawi) around whom a group of relatives and friends gathered. . . . Frequently the Bawi owned the land on which the group lived and other lands as well, all of which was an important source of wealth. He [the Bawi] maintained his prestige by providing his followers with numerous 'big feeds' and by helping them in time of need."[15]

These formal cultural rules acted to ensure broad community access to the staples of the indigenous diet. Thus food sources were rarely raided or exhausted by a few families or powerful individuals. And each person could be trusted not to steal. In the 1870s Powers observed acorn caches established miles from permanent village sites, conspicuous but left unguarded. Samuel Kneeland, too, remarked that the Yosemite Miwoks' honesty was proven by the existence of acorn and nut storehouses that were respected as property and not broken into by others. Anna Gayton commented on the Wukchumni Yokuts' trustworthiness: "Food was rarely, if ever, stolen from these granaries [storehouses of dried meat, dried fish, seeds, shelled acorns, and other foods next to the home]. Should a family run short of provisions, more acorns would be bought from someone who had plenty or, were the family really poor, the chief would see to it that they were fed." There are many other examples in the literature. With little fear of theft from fellow villagers, individuals were in a world of minimal plant resource competition.[16]

Barrows said that the tendency for individuals to hoard foods while gathering was "not a strong instinct with the Indian." Thus the communal orientation to food dovetailed with the cultural rules enforcing sustainable harvests. These rules, described in Chapters 2 and 4, dictated that whole plants and parts of plants must be left to ensure the next year's harvest. They were often reinforced with legends that taught right and wrong behavior with regard to hoarding and overharvesting. For example, mythic animals that became greedy met untimely ends.[17]

Seeds, bulbs, and other plant parts were important trade items, and trading served as a social cement between tribes that lived side by side. If a tribe needed a plant resource that grew plentifully in another territory, the chief visited the chief in charge of the area for the acquisition of that resource, who either granted permission for the group to gather or instructed his people to gather a certain quantity and pile it up before the visitors. Frémont made this observation of California Indians in his memoirs: "The Indians of this region [along the Cosumnes, Mokelumne, Tuolumne, and Auxumne Rivers] finding their food where they lived were not nomadic. They were not disposed to range, and seemed unaccustomed to intrude upon the grounds which usage probably made the possession of other tribes." The Nomlaki said that borders were maintained, but if on friendly terms with neighbors, oak trees were given in time of need. Relationships among the Pit River, Northeastern Maidu, and Northern Yana were particularly friendly, and these tribes often allowed reciprocal use of their root-digging grounds, fishing places, and other resource sites. The Tolowa had deer hunting and acorn picking tracts, and the hereditary owner was the nominal owner only: his kin had a right to use the area and "only outsiders had to ask permission (which was ordinarily freely given)." Respecting resource territorial limits and asking permission for gathering the plant and animal resources of another tribe were additional factors that prevented overutilization of plant resources.[18]

FOOD PLANTS REFLECTED IN CULTURE

The plants that provided California Indians with food were integrated into tribal cultures in a variety of ways. That these plants were valued by tribal members is reflected in the naming of people and locations. Whole towns, portions of creeks, and other sites often were named after edible plants; examples include the Wappo names *Unutsa' wa-holma-noma*, meaning "toyon-berry-grove town," and *Oso' yûk-eju*, meaning "going-to-make-buckeye-mush creek." Gathering sites were also named for the edible plant harvested there. A Foothill Yokuts man told Frank Latta, "We called Eshom Valley *Chetutu*, or Clover Place. There was always a nice field of sweet clover there in the spring."[19] This kind of naming demonstrated the confidence that "the plant will be there next year." Humans were named after food plants, too. Examples include *Cheso,* a Sinkyone female name meaning "tarweed blue," and *Kusetu,* a Miwok female name meaning "wild potato sprouting."[20]

The ethnobotanist Gary Nabhan has observed that plants helped to "shape and succor" the native cultures of the Sonoran Desert; they did this not only by providing calories but also by becoming characters in tribal myths. The same is true of the indigenous cultures of California. A recurring motif in hundreds of Indian myths is that a food plant actually has a human face or origin. For instance, the Cahuilla believe that all food was once human and could speak. The god Mukat chose particular people in the beginning who were to become plants and be converted into food for human use. The opposite transformation also occurs. Many creation myths tell of humans springing forth from seeds. Cattail seeds, in a Washoe legend, were turned into people; some became Miwok, others became Paiute, and still others became Washoe. These myths instruct humans that plants and people are from the same source and are related.[21]

One cultural universal was that food was never taken for granted. A gatherer asked the plant for permission to gather its parts, and gathering was followed by prayers and offerings of thanks. First-food rites were often held for special plants. Atsugewi women threw away the first roots they dug in a season and prayed for good luck. Young Modoc girls strung the teeth of moles (superb diggers) to form bracelets that were worn until puberty to assure luck in subsequent root-digging activities. At the end of the harvest season ceremonies were often held to thank the rains for nourishing the earth, or the black oaks for bearing heavily, or the Canada geese for returning.[22]

Between Hunter-Gatherers and Agriculturists

The way California Indians are viewed today by both the general public and many academics was powerfully shaped by the writing and research of a variety of scholars, anthropologists, and journalists in the late nineteenth and early twentieth century.[23] Based in part on their observations of Indians gathering "wild" plants in a rich, diverse natural environment, they solidified a view of indigenous Californians as primitive humans who, as noted previously, exemplified the category "hunter-gatherer."[24]

This mischaracterization of California Indians is important to consider in greater depth because of the bearing it has on the use of native plants for food, the subject of this and the next two chapters. As demonstrated in previous chapters with regard to the plants that provided materials for basketry and other material culture objects, the Indians of California did not just gather plants from the landscape; they tended, manipulated, and cared for these plants in ways that came to resemble the cultivation of domesticated plants in agricultural societies. To fully appreciate the tending of food plants by California Indians, we need to examine the prejudices that are the legacy of the anthropologists of one hundred years ago.

In the last quarter of the nineteenth century, scholars in history, biology, and anthropology at major British and American universities subscribed to racist generalizations concerning the intellectual superiority of "civilized" society over native "savagery." It was believed that humans progressed linearly from a crude animal state of nature to a refined, cultured state. Typical of the time were the ideas of Lewis Henry Morgan, who in his 1877 book *Ancient Society* defined the three basic stages of cultural development as savagery, barbarism, and civilization. These stages formed a sequence of progress; people such as the California Indians were in the least advanced stage—that of savagery, "the lowest race of men." In an 1889 article in the *Star* newspaper, John Wesley Powell defined these three stages of culture according to fire-making methods: "Savages make fire with two pieces of wood; barbarians with flint and steel or pyrites; civilized man by chemistry."[25]

The anthropologists and other educated Westerners who first began to systemically describe and document the fast-disappearing indigenous cultures of California in the late nineteenth century were part of this intellectual milieu and embraced the basic beliefs of the times. They believed themselves to be, as the anthropologist John Lubbock put it, "men trained to habits of methodical observation" who were "in a position to watch and describe the infancy of mankind." They saw the indigenous people in California as representing this "infant" stage because they lived at the whim of nature and did not know how to improve the land.[26]

In one of the first voluminous histories of California, Theodore Hittel described the California Indians in the following manner: "All were what the Americans, when they came to the country, termed 'Diggers.' They did not cultivate the soil, but lived upon what they could dig out or gather on top of the ground, and ate everything and anything within easy reach that would support human life, not excepting carrion clover, grasshoppers and grub-worms." In his diary, written before the Gold Rush, George Yount wrote: "In their wandering for a place of encampment, the party first encountered that species of red man peculiar to California and the Sierra Nevada range. From their mode of living on roots and reptiles, insects and vermin they have been called diggers. In fact they almost burrow into the earth like the mole and are almost as blind to everything comely. At the time our trappers supposed they had found the lowest dregs of humanity."[27]

Journalists helped to perpetuate this stereotype. In the 1870s Powers wrote of California Indians: "In times of great scarcity they are driven by the sore pangs of hunger to test everything that the soil produces, if perchance they may find something that will appease the gnawings of appetite."[28]

The apparent ease with which the Patwin shot deer with bows and arrows, or the Kawaiisu dug tubers with digging sticks, made a deceivingly simplistic impression on early anthropologists. Lack of famine, highly populated villages, and a rich, diverse landscape translated into a mode of life that involved little skill, imagination, or human ingenuity. But the window into California's native cultures was clouded by anthropologists' own cultural biases and myths. In truth, gathering and hunting took multidimensional knowledge, so rich and vast that it defied a comprehensive description. Unlike native dwellers on the land, anthropologists did neither hunting nor gathering for the food they required. Alfred Kroeber spent hundreds of hours among the Karuk, Yurok, and Yahi, yet the knowledge required for hunting and gathering remained intellectually inaccessible and unfathomable to him. His lack of understanding is demonstrated in a telling quote from his 1925 *Handbook of California Indians:*

> The securing of plant foods in general is not separated by any gap
> of distinctive process from that of obtaining grasshoppers, caterpillars,
> maggots, snails, mollusks, crawfish, or turtles, which can be got in
> masses or are practically immobile: a woman's digging stick will pro-
> cure worms as readily as bulbs. Again, it is only a step to the taking of
> minnows in brooks, of gophers, or lizards, or small birds: the simplest
> of snares, a long stick, a thrown stone even, suffice with patience, and a
> boy can help out his grandmother. The fish pot is not very different from
> the acorn receptacle, and the weirs, traps, stiff nets, and other devices
> for capturing fish are made in the same technique of basketry as the
> beaters, carriers, and winnowers for seeds. . . . Ducks were snared and
> netted, rabbits driven into nets, even deer caught in nooses and with
> similar devices. There is nothing in all this like the difference between
> riding down buffalo and gathering wild rice, like the break from whale
> hunting to berry picking, from farming to stalking deer.

He went on to say:

> The California Indian, then, secured his variety of foods by techniques
> that were closely interrelated, or, where diverse, connected by innumer-
> able transitions. Few of the processes involved high skill or long expe-
> rience for their successful application; none entailed serious danger, ma-
> terial exposure, or even strenuous effort. A little modification, and each
> process was capable of successful employment on some other class of
> food objects. Thus the activities called upon were distinguished by pa-

tience, simplicity, and crude adaptability rather than by intense endeavor and accurate specialization.[29]

Kroeber had succeeded in homogenizing the diversity of approaches to land use practiced by a hundred or so Indian nations. He and his colleagues bestowed the term "agriculturist" on the Mojave, Halchidhoma, and Quechan because they grew domesticated crops in the fertile floodplains of the Colorado River in the southeastern part of the state. But the rest of native California was stuck in a less progressive state of mankind—hunting and gathering.

Implicit in Kroeber's critical judgment is the idea that native people had not learned how to improve the land. Furthermore, there is an idealization of agriculture that places it at the pinnacle of human achievement. Learning how to produce food by intention, instead of harvesting it from nature, is frequently considered by archaeologists the most fateful and portentous development in human history.[30] Agriculture enabled surplus accumulation, stable settlement, and larger population concentrations, thus creating the foundation of civilization. Civilization, in turn, enabled the sophisticated and complex development of different realms of culture—the arts, religion, politics, and written language.

Looking for agriculture in terms of civilization, early anthropologists were blind to wildlands shaped by centuries if not millennia of in-depth knowledge and careful management. These activities were swept under the encompassing label "hunting and gathering." Thus a major historical distortion was created.

Anthropology has changed significantly in the past one hundred years. It has become much more sensitive, for example, to the issue of the inherent biases of the observer, but the basic elements of the nineteenth-century view of California Indians are still with us. The term "hunter-gatherer" is still used and still implies an evolutionary sequence of progress. The notion of the evolution of human cultures remains implicit in the layout of many current human ecology and anthropology textbooks and is explicit in recent anthropological journal articles that refer to this progression as "the ascent of man."[31]

The dichotomy between hunting and gathering and food production has tended to disguise the existence of a rich continuum of human–plant interactions ranging from true gathering to full domestication. In the words of the eminent geneticist Thomas Ledig, "The transition from foraging to farming is not a sharp break, rather it is a gradation" (pers. comm. 2001). Most if not all of the cultural groups in aboriginal California could claim an intermediate spot in this gradation because they enhanced and intensified food resources by practicing various forms of resource and land management.[32] Many of these management practices, along with selective harvesting strategies, were the same as those utilized in early agriculture to increase yields of the edible parts of domesticated plants. The protoagricultural techniques used by native people altered natural environments enough to put artificial selective pressures on many species of desirable plants, setting them on the path to domestication. In a few cases (e.g., devil's claw, discussed in Chapter 5), domesticated varieties of desirable plant species may have actually been created by the time Euro-Americans began to settle California, and many other species were arguably in a state of incipient or intermediate domestication.

Some scholars have speculated that the sorts of protoagricultural practices employed by the indigenous people of California are not only very ancient (perhaps thirty thousand to fifty thousand years old in Africa, Europe, and Asia) but also nearly universal among human cultures. The widespread knowledge of how to augment populations of wild food plants is a likely explanation for why there are many and diffuse origins of agriculture throughout the world.[33] In California, it is possible that protoagricultural practices were so successful in meeting human needs that there was no motivation to develop more labor-intensive techniques for growing domesticated crops.[34]

When societies adopted agriculture, it triggered a trend toward the simplification of human relationships with nature. Today's industrial agriculture relies on fewer and fewer crops, monoculture, chemical fertilizer, and pesticides, creating homogenized landscapes in which everything is dead but the crop. Subsidized by massive inputs of fossil fuels and tending to degrade the soil over time, our agricultural practices are clearly unsustainable. Critics of industrial agriculture contend that sustainable agriculture can only come from considering agricultural systems as ecosystems, dependent on diversity, natural processes, and species interactions for healthy functioning. In such "agroecosystems," predatory and parasitic insects feed on insect pests, insects from outside the system pollinate the crops, soil fertility is maintained through nutrient recycling, and mutualistic interactions benefit the crop plants. Ironically, the practice of this type of agriculture has many similarities with the protoagriculture of indigenous people in California. In both cases, the production of food is part of bigger and more complex processes.[35] Thus awareness of how California Indians blurred the line between gathering and agriculture can be helpful in the design of sustainable agroecosystems, which resemble the managed "natural" ecosystems of the Indians.

Knowledge of native management practices can also be valuable for conservation and wildlife biologists, who are interested in managing habitat for rare and endangered species. Agencies, companies, and individuals that have heightened interest in the "minor" products of forests and rangeland will also find the way Indians managed for multiple wild plant species relevant for perpetuating populations of these species in situ.

In the next two chapters, we accompany California Indian women and men into the field to witness the gathering of seeds, grains, greens, fleshy fruits, bulbs, corms, and tubers and observe the variety of horticultural techniques used to maintain the fertility of traditional gathering sites.

CHAPTER 9

Plant Foods Aboveground

Seeds, Grains, Leaves, and Fruits

They burned under the white oaks *[Quercus douglasii]* and you would see all the beautiful wildflowers that grew after a fire. It gives nourishment to the plants and they grow. I remember as a child I'd go from here [North Fork] down to the valley. It was the most beautiful sight you could see—poppies, niggertoes *[Dichelostemma capitatum]* and snowdrops—all kinds of flowers would just cover the hills. Many of the wildflowers we'd gather for seed and we would grind them such as the snowdrops *[Plagiobothrys nothofulvus]* and farewell-to-spring *[Clarkia* spp.] flowers. That's why the Indians burned down to the valley to keep it restored. They probably burned in the fall because everything's dry then.

ROSALIE BETHEL, North Fork Mono (1989)

As discussed in the previous chapter, California Indians had remarkably diverse plant diets, each tribe typically using as food the seeds, fruits, corms, bulbs, rhizomes, stems, and leaves of several hundred plant species. The members of each tribe memorized every aspect of each food plant growing in their territory, including the plant's growth habits, life cycle, habitat requirements, uses, storage particulars, and preparation and cooking methods.

The seeds of wildflowers and pines, the grains of native grasses, and the dry fruits of oaks (acorns) were among the staples of most Indian diets. These seeds, grains, and dry fruits provided ample protein, often higher in value than our domesticated grains and pulses, along with fats and carbohydrates. The leafy greens and stems of miner's lettuce, fiddleneck, cow parsnip, clover, and other plants, along with the pomes, drupes, and berries of a variety of shrubs and trees, provided needed vitamins, minerals, and fiber.

Indigenous people devised management techniques and harvesting strategies that maintained, created, and sustained the diversity of plants with edible aboveground parts. Many of the same horticultural techniques used in early and modern agriculture to increase the yields of edible plant parts—including pruning or burning shrubs to increase fruit production, saving and sowing seed, and burning areas to fertilize the soil and reduce competition—were used by California Indians.

Wildflower Seeds and Grass Grains

Avid wildflower watchers who today marvel at the springtime color display in Bear Valley, on the Vina Plains of southern Tehama County, or on the Carrizo Plains of southeastern San Luis Obispo County are witnessing a spectacle that once occurred over large areas of California. The seeds of these wildflowers were an important food source for native groups all over the state. Popcornflower *(Plagiobothrys canescens* and *P. nothofulvus)*, which still blankets the Sierra Nevada foothills, balsam-root *(Balsamorhiza sagittata)*, which forms a sea of yellow in open montane forests, owl's-clover *(Castilleja exserta)*, which carpets valley grasslands, and tidy-tips *(Layia platyglossa)*, which grow profusely in coastal prairies, were all relished by California Indian tribes for their seeds and harvested in quantities that seem unimaginable today.[1] Other species harvested for their seeds were *Blennosperma nanum* and goldfields *(Lasthenia glabrata)* in vernal pools, mule ears *(Wyethia mollis)* in open red fir forests, farewell-to-spring *(Clarkia purpurea)* in chaparral, fireweed *(Epilobium densiflorum)* along streams, rancher's fireweed *(Amsinckia menziesii)* in valley grassland, and common madia *(Madia elegans)* in coastal prairies. These plants produced seeds ranging from slightly larger than the size of the period at the end of this sentence to the size of a small sunflower seed and came in an assortment of colors—white, tan, brown, red, gray, black, and speckled. "I saw the Sacramento Valley before the footstep of civilization had made an imprint there," wrote Will Green, surveyor-general of California, in 1895. "Flowers of several hues covered the plain, and the ripened seeds made rich food [for the Indians]."[2]

Growing along with the wildflowers, often in the same habitats, were numerous tufted and rhizomatous native grasses that also produced grains the native people gathered in enormous quantities. Among the grass species producing tasty, nutritious grains were members of the genera *Leymus, Achnatherum, Nassella, Festuca, Phalaris, Panicum, Eragrostis, Hordeum, Elymus, Deschampsia, Bromus,* and *Melica.* Yokuts women, according to Frank Latta, could gather "several hundred pounds of seeds" of wild rice in one day.[3]

The record of useful native grasses is incomplete, as the flora had significantly changed by the time ethnographers began recording information about Indian cultures. By then, many California native grasses had been replaced with exotic grasses, although the grains of these aliens were quickly adopted as foods by tribes and recorded in the white men's ethnobotanies. These exotic species included wild oats *(Avena fatua)* and ripgut grass *(Bromus diandrus).* These grasses probably had harvesting characteristics and responses to indigenous management similar to those of the native species they replaced.[4]

SEED AND GRAIN GATHERING

For eons, Indian women all over California bent over tawny grasses and withered wildflowers and thrust their seed beaters over the inflorescences of the plants to remove their edible seeds. The grains fell directly into a burden basket or an open-mouthed basket that was then emptied into a burden basket. The swish, swish, swish of seed collecting was a familiar sound in the grasslands, as common as a meadowlark's flutelike song. Before going seed gathering in the spring, Foothill Yokuts women would smear their forearms and legs with tobacco, as rattlesnakes were said to abhor the smell and stay away. Shiny ornaments, such as abalone shell pendants, were worn in the field so as to startle the snakes and make them rattle.[5]

Louise Amelia Knapp Smith Clappe, better known as "Dame Shirley," described the process of seed beating among the Maidu: "One of these queer baskets is suspended from the back and is kept in place by a thong of leather passing across the forehead. The other they carry in the right hand, and wave over the flower seeds, first to the right and back again to the left alternately, as they walk slowly along, with a motion as regular and monotonous as that of a mower. . . . The seeds thus gathered are carried to their rancherias and stowed away with great care for winter use. It was, to me, very interesting to watch their regular motion, they seemed so exactly to keep time with each other; and with their dark shining skins, beautiful limbs and lithe forms, they were by no means the least picturesque feature of the landscape."[6] (See Figure 33.)

The tiny jet black seeds of red maids *(Calandrinia ciliata)* were gathered "by the pounds" by the Nomlaki and many other tribes. Phil Orr, former curator of the Santa Barbara Museum of Natural History, wrote of a Native American burial on Santa Rosa Island embedded in about twelve quarts of carbonized red maids seeds. Buttercup seeds were gathered "in great quantity" by the Yuki and other tribes. H. H. Rusby describes the buttercup seed harvesting method in 1906: "In the low meadows of California, *Ranunculus Californicus* grows thickly, as does the common swamp buttercup in the East.

Figure 33. Cecilia Joaquin, a Hopland Pomo, collecting seeds with a seed beater, photographed before 1924. Beating seeds of wildflowers and grasses with a shallow basket into a collection basket for food was the first in a series of management steps. Afterward, areas were burned to reduce plant competition and recycle nutrients, and some of the seed was saved and broadcast in the burned area. These practices are recorded for tribes of California and the Plateau, Great Basin, Southwest, and Pacific Northwest. Courtesy of the National Anthropological Archives, Smithsonian Institution. Photograph by Edward Curtis, #75–14715.

In May the Indians sweep a large gourd or closely woven basket, mounted on a long handle, through the tops of these plants, so as to catch the ripe seeds."[7]

Elthea Embody recorded in 1907 that Paiute women and children gathered the seeds of chia *(Salvia columbariae)*, "a wild grain" that grew "in the greatest profusion" in mountainous terrain. Farewell-to-spring and red maids were plentiful in the Clipper Gap and Auburn area before 1900, enabling the Southern Maidu to make ten- to fifteen-pound loaves of mashed, cooked seeds. Seeds gathered from the freshwater marshes were just as prolific: Klamath Marsh alone contained fifteen square miles of solid growth of water lilies that contained pods of edible seeds. Modoc and Klamath women could gather as much as twenty to thirty pounds of these seeds in one day.[8]

Published accounts and unpublished field notes from the turn of the century or earlier are filled with references to the gathering of grains from grasses. John Muir described collecting the grains of grasses by the Paiute on the east side of the Sierra Nevada: "I came to a patch of *Elymus*, or wildrye, growing in magnificent waving clumps, bearing heads six to eight inches long. The crop was ripe and Indian women were gathering the grain in baskets by bending down large handfuls, beating out the seed. . . . The grains are about five eighths of an inch long, dark-colored and sweet. I fancy the bread made from it must be as good as wheat bread." This was probably giant wild rye *(Elymus condensatus)*. The Mono Lake Paiute also harvested the grains of *Achnatherum speciosum*. Dan DeQuille described grain gathering among the Washoe: "[H]ere and there we saw Indian women, with their conical baskets on their heads, gathering the harvest of grass seed, while the children were playing nearby in the shade of the pines."[9]

SEED AND GRAIN STORAGE AND PREPARATION

In most cases, each kind of seed and grain was stored separately, in its own larder. Caches or granaries used for acorn storage were so tightly constructed that they were also used for storing grass grains. Granaries were perched on platforms of tree trunks and branches, or, in the mountains, on the flat tops of high boulders. Sometimes seeds were stored in baskets hung on walls in homes, or stored in pits lined with wild oats and cedar bark, as among the Washoe.[10]

In 1851 Bunnell described in his diary the food caches of the Yosemite Miwok in Yosemite Valley: "Their caches were principally of acorns, although many contained bay (California laurel), Pinon pine (Digger pine), and chinquepin nuts, grass seeds, wild rye or oats (scorched)."[11] During the

Tule Indian war of 1856, the military stumbled upon the stored food supplies of the Yaudanchi Yokuts, which contained "a great quantity of dried beef made from Packwood's fine cattle, stores of pine nuts, arrows, grass seed, and grasshopper cheese."[12]

To prepare seeds for consumption, many groups lightly roasted or parched them in a tightly woven winnowing basket with live coals to enhance their flavor and remove the husks or chaff. Miguel Costansó described this process among the Chumash in summer 1769: "In order to eat the seeds which they use in place of bread, they toast them first in great trays, putting among the seeds some pebbles or small stones heated until red. Then they move and shake the tray so it may not burn."[13] The seeds were then skillfully tossed in the air in regular, even motions with a shallow basket, allowing any excess chaff to be carried off in the breeze. Further roasting of the cleaned seeds often intensified their flavor. Henry Akers reported to Frank Latta that Yokuts women could clean wild seeds so well "they would almost shine."[14]

After winnowing, the parched seeds were sometimes eaten dry with the fingers, but more commonly they were pounded or ground into a flour. Unparched seeds were also ground. The Western Mono and the Sierra Miwok used deep mortar cups (up to one foot long) to pound small wildflower seeds into a flour. (The mortar cups used for acorn pounding were wider and shallower.) According to archaeologists, the earliest seed processing equipment used by California Indians was the milling stone (called a metate by the Spanish), a portable flat dish of stone. Edward Curtis described the preparation of seeds among the Mono in the central Sierra Nevada: "The bunch-grasses, wild oats, the tarweeds, and sunflowers were important sources of edible seeds, which, parched in the harvesting, were ground with metate and muller."[15]

Grinding slicks, which are smooth, shiny surfaces on flat, stationary boulders, were used to mill grass grains as well as other plant foods such as fleshy fruits and greens. An oblong, cylindrical hand stone was rolled over the grains on the slick, breaking and removing the chaff. Rock outcrops on ranchlands in Riverside County and many other areas throughout California are riddled with these polished, shiny surfaces. Archaeologists tell us that they are ancient kitchen counters and that the constant rubbing back and forth of a grindstone across plant foods to break the chaff smoothed these rock surfaces over time.

After grinding or pounding, the flour was often sifted in a basket. This was done by tapping the basket plate with the fingers so that the fine meal worked to the far edge and fell into another basket below. The coarse particles were then pounded further.

Seed meal or flour was prepared as food in a variety of ways. It could be moistened and eaten with the fingers, moistened and made into a seedball, eaten as a gruel without additional preparation, baked into a bread, or boiled into a mush or soup. The Southern Maidu sometimes mixed buttercup seed meal with manzanita cider. Each kind of seed had its own distinct taste, and often seed types were prepared and eaten singly. In some instances different kinds were combined to produce a special flavor. Small quantities of yampah (*Perideridia* sp.) seed, for example, were added by the Pomo to pinoles made of other seeds to give them added zest.[16]

The word *pinole* is often used in reference to wildflower and grass seed meals and flours. It is a Spanish term, derived from the Aztec word *pinolli*, that originally meant a meal made of ground corn or wheat and mesquite beans. The definition evolved to become more generic in California, encompassing all flours in the Indian diet made by pounding the roasted seeds of wildflowers and grasses. The flour was eaten dry or moistened and shaped into cakes or balls. If the flour was used in soup, gruel, mush, or baked bread, it was no longer considered pinole. Soups, gruels, or mush were boiled and called "atole" by the Spanish. In the making of some atoles, the coarser flour was used, and sometimes handfuls of finer meal were stirred into the atole to thicken it.[17]

Often the name of the plant from which the seed came was used to particularize the type of pinole. So there was buttercup pinole, red maids pinole, chia pinole, California brome pinole, and so on. V. K. Chesnut recorded more than thirty kinds of seeds gathered for pinole by the Yuki, Pomo, and Wailaki in Mendocino County. Harrington recorded twenty-nine kinds of seeds harvested for pinole by the Kitanemuk on what is now Tejon Ranch.[18]

A common way to eat pinole was to moisten it and make it into a ball that varied in shape from oval—like a small pear—to round. Round seed meal balls ranged from the size of a big marble to as large as an orange. This process was described by Father Francisco Palóu in 1776 of the Ohlone Indians: "The seeds they grind, and of the flour make gruel; and sometimes a kind of pudding or dough, which they form into balls of the size of an orange." Some seed meals such as those of the tarweeds were so oily that the particles adhered when pressed together with the fingers without adding water.[19]

BROADCASTING SEEDS AND GRAINS

When Indian women gathered seeds and grains with a seed beater and gathering basket, some seeds would fly too far or bounce out of the basket. They were not concerned about this loss because they recognized that some of the seeds needed to fall to the ground to produce the next year's crop. The incidental scattering of seeds during gathering, however, might not be sufficient to sustain the population of the plant over time. For this reason, the indigenous people in many parts of California deliberately saved seeds and grains and sowed them, usually by simply broadcasting the seeds in the appropriate habitats. This simple type of sowing, repeated annually for untold generations, likely led to the domestication of cereal grains in other parts of the world.[20]

Records exist of seed sowing by many native groups. Harold Driver and William Massey, in their volume *Comparative Studies of North American Indians*, reported the broadcasting of seeds by seven tribes: the Modoc, Eastern Shasta, Pit River (Achumawi), Northeastern Maidu, and Nisenan of northern California and the Mojave and Quechan of southern California. It is known that the Wappo increased the yield of certain plants by scattering their seeds, that the Chumash scattered seeds in cleared areas around their villages, and that the Southern Paiute sowed the grains of Indian ricegrass *(Achnatherum hymenoides).*[21] The anthropologist Julian Steward recorded seed sowing among the Paiute on the east side of the Sierra Nevada: "The brush in basins in the hills near the winter villages was burned and *Mentzelia* and *Chenopodium* seeds were broadcast. There is no question that this practice was native."[22] Alfred Kroeber reported that the Mojave planted wild seeds: "[T]he Mohave planted several wild herbs or grasses in their overflowed lands and gathered the seeds. These they call *akatai, aksamta, ankithi,* and *akyesa.* They are unidentified except for the last, which appears to be a species of *Rumex.*"[23]

BURNING FOR INCREASED SEED AND GRAIN PRODUCTION

California Indians knew that particular seeds and grains readily germinated and flourished after the land was burned. Burning for this purpose was commonplace, as reported by Kroeber: "The usual California practice was followed of burning the country over in order to clear out the underbrush for facilitating acorn gathering, and to foster the growth of seed-bearing annuals."[24]

Long before anthropologists began to study indigenous burning practices, the Spanish military and missionaries noted these habits. Consider the diary entry of Captain Fernando Rivera y Moncada at Monterey on October 3, 1774:

> There came a fire from the west that was burning the forage of the countryside, and as it neared the Presidio, the soldiers, servants and even I went out to fight it, not because of the danger to the houses but to preserve the grass for our animals. . . . The heathens are wont to cause these fires because they have the bad habit, once having har-

vested their seeds, and not having any other animals to look after except their stomachs, they [the Ohlone] set fire to the brush so that new weeds may grow to produce more seeds, also to catch the rabbits that get confused and overcome by the smoke.[25]

Driver and Massey also document the widespread practice of burning and describe its effects on perennial plants:

Burning to produce a better wild crop the next season was practiced over a large area in the west. . . . This treatment was applied to grasses and to small plants such as tar weed and wild tobacco. This method is especially effective for perennials because the roots beneath the surface are unharmed and the elimination of surface coverage gives the new sprouting crop the maximum of sunlight and the minimum of competition from other species.[26]

The grassland and woodland fires were frequent enough that they were rarely severe, and sometimes they were light and slow enough that they could be put out by hand: "J.R. [Chukchansi Yokuts/Miwok] claims that when he was a boy the Indians throughout this region [Sierra foothills] set fire to the brush after the seeds had been gathered (about July). The men started the fire and the women watched to see that it did not approach the houses. When it did, it was beaten out."[27]

George Gibbs reported that on the Eel River below Van Duesen's Fork the Indians lived on various seeds, "which, like Indians in the southern valleys, they collect after burning the grass." Driver recorded burning for better wild seed crops among the Yurok, Wiyot, Hupa, Chilula, Nongatl, Mattole, Sinkyone, Cahto, and Yuki of northwestern California. Harrington took copious notes on the culture of another northwestern tribe, the Karuk, and published the following: "And the wild rice plants also they burn, so that the wild rice will grow up good. They burn it far up on the mountains."[28]

According to Peri and Patterson, fires were set by the Pomo in grassy meadows or valleys after the summer grain harvest to encourage the growth of wild oats and tarweed (*Madia, Hemizonia,* and *Holocarpha* spp.) for food. The anthropologist Erminie Voegelin recorded burning for better wild seed crops among eleven tribes: the Klamath, Modoc, Eastern Shasta, Western Shasta, Atsugewi, Western Achumawi, McCloud River Wintu, Mountain Maidu, Northwestern Maidu, Foothill Nisenan, and Southern Mountain Nisenan. The Washoe burned areas around Lake Tahoe in fall to encourage the proliferation of grass and wildflower seeds. Florence Shipek learned from the Kumeyaay elders of San Diego County that after a certain unidentified grain was harvested in June or July, the Kumeyaay burned fields and broadcast seed over them. Philip Drucker, fifty years earlier, recorded

that the Kumeyaay deliberately set grasslands on fire for the purpose of increasing seed yields of a wildflower chia. This was a common practice among
the Cupeño and Cahuilla as well.[29]

We can reasonably assume that the combination of burning and sowing
of seed crops was understood and employed by most tribes because we have
extensive documentation of burning areas, sowing, and tending of native tobaccos. The anthropologist Omer Stewart describes the transference of these
practices across different species for different uses: "When the grasslands,
with their weeds and herbs, dried in the late fall they were set on fire nearly
every year, because the Achumawi recognized that burned-over plots produced tobacco and wild seeds more abundantly than the areas not burned."[30]

After fire exclusion policies were enacted by the federal government, Indians could no longer burn for seed production, but the connection between
fire and better seed growth is still remembered by elders in the Sierra Nevada
and in the grasslands of the Central Valley. Female elders of the Yokuts, Western Mono, and Miwok tribes remember their mothers, grandmothers, and
great-aunts going to areas after a fire to collect seeds. Beverly McCombs,
Chukchansi Yokuts/Miwok, said, "There is a plant with a little yellow flower,
kind of like a tarweed. My mother collected seeds from this plant and they
have an almond flavor. This plant comes up better after a fire" (pers. comm.
1989). Several Western Mono elders remember specific food species that
were plentiful after Indian-set fires—tarweeds *(Madia elegans)*, farewell-
to-spring *(Clarkia* sp.), snowdrops *(Plagiobothrys nothofulvus)*, chia *(Salvia
columbariae)*, and sunflower, or Hall's wyethia *(Wyethia elata)*. (See Table
3 in Chapter 4.) "A burn makes these plants come out more," said a North
Fork Mono elder.[31]

HARROWING

Some tribes prepared the soil before or after sowing seeds, often using a simple harrow made of brush. Sometimes the area was burned first. The process
of harrowing among the Karuk was described by Harrington: "Then they
break them open, when they sow, they scatter them over the ashy place. . . .
After they scatter the seeds, then they hunt a bush, then they drag the bush
around over it, so that the seeds will go in under the ground. Or they merely
sweep on top of it with brush. They sweep over where they have sown. It
is soft ground, it is ashes, where they burned the logs."[32] Although the
process Harrington described involved tobacco, a similar practice was probably used for the seeds of edible plants in this and other parts of California.
There are several records of tribes practicing harrowing in some form. For

instance, Central Sierra Miwok people at Groveland told John Hudson that they cultivated six kinds of seeds: *nu'wa si, ta'la gu, muk'si tana, cu'lu, si'hi ya'*, and *noco'pai yu*. (*Nu'wa si* is *Clarkia purpurea* subsp. *viminea* and *noco'pai yu* is *Wyethia helenioides*, but the others are as yet unidentified.) Hudson says that the cultivation of these seeds was the same as that for tobacco: "Burnt rich spot chosen and scratched with brush harrow, hauled by squaws. Seed sown broadcast."[33]

Saving seeds, scattering them on burned-over ground, and using a harrow to prepare the soil are practices that encompass many of the steps that anthropologists define as leading to domestication. The ethnobiologist Bruce Smith hypothesizes that "the first experiments with planting could well have consisted simply of efforts to enlarge stands of wild or camp follower food plants by broadcasting some of the harvested seed over a wider area, as the Kumeyaay did." He goes on to say, "[P]lanting might seem a simple and logical extension of hunter-gatherers' manipulation of food plants. In fact, however, it marks an essential change in the relationship of human societies to their environment."[34]

LASTING ECOLOGICAL EFFECTS OF INDIGENOUS SEED AND GRAIN MANAGEMENT PRACTICES

Many of the vast seed-bearing tracts that nineteenth-century Europeans wrote so poetically about were in part the products of indigenous seed beating, sowing, and burning. Indigenous burning helped to maintain the productivity of the grassland ecosystems by removing dead material and recycling nutrients. (Recent studies indicate that in the absence of fire, grassland ecosystems become choked with detritus, lowering productivity and reproduction, and that seed production dwindles in the absence of some kind of intermediate disturbance, such as herbivory, fire, or flooding.)[35] Burning may also have been responsible for the relative abundance of particular species, and thus the colors that Europeans viewed with awe. Repetitive burning would have selected for wildflower and native grass species that responded well to human disturbance. Species that did not vigorously resprout, in the case of perennial wildflowers and grasses, or could not germinate and grow in bare mineral soil, in the case of annuals, would be severely reduced or eliminated in the areas that were burned repeatedly. The density of the wildflower stands, and the size of the areas they encompassed, were likely due in part to the scattering of seeds during seed beating and the deliberate sowing of seeds. These practices may also have expanded the ranges and distributions of favored plant species.

Many of the native wildflowers and grasses grew in patches of a single species. This type of distribution may have been another artifact of native management. In 1868 Titus Fey Cronise wrote of California's wildflowers: "The different classes and genera do not usually intermix, but grow segregated in patches, some of which cover acres, and sometimes even square miles of space."[36] Jeff Mayfield also observed monocultural patches as he and his family crossed the San Joaquin plains in spring 1850: "As we passed below the hills the whole plain was covered with great patches of rose, yellow, scarlet, orange, and blue. The colors did not seem to mix to any great extent. Each kind of flower liked a certain kind of soil best, and some of the patches of one color were a mile or more across."[37]

Were these patches natural or anthropogenic? Judith Lowry, a native plant horticulturist who cultivates many of these edible-seed plants, says that "wildflowers such as red maids, globe gilia, tidy-tips, farewell-to-spring, and goldfields are not particular—they grow in many different soil types." This suggests that the observed patchiness of their distribution was not caused by unseen differences in soil or other environmental factors. "Today," says Lowry, "many wildflower fields contain a mixture of species, a change that could be attributed to the lack of indigenous management."[38] Other evidence for the anthropogenic hypothesis is the existence, in museum collections, of old pint-size jars of minute edible wildflower seeds and grass grains that were bought by or given to anthropologists who interviewed various members of Indian tribes in the nineteenth and early twentieth century. Each jar contains thousands of seeds of only one species, and the seeds are clean and free of chaff. If native gatherers had to collect these seeds from mixed stands, it would have taken enormous amounts of time to separate them by type. We have to conclude that they were gathered from monospecific patches. "If there is one seed type per jar and the seeds are clean," says Paul Kephart, a restoration practitioner, "then they [California Indians] were obviously practicing methods of agriculture such as the sowing of seed" (pers. comm. 2002).

Greens: Leaves and Stems

California Indians ate the leaves, buds, and stems of a great assortment of plants, including wildflowers, tules, cacti, and trees. These greens were eaten raw, stone-boiled in a basket, or steamed as potherbs in an earth oven. The leaves and tender tops of some greens were pulverized in bedrock mortars and eaten uncooked.[39] Leaves, stems, and young shoots were eaten as salads, added to acorn mush, or served as accompaniments to soups and stews. The Sierra Miwok alone used forty-eight types of greens; by comparison, we commonly eat only about a dozen domesticated greens today, in salads, as steamed vegetables, and in baked casseroles.

Native women harvesting, preparing, and eating delectable greens in late winter or spring were commonplace in various landscapes. At Brown's Flat, John Muir observed an elderly Indian woman with a basket on her back: "She was on her way, I suppose, to some wild garden, probably for lupine and starchy saxifrage leaves and rootstocks."[40] Jean-Nicolas Perlot, a Belgian argonaut in the Sierra Nevada, observed Miwok women in the early 1850s: "The Indian women had been cooking, since dawn, some wild clover, a sort of cress, and wild corn salad."[41]

Wild greens are packed with nutrition and are attractive for their vitamins—especially vitamins A and B, including thiamine, riboflavin, folic acid, niacin, and C and K. They are also rich in minerals, such as calcium, phosphorus, potassium, iron, chlorine, cobalt, and manganese. Leafy material and fibrous stalks also contributed roughage to the diet of California Indians, enhancing digestion.[42]

Different types of wildflowers with edible stems and leaves blanketed the hills, valleys, and meadows of many parts of California. Those with yellow and orange flowers were the monkeyflowers, California poppies, and fiddlenecks (*Amsinckia* spp.). The whites were the popcornflowers, yampahs, lomatiums, and angelicas; the pinks and purples were the thistles, shooting stars, bleeding hearts, and phacelias. Many kinds of clovers were gathered and eaten, as well as the leaves and flowers of various species of violets.

Since greens are highly perishable and have no feature that resists decay, they are rarely preserved in the archaeological record. Yet their importance in the traditional indigenous diet should not be underestimated. The leaves of *Phacelia cicutaria* var. *hispida*, for example, were said to have saved a certain tribe from starvation in Tulare County. Greens were prominent enough in the indigenous diet to warrant the marking of their special place in some aboriginal calendars, and they are featured in special ceremonies, such as clover dances, as well as many legends, throughout indigenous California.[43]

GREENS IN THE DIET

The consumption of greens by California Indians was largely seasonal. In spring, fresh, succulent greens were a welcome change after a long winter of dried foods. The emerging shoots pushing up through the earth, the new uncurling leaves, the bursting buds—all had a fresh, juicy green taste, without the bitter overtones or blemishes of old age. The emergents were dozens of shades of green, with subtle yet distinct differences in leaf forms visible to the well-trained human eye. Identifying plants at this stage—before flowering or fruiting—is difficult, yet California Indians knew exactly what they were eating.

The uncurling fronds of bracken fern were pinched and eaten raw. The new basal rosettes of leaves emerging next to the old russet brown seed stalks of docks were eagerly harvested and eaten. One of the first plants to be eaten in late winter or early spring was alumroot *(Heuchera micrantha)*, which grows in the cracks and crannies of granite cliffs in the Sierra Nevada. Another plant whose leaves were widely eaten in early spring was nude buckwheat *(Eriogonum nudum)*. The foliage of sun cups *(Camissonia ovata)* was eaten by the Ohlone.[44]

The young shoots of many kinds of plants were gathered, including four species of mule ears *(Wyethia glabra, W. angustifolia, W. helenioides, W. mollis)*, two kinds of soaproots, and white sage *(Salvia apiana)*. The young stalks of yampah, also referred to as sweet anise *(Perideridia kelloggii)*, were eaten by the Wiyot. The young, tender stalks of cow parsnip *(Heracleum lanatum)* were peeled and eaten raw by the Kashaya Pomo, the Coast Miwok, and many other tribes. The tender cores of the young shoots of cattails and tules, just beginning to emerge from the rootstock in late winter or early spring, were eaten raw like any young vegetable.[45]

The young leaves of many other plants were gathered before flowering and eaten after boiling; these included sweet cicely *(Osmorhiza chilensis)*, monkeyflowers *(Mimulus guttatus, M. moschatus)*, western larkspur *(Delphinium hesperium)*, and crimson columbines *(Aquilegia formosa* var. *truncata)*.[46] Other greens were gathered after flowering, because that was when an undesirable property disappeared; redwood sorrel *(Oxalis oregana)*, for example, was too sour if eaten young.[47]

The leaves of wild onions were gathered from little wet pockets under overhanging rocks or from crevices next to creeks. They were torn by hand one to two inches from the ground or pulled up and washed off in the creeks. In the higher elevations, often the bulbs were too small to bother with and the odoriferous stems and leaves were harvested before flowering. They were used as a culinary flavoring, much as we would eat chives. Today they are still gathered, and they can be eaten raw as a snack, rolled into balls and sprinkled with salt, or used as a delicious herb to accompany acorn mush and various kinds of meat. The leaves are eaten fresh, never cooked or stored for long periods.

One of the most important greens, a staple food of south-central and southern California, was the Spanish bayonet *(Yucca whipplei)*. The young flower stalks and the basal portion of the plant, with leaves removed, were eaten after being roasted in a pit oven with hot stones. Cutting the flower stalks before flowering may have caused hormonal changes in the plants, forcing them to produce "pups"—small, genetically identical plants adjacent to the parent plant.[48]

Many kinds of leaves and stems, on steeping, made a refreshing tea, drunk as a beverage, such as Sierra mint *(Pycnanthemum californicum)*, western Labrador tea *(Ledum glandulosum)*, and Mormon tea *(Ephedra* spp.).

CLOVER

Perhaps more widely eaten than any other green, clovers were a mainstay in the diet of many groups. Of the thirty-one native clover species in California, at least fifteen were prized as food. The Sierra Miwok alone ate at least eight types of clovers.[49] After gathering clovers in the fields, often Indians would set the fields afire and then broadcast seed. Because clover is a nitrogen-fixing plant, encouraging its growth by burning and scattering seed replenished soil nitrogen, as does the alfalfa a farmer plants as a cover crop in orchards. The cattle of Spanish and American settlers grew fat on these Indian-enhanced clover fields, and the ranchers grew rich.

Clovers were eaten fresh or steamed in an earth oven and dried and stored to be used in the winter for making soup. The Choynumni Yokuts yanked up handfuls of the plant, rolled them into balls between the palms of the hands, and then put them into their mouths. After thoroughly chewing the balls, they added salt to the mass by sucking on a stick of saltgrass *(Distichlis spicata)* crystals.[50]

Powers described the clover harvest among the Wintun (Wintu): "Clover is eaten in great quantities in the season of blossoms. You will sometimes see a whole village squatted in the lush clover-meadow, snipping it off by hooking the forefinger around it and making it into little balls." After Euro-American settlement, clover was still gathered in abundance: "Up until the late 1920s the Nisenan from Clipper Gap and Auburn trekked annually to the vicinity of Rocklin to gather wild onions and clover . . . bringing back barley sacks full of onions and clover."[51] Curtis reported of the Maidu: "The subterranean stock of the tule was a fairly important article to the valley bands, and young clover and pea-vines were devoured in the field with gusto."[52]

The preference of California Indians for clover and its heavy consumption were so evident that Euro-Americans often likened indigenous people to livestock grazers:

> Early travelers and explorers speak frequently of the fondness of the
> Maidu and neighboring tribes for fresh clover and a variety of wild pea,

and describe them in the Sacramento Valley as getting down on hands and knees in the fields, and browsing like so many cattle.[53]

In the clover season, when the meadows were bright with pink and white blossoms, whole rancherias [Consumnes tribe in the Sacramento basin] went out literally to graze, and the Indians might be seen lying prone in the herbage, masticating the clover tops like so many cattle.[54]

PREPARATION AND PALATABILITY

California Indians had numerous techniques for preparing greens for consumption. Although many greens were eaten fresh, others were cooked, soaked, leached, cured in the sun, dried, or freeze-dried to enhance the flavor or remove toxic or distasteful components. Some leaves—such as those of the checker mallow *(Sidalcea malvaeflora)*, eaten by the Coast Miwok and many other tribes—were steamed in an earth oven. The Nisenan steamed dandelion greens along with meats in an earth oven. The Kawaiisu gathered the leaves and stems of phacelias *(P. distans* and *P. ramosissima)* before flowering, laid them on a flat rock, placed hot rocks and another layer of stems and leaves on top of them, and then sprinkled water on the hot rocks to steam-cook the mass.[55]

Native people processed wild plants in ways that removed their toxicity, yet maintained their nutrition. The toxic alkaloids and protease inhibitors in lupine leaves—greens important to the Western Mono, Sierra Miwok, and other tribes—were eliminated by boiling the leaves.[56] Many tribes, such as the Wappo and Sierra Miwok, boiled the leaves and roots of milkweed in two changes of water; once the water was clear, the greens were ready to eat, and were considered a delicacy. The leaves of the California poppy were eaten by the Nisenan either boiled or roasted with hot stones and then laid in water. The basal leaves of prince's plume *(Stanleya pinnata, S. elata)* and jewelflower *(Caulanthus crassicaulis)*, also known as thick-stem wild cabbage, are highly emetic when fresh, but Great Basin tribes cooked them in several waters, each poured off, to make them edible. The Panamint put the leaves of prince's plume into boiling water for a few minutes, washed them in cold water five or six times, and then squeezed them out. Frederick Coville described these large crucifers as having the taste of cabbage. "The operation of washing," he wrote, "removes the bitter taste and certain substances that would be likely to produce nausea or diarrhoea."[57]

STIMULATING GROWTH BY PICKING AND PINCHING

Stripping stalks of leaves or cutting off young shoots may seem like rough ways to handle plants, but these actions were very similar to those of grazing tule elk or pronghorn antelope. Since many native plant species are well adapted to regular herbivory, able to quickly compensate for the loss of biomass, both human harvesting of greens and the munching of ungulates set plants back temporarily but had the longer-term effect of stimulating their growth. Herbivory studies have shown that concentrated grazing followed by rest periods for native herbs does not harm them and that in some cases plants browsed by mule deer and elk produce significantly higher numbers of flowers and fruits than plants that are not eaten. Plant ecologists call the phenomenon of increased growth after herbivory overcompensation.[58] Stephen Edwards, director of the East Bay Regional Parks Botanic Garden, has hypothesized that the now-extinct megafauna in California provided constant disturbance through grazing over millions of years, creating an environment favoring overcompensation and other adaptations to intense browsing.

Gardeners have a direct understanding of how a modest level of defoliation can stimulate the vegetative and reproductive growth of plants. Pinching back basil or cutting back chard gives the gardener a large handful of leaves for cooking and delays flowering. Where the plants are pinched off, axillary buds elongate, sending out new branches. This makes the plant bushier, encourages multiple flowering stalks, and stimulates a tremendous new growth of leaves. With some plants, such as the sour docks, indigenous people's pinching of leaves in spring enabled a second harvest of leaves in summer or fall.[59]

We know that native people in some parts of California were well aware that pruning leaves can encourage new, larger leaves because of the existence of good descriptions of the cultivation of tobacco, such as this one by Chalfant in 1933: "The Eastern Mono used a species of wild tobacco which they assisted by burning off the land each spring and trimming away the poorer leaves on the plant in the summer so as to favor the larger leaves."[60]

The gathering of greens was carried out with an eye to sustainable harvests. In general, some greens were always spared. Specific plants required specific methods of harvest. In the case of wild onions, for example, the Washoe were careful not to uproot the bulbs when harvesting the leaves, to make sure the plants would come up again and again. Fowler has documented among the Timbisha Shoshone of Death Valley the routine pruning of Panamint prince's plume *(Stanleya elata)* and desert prince's plume *(S. pinnata)* in spring:

> As the new growth was removed to be used as a green, people broke off last year's flower stalks and any dead leaves on these perennials, thus cleaning them up to make ready for continued new growth. . . . People therefore harvested only young and tender leaves in the early spring, giving the plant ample time to put on additional leaves to carry it through the late spring to early summer bloom. The pruning and cleaning promoted healthy growth for next year, according to what people were taught by their elders.[61]

Harrington recorded among the Kitanemuk of Tejon Ranch the harvest method for an unidentified plant used for salt: "Quite a quantity of this plant or small bush is gathered. When picking the plant, only the leafy upper part is taken. The root and stiff stem are allowed to remain, so that the plant will grow up again it is said."[62] From early accounts of tobacco planting, we know that many tribes gathered leaves in ways that ensured the plant would produce seed. For instance, the Maidu would pinch off the leaves, to use for smoking in stone pipes, and leave "the stalk to mature" so that "the seeds from it" could be "replanted the next year."[63]

Southern California Indian tribes harvested the leaves and flower stalks of agaves for food, which may have kept the plants productive for longer periods and delayed the flowering process, which kills the plant. Indians were well aware that if the flower stalk was harvested at the right time, the agave responded by sending out shoots from its base that take root and create new plants.

FIRING TO PROMOTE GREENS

It was a common practice up and down California for native people to set fires to stimulate the growth of plants important for greens. Fire was another form of disturbance that, like harvesting, benefited these edible plants.

Grasslands and chaparral areas in what are now the Napa and Alexander Valleys were burned by the Wappo to keep trails open and encourage the lush growth of clovers—a favored delicacy. The Maidu burned areas to encourage the growth of bulbs and greens. Burt Aginsky records the "burning of herbage for better wild crops" among the Valley Yokuts, Chukchansi Yokuts, Western Mono, and Southern, Central, and Northern Sierra Miwok.[64]

According to Peri and colleagues, the Pomo burned areas in which various edible clovers and bracken fern grew. Stewart recorded in his unpublished field notes the words of a Pomo informant: "Land burned every year, to make clover and wild oats grow. Kept down brush. Clover is eaten without cooking. Boo-te tops are eaten as greens as well as potatoes. These leaves are cooked with bull pine leaves. . . . When grass is burned upon the hills it was slow 'pickins.' Boo-te continued to be edible even when the top of plant burned. The women would just start to dig any place they thought might yield anything." Burning for grasshoppers probably also enhanced clovers. The Yana are known to have gathered clover, edible roots, and seeds in Basin Hollow and to have burned out grasshoppers there as well.[65]

The naturalist José Longinos Martínez, in his 1792 journal, reported indigenous burning to encourage different kinds of young shoots used for greens: "In all of New California from Fronteras northward the gentiles have the custom of burning the brush, for two purposes; one, for hunting rabbits and hares (because they burn the brush for hunting); second, so that with the first light rain or dew the shoots will come up which they call pelillo (little hair) and upon which they feed like cattle when the weather does not permit them to seek other food."[66]

Native people today remember that edible greens were not "naturally" productive continuously, over many years, but required burning to maintain their quality and quantity each year:

> The only plant that I know benefited tremendously from fire in the high mountain meadows was the clover. Clover patches were burned. In Squaw Valley, my dad said that the Indian people burned clover patches right by the Mission. There's not that much clover left now because of all the buildings, the houses. They've destroyed a lot of the clover. There used to be sloping meadows of clover. The clover was gathered in March and April. When it got older, it got tougher, and it didn't taste good. You ate it when the leaves were tender. I remember the people wouldn't uproot the whole plant. They would pinch it off. My dad said that they burned it after it produced seeds and after the seeds fell. (Lalo Franco, Wukchumni Yokuts, pers. comm. 1992)

> They used to burn areas for the clovers. It would increase the quality— the clovers would be young and tender and increase the amount. (Rosalie Bethel, North Fork Mono, pers. comm. 1989)

> Miner's lettuce comes up good after a fire. (Beverly McCombs, Chukchansi Yokuts/Miwok, pers. comm. 1989)

Indigenous burning favored a variety of leguminous forbs, many of which have long-lived seeds that need scarification from fire to germinate. It is probably no coincidence that the seeds and leaves of these legumes found their way into the diet of native peoples. By favoring and tending these nitrogen fixers in the landscape, Indians benefited many associated species as well. Nitrogen-fixing plants important for their edible greens included wild peas *(Lathyrus vestitus, Lathyrus graminifolius),* meadow hosackia *(Lotus oblongifolius),* and some of the lupines *(Lupinus bicolor, L. succulentus, L. microcarpus* var. *densiflorus, L. polyphyllus, L. latifolius).*[67]

Fleshy Fruits

California is well endowed with wild, succulent fruits: more than one hundred species of trees, shrubs, vines, and cacti produce edible fruits in the form of berries, drupes, and pomes. These fruits were gathered by California Indians from deserts, chaparral, oak woodlands, evergreen forests, and a variety of other landscapes, from sea level to the higher elevations of various mountain ranges.[68] Like today's domesticated varieties, many of the fruits eaten by California Indians were from plants in the rose family: these included wild cherries, wild plums, wild strawberries, wild raspberries, wild blackberries, wild apricots, and thimbleberries. (Wild strawberries—three species of which grew profusely in habitats as diverse as stream banks, coastal prairies, sand dunes, Monterey pine forest, and under giant sequoias— probably encompassed more acreage than any of the other plants bearing fleshy fruits.) Other commonly eaten fruits were those of the wild grape, huckleberry *(Vaccinium* spp.), gooseberry *(Ribes* spp.), manzanita *(Arctostaphylos* spp.), California fan palm *(Washingtonia filifera),* sourberry *(Rhus trilobata),* prickly pear cactus *(Opuntia* spp.), and elderberry *(Sambucus* spp.). All these fruits were as high in vitamins and minerals as our domesticated fruits today, yet lower in sugars and calorie content. Elderberries, for example, are loaded with calcium, phosphorus and iron and contain three times the vitamin A found in peaches. Today native people still enjoy some of these fruits and add them to Western foods—as filling for pies and toppings for ice cream and to make jellies and jams and sauce for dumplings.[69]

GATHERING BERRIES, DRUPES, AND POMES

Indians gathered most fruits by hand, breaking off the single fruit or cluster. Some shrubs, such as gooseberry *(Ribes* spp.), manzanita *(Arctostaphylos* spp.), and buffalo berry *(Sheperdia argentea),* were shaken or knocked with sticks to make the fruit fall off into baskets placed below. With particular species of manzanita bearing very sticky berries, the Sierra Miwok would sweep under the bushes and knock the fruit to the ground rather than use

Figure 34. Angela Lozada, Maria Garcia, and Juana Encinas picking gooseberries at Tejon. Various tribes, such as the Maidu, Western Mono, Miwok, and Pomo, burned areas to foster the growth of shrubs and trees with edible fruits. Courtesy of the National Anthropological Archives, Smithsonian Institution. Photograph by John P. Harrington, #91–33552.

their hands. The Cahuilla detached clusters of California fan fruits with a twisting motion of a long, notched willow pole. To harvest blackberries from thorny bushes, Yokuts gatherers would latch long, hooked branches of cottonwood or willow onto thorny shoots to move them to one side in order to get at the berry-laden branches. Women in many tribes wove special berry baskets and sometimes hung them around their necks while gathering, leaving both hands free.[70]

Early photographs show large, concentrated patches of accessible berries from which a gatherer could gather a lot of fruit in a short time. (See Figure 34.) Latta mentions that he and several Wukchumni Yokuts people gathered fifty pounds of blackberries in one day.[71] What is clearly evident from diaries and anthropologists' notes is that many of these fruits were gathered in substantial quantity. Other descriptions report on the sheer abundance of fruit. Myriad grapevines, some six inches in diameter, grew along creeks and streams or damp arroyos, winding their way up old sycamores and alders and draping the branches with clusters of purpling fruit. Isaac Wistar, while on the Salmon River in Siskiyou County during late summer, discovered a thicket of wild plums covering less than an acre in which every tree was bending under the weight of the wild fruit. Barrows reported that the holly-leaved cherry *(Prunus ilicifolia)*, also called *islay*, in a single canyon often produced enough fruit to feed an entire Cahuilla village and that scores of such canyons could be found.[72]

O. B. Powers, charged with mapping the Calaveras Wagon Road Route from Carson Valley to Calaveras County, started out from Murphys and described the scenery approximately ten miles northeast of the Calaveras Grove:

> Continuing in the same direction four miles further we reached Black Springs; a beautiful cold spring, surrounded with aspens and plenty of grass for animals, in the immediate vicinity. Here we found gooseberries and raspberries in great abundance. . . . On the whole route there is plenty of excellent timber, plenty of good grass, and "any quantity" of raspberries, gooseberries, etc. Some of the gooseberries were as large as a pigeon's egg, and of a most excellent flavor, and so abundant that we could have gathered a bushel each in two hours.[73]

Berry gathering was so satisfying, and such an efficient means of procuring food, that families would often travel great distances to harvest berries: "The report of a plentiful supply of chokecherries would lead many [Washoe] families to travel 20 or 30 miles to gather them." The Indians of Mendocino County in the early 1900s would travel a dozen or more miles to gather California huckleberries *(Vaccinium ovatum)* on foggy ridge tops. Some of the long-distance traveling noted by white settlers reflected the displacement of native people and the clearing of native fruit trees and shrubs from their land.[74]

Although gathering fruit was simpler and less labor-intensive than gathering many other foods, it nevertheless required an intimate familiarity with nature and natural patterns. Many animals, including birds and grizzly bears, also ate the fruits and competed with Indians for this resource. For this reason, timing of the harvest could be crucial. Indians watched the other animals and linked their behaviors with the ripening of the fruit. Goldfinches, for example, would begin to whistle more frequently when it was time for the Foothill Yokuts to gather blackberries. Keeping a close watch on weather patterns was also important. For example, rosehips (the fruits from wild roses) tasted sweetest after the first light frost or cold nights of fall. The Karuk harvested California huckleberries after the first frost because that was when they were sweetest.[75]

PREPARATION OF FRUITS

Most fruits required little or no preparation before being cooked or eaten. The nasty spines of some fruits were only a minor hindrance to getting at the sweet, juicy innards. The Kumeyaay impaled the fruit of the prickly pear on a stick, for example, rubbed it in the dry grass to remove the spines, and ate it raw after the seeds were removed. The Pomo shook gooseberries back and forth in a winnowing basket with hot hardwood coals from the fire to singe off the bristles, or mashed the berries with a pestle in a mortar.[76] Central and southern California tribes prized holly-leaved cherry, not for its minimal fleshy fruit, but for its seed kernel or pit, which had to be subjected to a lengthy preparation and cooking process to remove the poisonous hydrocyanic acid it contained. The flour made from the processed pits was said to taste like beans or chestnuts.[77]

Some fruits—such as wild strawberries, thimbleberries, salmonberries, and raspberries—were eaten raw on the spot. Indian children often gathered these berries for snacks and returned home with berry-stained hands and lips. The sweet fruits of the fan palm, popped in the mouth, tasted much like the domesticated date. Many other fruits were commonly dried. Huckleberries and elderberries were gathered by the burden basketfuls, dried, and stored over the winter in baskets. The Kumeyaay dried wild grapes into raisins and then cooked them before eating. Fruits laid out to dry on cleanly swept ground was a common sight near Cahuilla homes in southern California. The fleshy green covering of California bay nuts was eaten raw by the Kashaya Pomo after it had been loosened and split while the nuts lay drying in the sun. Some fruits were pounded or mashed and then shaped into large cakes or loaves, which would then be dried. These loaves were broken into pieces and softened with water as needed. Slicks were used for mashing fruits such as manzanita berries and sourberries and removing the seeds.[78]

Some fruits were cooked before being eaten—either roasted, boiled, or baked in earth ovens. Some berries, such as toyon, were cooked on hot rocks until they bubbled, then mashed in the hands and eaten.[79]

Many California Indian groups used fruits for making beverages and soups. Puckery sourberries were pounded into a pulp, mixed with cold water, shaken into a sour fizzy drink, and drunk by the Sierra Miwok in hot weather. (Sourberries are still sold at acorn festivals and powwows, eaten at other tribal gatherings, and offered to visitors as afternoon snacks.)[80] One of Walter Goldschmidt's Nomlaki informants reported that the Nomlaki "put grapes into a basket and mashed them into a kind of soup, getting out what seeds they could by hand." This soup made "a good drink to wash down pinole with." Palm fruits, besides being eaten raw, were ground into a flour, mixed with other flours and water, and eaten as a mush.[81]

The fruit beverage that quenched thirst in more California Indian homes than any other was a type of cider made from the crushed berries of manzanita shrubs. The fruits were often mashed in mortars or slicks into a fairly dry pulp and set in open-work basket colanders. Water was then poured over the mass and the liquid drained into another watertight basket, producing a refreshing drink. Merriam described it as being "in color and flavor like the very best apple cider[,] . . . cooling and delicious."[82] Each woman had her own special manzanita shrubs that provided the best-tasting berries. Cider is still made by Indian families in many parts of California.

TENDING BRAMBLES, BUSH FRUITS, AND WILD ORCHARDS

Today all of our domesticated fruit-bearing plants require considerable human care to retain their productivity over time. They must be pruned annually to maintain an optimal canopy density, to encourage the growth of new, vigorous, fruit-bearing shoots, and to remove old, unproductive and diseased wood. The thinning of the canopy allows sunlight to reach more leaves in the interior. Left unpruned, a tree will soon produce most of its fruit at the top of the tree, where there is sufficient sunlight.

Wild species, many of which are closely related to our domesticated fruits, may respond similarly to human tending and intentional disturbance. In one study, fruit production in young gooseberry seedlings, established after a fire in the Stanislaus National Forest, began five years after the burn and increased each year thereafter for another five years. Other scientific studies have shown that many of our native shrubs with edible fruits increase fruit production following stand thinning.[83]

California Indians were aware of the relationship between disturbance and productivity in native fruit-bearing plants and used both pruning and fire to keep the favored plants healthy and productive. They knew that without fire, productive berry gathering sites would degenerate, as the density of branching per individual shrub or tree increased, decreasing light to the fruiting structures. Furthermore, surrounding trees might outcompete these shrubs, trees, and vines, further curtailing their fruit production. According to Lalo Franco, "Some of the other berry plants that benefited from fire were elderberry and chokecherry. If you look at chokecherry bushes, they can become really congested. When fires were set in these areas, the plants weren't as congested. You're able to reach in the bush and grab berries. Fires would make more berries and make the shrubs healthier" (pers. comm. 1991).

There are a number of references to burning for fruit production in the historical literature. Driver reported for the Wiyot that "[b]urning [occurred] every two or three years, to get better berry and seed crops, and to increase feed for deer." The Pit River burned fields and forests to drive game and stimulate growth of seed and berry plants. Peri and colleagues reported that the Pomo people burned manzanita shrubs and that the berries provided food, the leaves provided medicines, and the branches were used for clubs. They also documented the pruning of huckleberry and the burning and pruning of toyon, which were both used for their edible berries. Harrington noted that the Karuk in northwestern California burned huckleberry bushes so that they would "grow up good."[84]

The burning of chokecherry, manzanita, strawberry, and elderberry patches was recorded among the Foothill Yokuts, Western Mono, and Miwok. Burning was carried out to increase fruit production, thin dense shrub canopies, and reduce insect activity by eliminating old wood. Avis Punkin (Mono) said her grandmother burned in July for the manzanita: "It would come back better with more fruit" (pers. comm. 1991).

California Indian elders remember burning for wild strawberries. Lydia Beecher reported, "After they'd light the fires, the strawberries would come up better" (pers. comm. 1991). And according to Hazel Hutchins, "Fires help the berries too. There are some kinds of plants that grow up in the mountains and they set the fires for them" (Mono, pers. comm. 1992). Klamath River Jack, in a 1916 letter to the California Department of Fish and Game, noted, "Fire burn off old wood on berry bush make new wood grow and lots big berry come."[85]

Fire had other positive effects. It helped to control diseases and pests, a benefit that Indians were well aware of. "Indian have no medicine to put on all places where bug and worm are," said Klamath River Jack, "so he burn; every year Indian burn." Burning berry patches to kill insects was also documented by Donald Jewell among the Maidu.[86] Austen Warburton and Joseph Endert wrote of the Yurok: "In the fall of the year it was the duty of certain men to burn patches of oak, hazel, and huckleberry brush to eliminate fungus and insect damage and improve the crop in the next year. In the second year after burning there was usually a heavy increase in hazel nuts, acorns, and berries. In 1885–95, it was not unusual to see them bring in loads."[87]

Fire also helped to maintain an optimal spatial arrangement of individual plants, which was important for good berry production:

> One of the plants that really benefited from fire that I heard about from the old people was manzanita. The berries were made into cider. The leaves and bark were used for medicinal purposes. You can't have a lot

of manzanitas all cluttered. The tree has to be free. When they're not all cluttered, they get really, really big and that's when they were really beneficial. A fire clears out all the smaller bushes around the larger manzanitas. The small manzanitas—they had to keep those under control. Because then they wouldn't let this large tree benefit. (Lalo Franco, pers. comm. 1991)

California Indians' management of fruit-bearing native plants in many ways laid the foundation for domestication of some of the berries grown today. Picked from and tended over hundreds or even thousands of years, wild strawberries, raspberries, blackberries, and others had already become adapted to human cultivation when horticulturists began the selective breeding that led to modern varieties. Further, many of today's berries have benefited from crosses with wild genetic resources, which has conferred such benefits as disease resistance. Thus the stewardship and preservation of these wild plant resources by California Indians has tremendous value today for the berry farmer.

Pine Nuts

For native people in the Sierra Nevada, Great Basin, and elsewhere, pine nuts were an important food resource. Loaded with protein and unsaturated fatty acids and with a respectable carbohydrate constituent, pine nuts could be gathered in large quantities. They were also stored for long periods. A wide variety of pines produced nuts that were eaten as food, for example, the Torrey pine *(Pinus torreyana)*, Coulter pine *(P. coulteri)*, Monterey pine *(P. radiata)*, Pacific ponderosa pine *(P. ponderosa)*, and Jeffrey pine *(P. jeffreyi)*, but the most highly valued nuts came from gray pine *(P. sabiniana)*, sugar pine *(P. lambertiana)*, Colorado pinyon *(P. edulis)*, and singleleaf pinyon *(P. monophylla)*. Gray pines flourished throughout much of the California Floristic Province; sugar pines flourished in the middle elevations of the Sierra Nevada and Cascade Range and in the mixed evergreen forests of northwestern California; and pinyon pines prospered in the eastern Sierra, desert mountains, and Great Basin.

The delicious nuts were eaten raw, parched in a basket, or steamed in an earth oven. They were also pounded into a flour that was made into butter, soup, or bread; mixed with other dry seed meals; or mixed with dried salmon. The nuts of certain species, if gathered early enough, could be cracked open with the teeth; if gathered later, they were cracked with an oval nutting stone, with the nut placed on a type of anvil. The nuts were eaten as a snack between meals, and they were dried and stored, shelled or unshelled, for lengthy periods in large coiled baskets or underground granaries. They were offered to the hungry wayfarer and are still commonly eaten and offered to guests. Pine nuts were so important to the western branch of the Atsugewi that their name *Atsuge* literally means "pine-tree people." In many California indigenous cultures the word for "pine" is the same as the word for "pine nut," showing the importance of this part of the plant.[88]

California pine trees were more than just sources of food. Every part of the tree had a use. The branches and trunks were burned as firewood. The needles were used as tinder, made into torches, and burned to make a smudge applied to spider bites. The new, freshly scented needles were fashioned into pine spindle dollhouses for Sierra Miwok children and spread on the floor of earth lodges. The supple branches could be used as stirring sticks and the straight saplings for lifting hot rocks from the fire. Charred twigs of young pine were fashioned into ear plugs by the Sierra Miwok, and the large branches were made into bull roarers. Springy limbs were carved into Washoe bows. The oozing pitch from pinyon pines and ponderosa pines was melted and applied to baskets to make them watertight. This adhesive was also used in constructing Mono whistles, Paiute arrows, and Washoe toy tops. Pine roots were used in northern California basketry; the bark, for constructing conical houses, summer lean-tos, and the roofs of acorn granaries. The chocolate brown nuts were fashioned into necklaces and sewn onto fringed skirts and aprons. They were strung on strands and used as a type of women's money by the Modoc, Wintu, Tolowa, and other tribes.[89]

There is ample evidence that many California Indian tribes managed pine groves by pruning branches, knocking off cones, weeding, and burning. Management techniques varied by tribe and by pine species. The management of sugar pines in middle elevations of the Sierra Nevada, along with the collection and preparation of sugar pine nuts, is discussed in Chapter 5. Here I describe how California Indians managed gray pines and pinyon pines.

GRAY PINE

Big gray pines, with sparse foliage and springy limbs, could be climbed with relative ease and their large cones twisted off. The Sierra Miwok harvested the cones in spring when young and still green. When roasted for twenty minutes, they yielded a brownish, syrupy food that tasted like sweet potatoes.[90]

The cones were also gathered in early fall for their nuts. Muir made the following observation of the Yosemite Miwok harvesting pine cones: "Indians climb the trees like bears and beat off the ones or cut off the more fruitful branches with hatchets, while the squaws gather and roast them until the scales open sufficiently to allow the hard-shell seeds to be beaten out." According to Merriam, Indians in the Mariposa area put gray pine cones in the fire "long enough to burn off the thick resin with which they are heavily coated. This served a double purpose, getting rid of the sticky gum and at the same time toasting the nuts a little."[91]

Jepson reported trees as large as four feet in diameter and ninety feet high in the foothills. Most of these were cut during the Gold Rush to fuel the engines at the quartz mines. Jepson said that the nuts were the most useful, next to acorns, and noted that it was "no small wonder then that the Indian looked on in distress whenever the 'white man' cut down a Digger Pine."[92]

The Pomo, Ohlone, Sierra Miwok, Mono, Maidu, and Yokuts burned gray pine areas to keep down the thickets of manzanita, buck brush, and chamise that would otherwise grow around the widely spaced trees, making them inaccessible and more prone to destruction in wildfires (as they are today). Many accounts describe the openness of gray pine stands. C. Hart Merriam described the foothills and lower slopes of the Sierra Nevada and the Coast Ranges as being "carpeted with wild oats, interrupted by thickets of berry-bearing manzanitas and shaded by open forests of nut-bearing Digger pines [gray pines] and numerous kinds of oaks, which together furnished the principal food of the people." Fire return intervals before major American settlement are short in blue oak woodlands, and blue oak is often a major associate of gray pine. Fire-scarred trees sampled in a blue oak woodland 30 kilometers east of Marysville showed a fire return interval ranging from 8 to 49 years, with a median of 28.5, before 1848. Scott Mensing documented the fire frequency as fire scars in blue oaks on three sites in the Tehachapi Mountains. He found that the mean fire return intervals for the presettlement period ranged from 9.6 to 13.6 years.[93]

In 1934 old Pomo Indian informants and early settlers in Mendocino County told Omer Stewart that the hills and valleys of northern California were formerly free of brush: "The great redwood forests were open and clear. Oak and pine woods occurred, too, but had trees widely spaced so that grass grew between the trees and game could be seen at a distance. Indians and settlers agreed that the thick brush and undergrowth presently characteristic of these forests had been formerly kept in check and the forests kept open by frequent, almost annual, intentional burning-over by fires set by the Indians."[94]

James Hutchings described how the Ohlone set fires on Mount Diablo: "The cañons of this mountain are lined with stunted oak, and pines; and wild oats and chaparral, alternately, grow from base to summit. In the fall season, when the herbage and dead bushes are perfectly dry, the Indians have sometimes set portions of the surface on fire, and when the breeze is fresh, and the night dark, the lurid flames leap, and curl, and sweep, now to this side and now to that, and present a spectacle magnificent beyond the power of language to express."[95]

Gayton reported what she was told by a Yokuts and Sierra Miwok man, named J. R., about how the Indians set fires in blue oak–gray pine areas in the Sierra foothills: "It [the fire] burned the hills, all over, clean through to the next one. The trees, which were green did not ignite easily: however, dead trees and logs were all cleaned up that way."[96]

Thomas Ledig, a geneticist with the Forest Service, believes that the disjunct distribution of gray pine may be a result of accidental or deliberate seed dispersal by California Indians. There are three notable extensions of gray pine beyond the edges of the Central Valley, and all occur along old Indian trails: along the Pit River on the Modoc Plateau, along the Trinity River in and to the southeast of what is now the Hoopa Valley Indian Reservation, and along the Eel River in and south of the Round Valley Indian Reservation.[97]

SINGLELEAF PINYON PINE

The singleleaf pinyon pine's large, orange-red to chocolate brown nuts have been an indispensable food for California Indians east of the Sierra Nevada crest for millennia. Frémont noted in his journal on January 25, 1844, as he and his men crossed the Great Basin: "These [pine nuts] seemed now to be the staple of the country; and whenever we met an Indian, his friendly salutation consisted in offering a few nuts to eat and trade."[98]

Pinyon charcoal and seed coats have been uncovered in the firepits of Gatecliff Shelter in central Nevada, radiocarbon dated to about six thousand years of age. Archaeological excavations done in Owens Valley by Robert Bettinger reveal that pinyon exploitation in central eastern California is at least one thousand years old; the evidence includes the presence of milling equipment and circular floors within the pinyon-juniper zone. Archaeobotanical research shows that pinyon nuts were brought to winter villages on the Owens Valley floor from distant pinyon groves as early as 2000 B.P.[99]

Great Basin tribes traditionally harvested the cones using two major methods. One way was to gather the cones in late August or early September, while they were still green and tightly closed, using a hooked stick. The stick either snapped the cones off the limbs or was hooked around each flexible limb, pulling it down to the gatherer, who hand-twisted the cones from the branches. The other method was to wait until the cones opened and then whip the trees with a pole, removing the cones or knocking nuts out of the mature cones. The fallen cones and nuts were then collected from the ground. With either method, dead or dying branches were pruned back, a practice that the Timbisha Shoshone and Washoe say was "good for the trees." In addition, the Shoshone traditionally pinched off or broke the growth tips by hand on the lower branches. This practice, combined with the breaking of the growth tips higher on the trees during the whipping process, was also beneficial, according to the Timbisha Shoshone, as it fostered the production of one or two new growth buds.[100]

In 1940 Levi Burcham wrote about how widely spaced the pinyon trees were in the groves used by tribes for their nut production when white settlers first arrived. He said that this spacing "lessened the danger of damage from fire, insects, or disease."[101] This woodland structure in fact may have been human created, a result of native pruning, tending, and burning.

Aware that pinyon trees are not fire-resistant, Indians pruned back dead wood in the canopies and cut back low-lying limbs under the trees that could catch fire. Tribes also raked litter and duff from under pine nut trees and removed by hand any live growing shrubs that might act as fuel ladders. These practices protected the trees from the devastating effects of wildfires and, even more important, worked in conjunction with fires set intentionally.

Marshall Jack (Owens Valley Paiute/Central Miwok/Washoe) speaks of ancestral burning on the east side of the Sierra Nevada:

> My grandfather Jack Lundy . . . talked about burning. They burned to increase foods such as wild onions, elderberries, and caterpillars, and to clear out the underbrush to bring in the new growth for the animals. They burned at Paiute meadows, Sonora Junction and the Bishop region. Also Inyo Valley and Crawley Lake. Crawley Lake used to have a lot of pinyon pine trees. They'd burn every three years to increase cone production the following year and to decrease the grasses and duff from the pines. It would keep them from losing pinyons from a really big fire. (Pers. comm. 1990)

What would carry the fires? Evidence is mounting that many of the pinyon groves had grassy understories that would provide fine fuels to carry light surface fires. Fires, both wildfires and Indian-set fires, occurred every ten to ninety years in precontact times and were a major force in reducing tree encroachment and perpetuating the grassy understory in between the pines. Modern prescribed burning can reestablish understory species and produce up to a severalfold increase in herbaceous production.[102]

Early Euro-American settlers in the Great Basin logged pinyon trees to furnish charcoal and timber for silver ore mines and to supply ranches with fuel and rough fencing. Many of the groves that survived were overgrazed by sheep and cattle, causing soil erosion and a drastic reduction in native grasses and forbs. Later, fire suppression allowed more young trees to survive, increasing tree densities and further shading out perennial and annual herbaceous plants.[103]

By 1940, according to Burcham, mismanagement of pinyon woodlands had led to widespread infection by mistletoe, insects, and pathogens, decreasing the productivity of the woodlands. Burcham suggested that if managed properly the pinyon tree could serve two compatible purposes, nut production and watershed protection—goals that Native American tribes had fostered for many generations.[104] Public lands agencies, however, did not heed Burcham's message; they chained large tracts of pinyon trees and reseeded areas with non-native grasses such as Asian crested wheatgrass *(Agropyron cristatum)* to "improve" lands for livestock.

The nuts of pinyon pine are still gathered by the Washoe, Paiute, Mono, Shoshone, and other tribes. Nuts are eaten raw or roasted in their shells and eaten as a snack. They also are parched, shelled, and parched again, pulverized into a flour and made into cakes or soup. The Washoe eat pine nut soup with beef, rabbit, or deer meat. Nuts are also boiled like beans. Lorraine Cramer (Southern Sierra Miwok/Mono Lake Paiute) says, "I have to hide them from the kids, they're just like sunflower seeds so they don't taste bad raw. I cook them like acorns. Add water and it gets thick and gravylike. It's very rich and you can't eat very much of it."[105] Washoe elders look for environmental cues to forecast the ripening of the nuts. "When the fruits of the wild rose are red the pine nuts are ready," says one Washoe elder. Families rely on several groves, rotating their harvests to accommodate the periodicity of the nut crops.

Acorns

Large, venerable oak trees growing in valleys, foothills, and mountains produced acorns that were the staff of life to indigenous peoples in California for millennia. The anthropologist Edward Gifford called balanophagy (acorn eating) "the most characteristic feature of the domestic economy of the California Indians." Among the more than a dozen species of oaks that provided acorns were California black oak *(Quercus kelloggii)* in the Sierra Nevada, eastern California, parts of central coastal California, and southern California; tan oak *(Lithocarpus densiflorus)* in northwestern California and parts of central coastal California; and coast live oak *(Quercus agrifolia)* along parts of central and southern coastal California.[106]

Acorns were second only to salt as the most popularly traded item of the indigenous people of California. The eastern slope Paiute would return from the west side of the Sierra Nevada with immense loads of California black oak acorns on their backs—a round trip of one hundred miles. Today many tribes still highly value the acorns of this and other species of oaks, and acorn collection provides continuity with their past and long-term association with places, strengthening tribal ethnicity.[107]

Widely distributed and widely used, California oaks provided a fountain of resources to California Indian people. In addition to food, they yielded medicine, dyes, utensils, games, toys, and construction materials. Oak bark was boiled down until black and used as a dye for basket wefts by the Western Mono and Foothill Yokuts of Squaw Valley. Wooden mortars were carved from gnarly oak burls, giving special flavor to pounded acorn or ground tobacco. Oak made excellent wood for Sierran sweathouse fires, and the hair of Miwok mourners was singed short with oak coals. Acorn musical string toys, tops, and buzzers kept children entertained, and acorn dice games kept adults enthralled for hours. Western Mono and Foothill Yokuts acorn mush stirrers were made of looped green oak shoots; tongs for lifting hot rocks from the fire to heat acorn mush were made of older oak branches. Miwok treks through snowy mountain passes were made on snowshoes of oak. And Yosemite Miwok warriors fought dressed in rod armor of young oak sprouts, carrying wooden bows of oak. Sharp oak twigs performed as needles for the Tcainimni Yokuts, and oak galls served as an emetic. Mighty oak trunks served as Miwok earth lodge support beams and hollowed foot drums. Large quantities of young, supple sprouts were utilized for myriad items.

To compete successfully with animals such as rodents, most tribes collected acorns before they fell, knocking them out of the trees with long poles or climbing the trees and whipping the acorns down. Other competitors, however, also harvested acorns while they were still on the trees. Studies have shown that jays can harvest as many as four hundred acorns per hour from tree canopies, and acorn woodpeckers will store thousands of acorns in old fence posts, snags, or living trees. (In times of famine some Sierran tribes would raid the stores of acorn woodpeckers, using a deer antler to pry loose the acorns from their little holes.) Bears were also formidable acorn eaters. The Reverend Walter Colton, in *Three Years in California* (1860), described in the late 1840s how grizzly bears would approach the harvesting of acorns:

He is an excellent climber, and will ascend a large oak with the rapidity of a tar up the shrouds of his ship. In procuring his acorns, when on the tree, he does not manifest his usual cunning. Instead of threshing them down like the Indian, he selects a well-stocked limb, throws himself upon its extremity, and there hangs swinging and jerking till the limb gives way, and down they come, branch, acorns, and bear together.[108]

Even with the stiff competition, Indians did not harvest all acorns but left some for the wildlife.

The archaeobotanist Eric Wohlgemuth says, "The earlier direct evidence of acorn use in California is charred acorn nutshell found in archaeological sites dating to 7400–9900 B.P. east of Mt. Diablo and 7800–10,200 in the central Sierra foothills. Acorn nutshell is found in archaeological sites dating to all subsequent periods and is the most abundant charred plant food residue in all regions of central California, supporting the ethnographic information that acorns were the principal staple of California Indians" (pers. comm. 2004). Estimates of the beginning of intensive harvesting of acorns date to at least 2500 B.P. in central California and the Sierra Nevada. Mortar cups—small pockets in granite outcrops for pounding acorns and other foods—are ubiquitous in the Sierra Nevada, attesting to the widespread popularity of acorns.[109]

MANAGING OAK STANDS WITH FIRE

Many tribes in California used fire to ensure continual yields of high-quality acorns. The anthropological literature contains records of a variety of tribes across the state burning under the oak species growing in their territories. These include the Dry Creek and Cloverdale Pomo people, who burned under the tan oak, the valley oak, the Oregon oak (Quercus garryana), and Nuttall's scrub oak (Q. dumosa); the Kashaya Pomo of Redwood Valley; the Wappo; the Yurok and Tolowa of northwestern California; the Luiseño of southern California; the Maidu of the northern Sierra foothills; and the Ohlone of central coastal California.[110]

In addition, many individuals from various tribes have documented the use of fire by their ancestors for management of the acorn crop, including Lydia Beecher (Mono), Dave and Ed Bowman (Wobonuch Mono), James Rust (Southern Sierra Miwok), and Ron Wermuth (Tubatulabal/Kawaiisu/Yokuts/Miwok). Wermuth says, "[T]hey burned in the higher elevations [of the Sierra Nevada] under the oaks to keep the weedy stuff out and it kept the oak limbs off the ground and it made a better acorn crop. They burned in late fall, about October, when it starts to chill" (pers. comm. 1992).

The anthropologist John Duncan's account of the Maidu's burning of oak woodlands lists its many beneficial effects:

> There was considerably less chaparral and underbrush [in aboriginal times], due to the Maidu practice of burning off the areas near where they lived each fall and winter. They preferred an open, grassy, oak savannah habitat for several reasons. Open country is much easier to travel in than country with thick underbrush; it is easier to find game and harder for enemies to sneak up on the camp. More bulbs and greens grow in such an environment, and it is easier to gather acorns on bare ground. Since the fires were set every year, the deep detritus of our present forest floors did not have a chance to accumulate, with the consequence that only the grass and small brush would burn, leaving the larger trees to survive the flames.[111]

It is clear that oak stands were burned to achieve a variety of goals. Following are ways in which California Indians used this management tool:

Facilitate acorn collection. Regular fires reduced undergrowth and the accumulation of duff, making movement easier and fallen acorns easier to spot. The anthropologist Fred Kniffen, in his notes on the Kashaya Pomo of Redwood Valley, wrote that dry weeds and brush were burned every year after seed gathering was finished. In Driver's account of the Yurok, he mentions specifically that burning got rid of the leaves and enabled fallen acorns to be seen.[112]

Suppress diseases and acorn pests. Burning increased the quality and quantity of acorn production by helping to control the insect pests that would otherwise consume a large percentage of the crop and by controlling oak parasites. Driver recorded that the Tolowa burned under oaks to kill parasites on or underneath the trees.[113] As discussed in Chapter 4, regular burning in fall helped to break the life cycle of the filbert weevil and the filbert worm, ensuring a nonwormy acorn crop.

Stimulate the production of epicormic and adventitious branches for making cultural items. Following a fire, suppressed buds along the base of the oak trunks or on the root crown were released, sending up long, straight, flexible shoots useful for making many items. (See Chapter 7.) Deliberate burning under the oaks also fostered the growth of the significant quantities of oak shoots needed for certain types of baskets.

Decrease the likelihood of major conflagrations that would destroy the oaks. Regular fires kept fuels, in the form of shrubs, young conifers, leaf litter, and dead branches, from accumulating. Without these fuels, fires were much less likely to climb into the canopies of oaks and damage or kill them.

Encourage the growth of edible mushrooms. The black morel *(Morchella elata)* amasses dense colonies in recently burned areas in black oak–ponderosa pine forests. This may be due to elimination of competition from higher plants, an increase in nutrients from ash deposition, or increased soil temperatures. Called *cuyu* by Western Mono families, the morel is a delicacy and still frequently harvested in June or July in the mountains.

Increase edible grasses and other seed-bearing herbaceous plants under the oaks and in the surrounding forest. As discussed earlier in this chapter, oak woodlands would produce good crops of herbaceous plants and grasses bearing edible seeds and grains if they were burned regularly.

Enhance the growth of deergrass. Deergrass *(Muhlenbergia rigens)*, one of the plants most prized for basketry materials, grows in association with oaks, and burning favored the growth of individual plants and increased the size and vigor of populations. (See Chapter 6.)

ECOLOGICAL EFFECTS OF BURNING UNDER OAKS

California Indians greatly modified the forest and woodland communities in which oaks were dominant species through their regular use of fire. These communities were much more open, ecologically diverse, productive, and resistant to catastrophic fires than they would have been in the absence of human management. Indian practices favored the largest oaks (from 80 to 500 years old), which also produced the largest number of acorns, and increased their longevity. Bear clover, chamise, manzanitas, and other highly flammable shrubs or trees did not grow under the oaks at favored collection sites. Early written descriptions tell us that the oak understories in certain areas were largely composed of grasses and broad-leaved herbaceous plants. Indian elders confirm that this was the case. For example: "There was nothing in those days to burn. The fires set wouldn't get away from you or take all the timber like it would now. It wouldn't burn the black oaks, only the grass underneath" (James Rust, Southern Sierra Miwok, pers. comm. 1989).

Even as it protected the existing large trees, frequent burning under oak trees may also have promoted successful oak seedling and sapling recruitment. Recent studies have linked successful oak regeneration with periodic fire. According to the fire ecologists Boone Kauffman and Robert Martin,

"[S]uccessful germination of acorns is enhanced in areas where duff layers are weakly developed. These are probably the seedbed conditions in which the black oak evolved. Accumulations of forest floor fuels apparently represent degraded seedbed conditions for black oak."[114]

Food from the Land

It is difficult for us today to understand and appreciate the relationships that California Indians had with their food resources. We buy our food at the supermarket, much of it processed and therefore showing little evidence of its source. We can count on its always being there, and the knowledge required to procure it is no more complicated than knowing the location of a food store and being able to count change. Even when we venture into the wilderness, we are tethered to a food production system that we participate in only as consumers, since all our food is carried on our backs. For this reason we can only be visitors.

An indigenous person in aboriginal California, in contrast, had a direct, personal relationship with every morsel that filled his or her belly. He or she knew where it came from, what was required to gather, process, and store it, and how the plant or animal that provided it had to be treated and tended to ensure a continuing supply. Each person was dependent on nature, both vulnerable to its vagaries and benefiting from its remarkable abundance. Food was inherent in the environment and present everywhere in the landscape, thanks to natural processes, the human ingenuity behind management practices, and social cooperation. When one moved across the landscape, one gathered food, whether it was fungus coming up after the first fall rains, roots waiting in the cold winter soil, greens sprouting in the spring, or berries ripened by the summer sun.

Gathering edible plants carefully and respectfully was, and continues to be, viewed by indigenous people as both good for the gatherer and good for the plants. When humans collect and eat native foods, they become part of the localized food web, full participants in the places where they live. In *The Flavors of Home*, Margit Roos-Collins tells us, "Observing nature closely can be a grand passion, but tasting moves that relationship beyond the platonic. . . . Through the medium of wild plants, the minerals of the places I love have been knit into my bones."[115]

CHAPTER 10

Plant Foods Belowground

Bulbs, Corms, Rhizomes, Taproots, and Tubers

After they [the Yurok] got done collecting [bulbs and corms], they would go through and take all the little ones off and replant them . . . so there would be a continual supply.

KATHY HEFFNER, Wailaki (1992)

On the morning of March 28, 1772, Captain Pedro Fages and his men of the Spanish Cavalry, accompanied by Fray Juan Crespi, arrived at a northwestern Yokuts village and were graciously received with gifts—goose decoys, soaproot, and a curious, round food resembling a miniature baked potato. Crespi called it "cacomite."[1] It tasted sweet and pleased the palate of the Spanish soldiers. This round, white vegetable was probably a kind of brodiaea, or grassnut. One hundred years later, early settlers observed Yokuts people gathering brodiaeas along the sand sloughs and rivers of the San Joaquin Valley. The plants were said to grow so thickly that the gatherers would retrieve hundreds of them by running wicker sieves through the sand.[2]

Brodiaeas were just one of the many kinds of perennial wildflowers with underground storage organs that grew all over California in a variety of communities, from the salt marshes to the subalpine forests. Many of these plants, called geophytes, were edible, and their underground storage organs—bulbs, corms, rhizomes, taproots, and tubers—served as an important food source for nearly every group of California Indians. The underground organs are an adaptation to the long, dry summers typical of California's Mediterranean climate. Whereas annual wildflowers survive from year to year as seeds, geophytes store water and nutrients in their underground organs in spring, allow their leaves and stems to die back in summer or fall, and then grow new stems and leaves the following year. This strategy also happens to be a good adaptation to poor soil conditions, herbivory, and fire. (See Figure 35.)[3]

The geophytes gathered for food in aboriginal California included wild

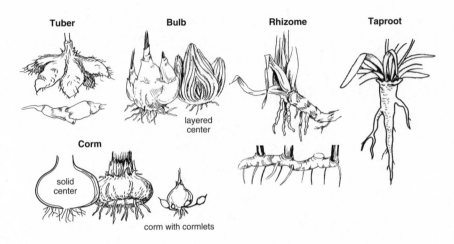

Figure 35. Many of the plant species used by California Indians have excellent vegetative reproduction by tubers, bulbs, rhizomes, or corms, or plants regenerate from taproots. Vegetative reproduction enables the plant to survive and reestablish itself in place after human disturbance and often expands the portion of the site the plant occupies. A tuber is a thickened, short, underground stem with numerous buds or eyes; a bulb is a short, basal, underground stem surrounded by thick, fleshy, modified leaf bases; a rhizome is an underground swollen stem, growing horizontally and rooting at the nodes; a corm is the enlarged base of a stem, bulblike but solid, not layered like an onion; a taproot is a large, central root that goes down.

onions (*Allium* spp.), lilies (*Lilium* spp.), sanicles (*Sanicula* spp.), common goldenstar *(Bloomeria crocea),* fawn lilies (*Erythronium* spp.), fritillary (*Fritillaria* spp.)., mariposa lilies (*Calochortus* spp.), brodiaeas (*Brodiaea, Dichelostemma,* and *Triteleia* spp.), yampahs (*Perideridia* spp.), soaproot *(Chlorogalum pomeridianum),* purple amole *(Chlorogalum purpureum* var. *purpureum),* lomatiums (*Lomatium* spp.), camas *(Camassia quamash),* and others. Also, the roots of certain species were used, for example, the taproots of bitter root *(Lewisia rediviva)* and the short rootstocks of cattails *(Typha* spp.).[4] California Indians use the generic terms "Indian potatoes" and "root foods" to refer to the bulbs, corms, rhizomes, taproots, and tubers of these plants.[5]

Women of the various tribes pried loose from the earth these dozens of root foods with digging sticks made of mountain mahogany, serviceberry, or buck brush. These sticks were often fire-hardened to increase their durability and are markedly similar to those used throughout the world, such as in Africa and Australia. (See Figure 36.)[6] Digging sticks in the anthropological collections of the Phoebe Hearst Museum at the University of California, Berkeley, are well worn, polished by the constant buffing of hard-

Figure 36. Rosa Charles and Billy George, a Wintu couple, digging for yampah (*Perideridia* sp.), 1931. The digging of many different kinds of bulbs and tubers with a hardwood digging stick, replanting propagules, and burning over areas to increase numbers, densities, and size of subterranean organs of wild plants for food were common practices throughout the West. Courtesy of the Santa Barbara Museum of Natural History. Photograph by J. P. Harrington, JPH-CA-WT-201.

working human hands. The soil in certain meadows and prairies and in the grassy understories of woodlands and forests was soft and pliable from centuries of women digging with their sticks in the same gathering places as their mothers, grandmothers, and great-grandmothers.

The indigenous people of California actively managed the populations of plants that bore edible corms and bulbs. To be able to harvest large quantities of these food plants and at the same time perpetuate their populations, they developed a variety of harvesting and management techniques that not only ensured the reproduction of the target species but often enhanced the vitality and size of their populations:

- spare individual plants to allow for future production;
- harvest after seed-set to ensure that seeds fall at the site;

- dig bulbs and corms so as to disperse the bulblets and cormlets that will mature into full-fledged plants;

- replant cormlets and bulblets;

- deliberately break the stem above the root crown or bulb, leaving some stem and root tissue;

- leave a lower section of the tuber;

- weed around favored plants; and

- burn areas in which the plants grow to decrease plant competition and recycle nutrients.[7]

It is widely known that gold miners and settlers derisively dubbed California Indians "diggers" after they witnessed Indian women digging up root foods, or walking along foothill trails, carrying burden baskets full of bulbs and corms, digging sticks in hand. Little did they realize that the digging they sneered at was to a considerable extent responsible for the gardenlike, paradisiacal quality of California they so admired. Essentially, the women who churned the soil year after year with their digging sticks were gardeners.

Indian Potatoes in Indian Cultures

Tasty, abundant, and nutritious, the various kinds of tubers, corms, and bulbs were a culinary staple for most California Indian tribes. Although the remains of root foods are underrepresented in the archaeological record because of their lack of hard parts,[8] ethnobiologists argue that historically, and perhaps prehistorically, Indian potatoes were in fact a dietary mainstay in some regions—as important as fish and game in parts of the Pacific Northwest and rivaling acorns and other seed crops in parts of California.[9]

John W. Harshberger, who coined the term "ethnobotany," wrote to John Hudson of Ukiah in 1928 to ask for photographs of California Indians digging bulbs and roots for foods so that he could use them in various lectures on the history of botany at the University of Pennsylvania. In the letter he made this comparison: "In one of the lectures of this course, I have drawn a parallel between the early Greek Rhizotomi (root diggers) and the California Indian who, instead of cultivating maize and other crops in ancient times, dug up the roots of wild plants, [and] made use of bulbs, seeds and fruits as vegetable food."[10]

The importance of root foods in Indian cultures is reflected in a variety of ways. The Yokuts called one special gathering patch *Kawachu*, meaning "place of the grassnuts," and gave the name *Kawachumne*, "the people of the grassnut place," to the resident Yokuts tribe.[11] Legends hint of the abundance of Indian potatoes as a food source: in a Sinkyone myth Coyote stole the sun, but the people overtook him, so he threw it in a gulch. Then he made an old basket and dug Indian potatoes. He dug them so fast that he covered the sun with them. Some tribes had a generic name for bulbs and corms as a distinct plant part. The Hupa, for example, called bulbs *yinetan*. V. K. Chesnut recorded that the Indians in the Mendocino area appended the word *bo* to the names of plants to denote the edible bulb, corm, or tuber of that plant.[12]

NUTRITIONAL VALUE

Root foods are high in carbohydrate content and were second in importance only to seeds in the California Indian diet. In some cases, bulbs were important emergency foods when acorn crops failed. Interestingly, the protein content of some of these bulbs is quite high, higher than that of acorns and important edible seeds. A biochemical analysis of the bulbs of camas performed in the 1950s showed 7 percent protein, which compares favorably to the 2.9 percent to 6.3 percent protein of acorns.[13]

A Wukchumni Yokuts woman described a tall grassnut with a yellow flower and large bulb as "rich and fattening—good for our blood." When a Pomoan child was first born, the mother went on a vegetable diet made up of acorn mush, salads, and baked bulbs. Certain Indian potatoes had effects that caused native people to avoid eating them at specific times. For example, breast-feeding Pomo mothers stayed away from griping onions *(Allium bolanderi)* and the corms of white brodiaea *(Triteleia hyacinthina)* because they caused infantile diarrhea.[14]

PREPARATION AND STORAGE

Sometimes bulbs were eaten raw, but mostly they were cooked by boiling, steaming, roasting, or baking in an earth oven. Chumash women on Santa Cruz Island had specially designed earth ovens as large as six feet square in which they baked huge quantities of brodiaeas *(Dichelostemma capitatum)*.[15] Some bulbs had to be cooked before being eaten because they contained toxins or other compounds with adverse effects (e.g., camas contains inulun).[16]

Numerous tribes stored tubers, bulbs, and corms over the winter, after baking, roasting, or drying them. Patrick Breen, a member of the Donner party during the harsh, stranded winter of 1847 in the Sierra Nevada, wrote in his diary on Sunday, February 28: "Solitary Indian passed by yesterday come from the lake had a heavy pack on his back gave me 5 or 6 roots resembling onions in shape, taste some like a sweet potato, all full of little tough fibres."[17] Underground swollen stems of many different kinds were strung on cords or kept in baskets or sacks. Camas is reported to keep indefinitely when stored. The Modoc roasted camas bulbs in the ground, kneaded them into cakes, and dried them. Joaquin Miller reported, "In this state if kept dry it will retain its sweetness and fine properties for months."[18]

Growing "as Thick as the Grass"

Much historical evidence suggests that geophytes and other root-food plants were amazingly abundant in parts of California. According to Frank Latta's consultant Mrs. Francisco Dick (Bankalachi Yokuts), *koo-nuk*, a type of brodiaea with waxy yellowish white flowers (probably the golden brodiaea, *Triteleia ixioides*) grew abundantly in the Central Valley—"as thick as the grass." In the early 1900s Chesnut noted that a patch of *Triteleia laxa*, a related species dug by the Indians of Mendocino County, was densely covered with two hundred plants per square foot. A Yurok woman said that a mountain valley north of the Klamath River was so crowded with brodiaea it reminded her of a blue lake. Arrowhead *(Sagittaria latifolia)* tubers were gathered by the Klamath in northern California and southern Oregon "by the canoe load." Saxton Pope reported of the Yahi that an Indian could go out on an apparently barren hillside and with a sharp stick dig up enough brodiaea bulbs in an hour to furnish food for a good meal. O. B. Powers, a road surveyor, in 1855 found "heaps of wild onions" that the Washoe had been gathering and piling up in Indian Valley about four miles from Mokelumne Valley.[19]

Mrs. M. H. Manning observed that at Fort Bidwell in 1903, "edible tubers of *Perideridia* [yampah] highly esteemed by the Indians were dug by the sackfuls before the blossoms come and kept on hand for winter use. They were commonly occurring as plentifully as if planted in favorable localities." The ethnographer John Hudson recorded in Paiute territory: "Seed and nut food were very abundant along the lower riverbank of the Owens River and lake. Chia abounded. Grassnuts covered the ground were large and very nutritious."[20] According to the historians Odie Faulk and Laura Faulk, the Modoc could gather about one-half bushel of *ipos*, or yampah *(Perideridia oreganum* and *P. gairdneri)*, in one day.[21]

Based on an interview with non-Indian Henry Akers in 1933, Frank Latta reported that the Indians could gather three or four barley sacks full of grass-nuts along the Kings River near Centerville in a few hours. Akers said, "I ate lots of them myself. They were real good. The hogs got just as fat as butter on them. You could see acres of sandy ground rooted up a foot or more deep where the hogs had been eating grass nuts."[22]

The sheer numbers of bulbs, tubers, and corms of different types allowed a flourishing horticultural trade in the late 1800s. Carl Purdy, for example, harvested (with hired help) as many as 75,000 bulbs of Diogenes' lantern *(Calochortus amabilis)* per year to send to New Jersey, receiving $1.50 per 100 bulbs. Purdy could harvest 4,000 bulbs (of different kinds) per ten-hour day—or 400 an hour and 6 a minute—and he was collecting from May until late November.[23]

By the late 1800s, however, the numbers of geophytes had begun to decline as a result of ranching and farming that disrupted the long-standing indigenous relationships with subterranean plant parts. The farmer's plow and the trampling and grazing of cows and sheep destroyed the plants' habitats. Hogs ate bulbs and tuberous roots, leaving a wide trail of uprooted ground a foot or more deep wherever they feasted. Tuber fragments and bulbs left behind dessicated in the sun. In 1938 the anthropologist Gladys Nomland reported that the Indians could no longer gather camas because of cultivation of fields and sheep and cattle grazing.[24] Store-bought potatoes took the place of Indian potatoes in the Indian diet.

During the twentieth century, a variety of factors led to further losses in bulb and tuber populations: destruction of habitat for housing and agriculture, fire suppression, reductions in the numbers of native animals that coevolved with geophytes, and harvesting for the horticultural trade. Although much of the decline of geophytes across California can be attributed to these direct effects, the decline could also be tied to the absence of Indian management of this resource. Paradoxically, geophytes respond well to digging.[25]

Digging: Assisting Reproduction

Decades ago, Harrington asked several Karuk elders about digging Indian potatoes and jotted down these notes: "[These are] practices bordering on a knowledge of tillage. But they knew indeed that where they dig cacomites all the time, with their digging sticks, many of them grow up the following year. . . . They claim that by digging Indian potatoes, more grow up the next year again. There are tiny ones growing under the ground, close to the Indian potatoes."[26] The "tiny ones" Harrington mentioned are referred to as bulblets or cormlets by botanists and as offsets by horticulturists. Clinging tightly to the base of the large, marble-sized mature bulb or corm, they are an important means of asexual reproduction. As many as fifty cormlets can be found on one "mother" corm. They often resemble grains of rice, and most are no bigger than a ladybug. The edible geophytes bearing cormlets or bulblets include the brodiaeas (*Brodiaea, Dichelostemma*, and *Triteleia* spp.), some of the true lilies (*Lilium* spp.), some of the mariposa lilies (*Calochortus* spp.), the onions (*Allium* spp.), and others. The growth of bulblets and cormlets in some species is slow or suppressed until they are detached from the parent.

When California Indians dug up the parent bulbs or corms, the bulblets or cormlets—many of which were tenuously attached to the parent—were detached and remained in the soil to grow. Those that were not knocked off in the digging process were intentionally removed. The digging sticks churned and broke up the soil in the collecting patch, aerating it and dispersing the detached bulblets and cormlets. Thus the little offsets found themselves in ideal conditions for rapid growth into mature plants. This process, carried on for millennia, ensured each successive year's harvest.

A variety of other techniques also helped the geophytes to reproduce. The Cahuilla Indians of southern California, for instance, harvested only the large corms and replanted the smaller ones. Elders instructed the young girls collecting bulbs not to take everything. Through proper harvesting or dividing, an individual plant could be multiplied or regenerated indefinitely.[27] It was also a common practice to time the digging of geophytes to allow them to accomplish their sexual reproduction cycle by seed. The Chumash, Yurok, and other groups often gathered plants after the seeds had set, leaving the seed stalks behind and taking only the swollen subterranean parts. The seeds would fall into the cultivated ground, where they would have an increased chance of germinating. "Delaying the harvesting of bulbs until late in the season would also be the time when carbohydrate contents would be highest," according to the physiological ecologist Philip Rundel (pers. comm. 2004).[28]

California Indians have provided a number of accounts of how geophytes were formerly harvested and cultivated. According to Lalo Franco (Wukchumni Yokuts):

> We used to go up to the mountains and my dad would spot the brodiaea. We'd dig around and there would be a whole cluster of them in an area. . . . They would only take half of the plants in a cluster. They would never pick them all in one area. You only take what you need. The babies were taken off and put back in the ground, along with the little roots.[29]

Stanley Castro (Southern Sierra Miwok) said that his mother used to pick large bulbs of a bluish flower [probably *Dichelostemma capitatum*] to eat. The bulbs had "babies" attached. "She would leave the babies behind to make more," he said (pers. comm. 1990). According to Virgil Bishop (North Fork Mono):

> We gathered Indian potatoes [possibly *Sanicula tuberosa*] in May or June when the leaves are green and when in flower with digging sticks. The stem and leaves are small like celery. They grow in grassy areas near ponderosa pines. You boil them just like a potato and they're eaten plain. We'd go back to the same area and gather them. My mother and grandmother would only take the best and the biggest. They wouldn't harvest the smaller ones. They also gathered two or three kinds of wild onions in the foothills along streams. The plants were harvested in spring before flowering. They never cleaned everything out. They would always leave some behind.[30]

Some Western Monos and Miwoks still deliberately break off a portion of the fist-size bulbs of soaproot, leaving behind the root crown, which will, they assert, grow into an entirely new bulb in a few years. Barbara Bill (Mono/Yokuts) said, "We harvest the soaproot and break them off at the roots so the roots grow into new plants. I've noticed they grow a lot more in the areas where we gather it" (pers. comm. 1991).

Bertha Rice, former director of the California Wild Flower Exhibit, wrote in 1920, "The Indians, for all their wild natures, exercised discretion and good sense in gathering the choice bulbs [of *Lilium washingtonianum*] from their wild gardens, taking care to leave stock for the next year's harvest."[31]

A number of records from the ethnographic literature describe collecting patches that resembled cultivated garden beds. Edith Van Allen Murphey reported of the Indians of northern California, for example, "The flowers [brodiaea] were beautifully marked and colored, and as the bulbs grew in beds, they were easily harvested." Of mariposa lilies, she commented: "While they may seem small for digging, they grow in beds, and one can do blind digging on them after locating even one seed stalk."[32] Frémont stated about a tribe in the Pacific Northwest that "large patches of ground had been torn up by the squaws in digging for roots, as if a farmer had been preparing the land for grain."[33]

The edible geophytes of California evolved in an environment inhabited by large herbivores that would have been capable of digging up large numbers of bulbs and corms. The now-extinct megafauna, particularly three species of ground sloths *(Megalonyx jeffersonii, Nothrotheriops shastensis,* and *Glossotherium harlani),* two species of bears *(Arctodus simus, Ur-*

sus arctos), and the peccary *(Platygonus compressus)*, were probably major herbivores of bulbs throughout the late Pleistocene from perhaps 300,000 to 10,000 years ago (Steve Edwards, pers. comm. 2000).[34] It is likely that lilies, brodiaeas, onions, and other geophytes evolved easily detached multiple propagules as an adaptation to the intense herbivory they faced from these animals; though the larger parent bulbs and corms might get eaten, the bulblets or cormlets would survive to carry on the plant's genotype. California Indians simply took advantage of these adaptations by filling the former ecological role of the large herbivores and adding a layer of purposeful management.

The indigenous people themselves, however, might have explained it somewhat differently, perhaps attributing their harvesting of bulbs and corms to the teaching of gophers, as revealed in a Karuk story:

> The people knew, that if a seed drops any place, it will maybe grow up. They knew that seeds are packed around in various ways. Sometimes they see at some place a lot of Indian potatoes and then they dig in under. Behold there are lots underneath. Something is doing that, is packing it around down under the ground. And in the myths Gopher did that same thing; he packed *upva'amáyav* [tubers] around; he packed them around. A'ikrêen [sugarloaf bird] brought them in from Scott Valley, he brought some in for his younger brother. He said to his younger brother: "Do not let my wife see you when you are eating the *upva'amáyav* [tuber], do not let her see you eating them." And that is why he used to eat it upslope, upslope then, Gopher. It came up, every place he went; those were the only places where there was *upva'amáyav*, the places where he went.[35]

Burning: Creating Favorable Conditions

The ubiquitous California Indian practice of burning was also applied specifically to edible geophytes to enhance their growth. "Where the ground has been burned," recorded the linguist Alice Shepherd among the Wintu, "wild potatoes grow in bunches and ripen big."[36] Fires recycled plant nutrients, cleared accumulated litter, eliminated competitive grasses and shrubs, and may have activated bulblet production.

The clearing away of competing shrubs and small trees was an important result of burning: studies have shown that some of the edible geophytes will not flower or will flower poorly if shaded, thus reducing the number of plants producing seeds. Also, shading by surrounding plants, even smaller ones, could limit the summer warmth that many of the native bulbs and corms need to initiate the formation of buds in their resting bulbs.[37]

Several individuals in the Western Mono, Sierra Miwok, and Foothill Yokuts tribes remember the former burning practices for such species as yampah, mariposa lily, and brodiaeas and connect that burning with better growth. James Rust, a Southern Sierra Miwok elder, lamented the loss of this indigenous tradition and its beneficial ecological effects: "Snowballs, niggertoes, poppies, mariposa lilies, and sunflowers hardly come up now. There are too many weeds, grass, and brush. They used to come up in the spring after a fire" (pers. comm. 1989). Ruby Cordero, a Chukchansi Yokuts/Miwok elder, remembers the relationship between fire and bulb abundance:

> At the Picayune Reservation they used to burn for wild potatoes. We used to go where it had burned and we'd find great big potatoes otherwise the potatoes are little. *Homogi* [*Perideridia* sp.] and *dana* [probably *Sanicula* sp.] grew together. When they'd burn for *dana*, they'd burn for *homogi* too. They'd burn different places and in the same spot every other year. They knew just where they grew and they'd burn there. They'd burn an acre. It wouldn't grow in the brush. The shorter ones grew in the flat open areas of white oak and bull pine, in the meadows, at lower elevations. The taller ones grew under ponderosa pines. They'd set the fires about now [August 12] or later in the fall about September. Then by spring the plants start coming back up. It would fertilize the ground—maybe the ashes do something. (Pers. comm. 1991)

Another North Fork elder remembers: "We ate the wild potatoes, brodiaea with a purple flower, tana with lacy leaves and little yellow flowers, and wild carrot. All of them came up better after a burn. These were gathered near Rediger Lake" (pers. comm. 1989).

Roger Raiche, a gardener, has suggested that bulbs can sit for a decade or more and wait for fire or other favorable environmental conditions before breaking ground. After a fire, bulbs are sometimes so abundant that they form a solid mass of purple, pink, red, yellow, or white. Long after California Indians discovered the benefits of burning patches of Indian potatoes, ecologists confirmed that periodic burning promotes the growth of a number of bulbous and tuberous plants and labeled some of these "fire followers."[38]

Burning and digging together most likely created population-level changes: denser patches of plants, larger numbers of plants, and larger gathering tracts. Many of the important geophytes are only found in the open places in conifer and hardwood forests, coastal prairies, valley grasslands, and montane meadows, and their populations would have dwindled without regular burning to create and maintain open areas and light gaps.

Burning to promote the growth of geophytes was practiced by indigenous groups on every continent. The San of South Africa traditionally burned the veldt at the end of the dry season, "in order that the edible bulbs, roots, etc., should come up better during the approaching rainy season." The Australian Aborigines of Victoria still dig up the tubers of murnong *(Microseris scapigera)*, and gathering areas historically were burned over to increase production. The Indians of the Pacific Northwest selectively harvest camas (*Camassia* spp.) bulbs, leaving the smaller ones behind for future harvests, and historically fired the prairies to improve the quality and numbers of the bulbs.[39]

Early Stages of Domestication

Long before the full development of root crop agriculture, humanity's relationship with many native plants with edible subterranean parts had already shifted from one of pure predation to one of mutualism. Over long periods, tillage, selective harvesting, and burning had subtle yet profound ecological impacts at the species, population, community, and landscape levels within a multitude of habitats in different parts of the world. Digging up the subterranean organs of wild plants for foods was perhaps the oldest form of tillage, one that became the precursory management technique and the ecological foundation for the development of root crop agriculture in some areas.

Through millennia of human selection, protection, and replanting of offsets, certain geophytic species underwent genetic changes. The recurring excavation of plants with vegetative reproductive parts and the replanting or breaking off of such parts selected for specific genotypes that thrived under human disturbance regimes. This process certainly occurred in many parts of the world before the actual domestication of such plants as the potato and was very likely under way in California as well.

Human selection would have favored corms and bulbs that produced the greatest number of cormlets and bulblets, because in a context of human gathering pressure, those genotypes producing the most offsets would leave the greatest number of offspring. Similarly, among the tuberous plants, human gathering would have selected for tubers that easily broke apart in response to harvesting attempts.

One California species whose traits may have been influenced by human selection is the soaproot *(Chlorogalum pomeridianum)*. A portion of the bulb breaks off easily just above the base when subjected to the light pressure of a sharp stick. This is the part that California Indians harvested, leaving behind the root crown to produce a new plant. (See Figure 37.) "It would not take very long for humans to select for this trait of breaking above the root crown," says the plant ecologist Michael Barbour.

Dig soaproot
summer–early fall

Leave or replant root crown

Put seeds into hole

Repeat harvest
after 3–5 year rest period

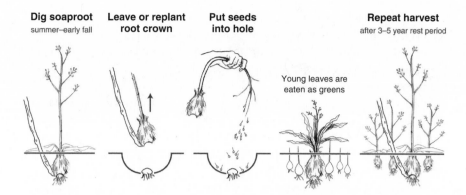

Young leaves are eaten as greens

Figure 37. Steps in the cultivation of soaproot *(Chlorogalum pomeridianum)*, an important plant to the majority of California tribes. The bulbs provided glue, fish poison, and food; the young leaves were eaten; and the old leaf sheaths that clothe the bulb were used as bristles for brushes. The plant is still gathered today.

Another example of a plant that may have been genetically influenced by human harvesting is Gairdner's yampah *(Perideridia gairdneri)*. It often has branching, spindle-shaped tuberous roots. In digging, these tubers break at the thinnest and weakest point. The remaining tuberous fragments are often composed of both root and stem tissue. According to the eminent botanist Lincoln Constance, who has studied the Apiaceae (the family in which yampah belongs), "roots of *Perideridia* when put in the ground reproduce tubers. . . . They're classified as 'tuberous roots' " (pers. comm. 1989). By gathering these subterranean parts before flowering and breaking them off to leave pieces behind, humans may have favored those tubers that leave the largest number of fragments.

Gathering as a Geophyte Conservation Strategy

Today 66 species and 102 taxa of California's monocotyledonous geophytes are listed as uncommon. An unusually high proportion of California's state and/or federally listed rare, endangered, or threatened plant taxa are monocotyledonous geophytes. Some of these plants were eaten by California Indians, and many were formerly more common. Their status is linked primarily to habitat fragmentation and loss. Many of them have also been diminished in numbers by gardeners and purveyors of rare garden plants who gather corms, bulbs, and so on, from the wild to sell to elite gardeners willing to pay a premium for unusual plants (e.g., western lily *[Lilium oc-*

cidentale] with its giant flower). Their declines may be due in part to the reduction of human and other mammal disturbance.[40]

Although little is known about the biology of rare geophytes, and the
different patterns of rarity may each require a somewhat different form of
protection and management, the cultural knowledge of native peoples may
be useful in restoring and managing some of California's rare and endangered geophytes. Given what we know about the biology of geophytes and
the harvesting strategies of native people, it may be that the kinds of disturbance brought about by former Indian harvesting should be a key part
of restoration strategies.[41]

Some species may require thinning for their populations to remain vital. The many tiny bulblets or cormlets they generate grow and reproduce
and eventually overcrowd the area and exhaust its nutrients unless the
stands are thinned, as they would be if harvested.[42] Other species may have
cormlets or bulblets that remain suppressed while attached to the parent,
and their growth is activated only when they are knocked off by some kind
of disturbance, such as gopher activity or human gathering. Lalo Franco links
the absence of gathering to the decline of one of the rare and endangered
brodiaeas (the Kaweah brodiaea, *Brodiaea insignis*): "There were three kinds
[of brodiaeas] that our people picked up here on Dry Creek in Tulare County.
The one that has a real small flower with a little bell—that's the Kaweah
brodiaea, the real rare one. We used to gather it. . . . I think that the plants
don't do very well and tend to die out if no one is digging."[43]

Particularly important for some genera, such as the true lilies (*Lilium*
spp.) and the checker lilies (*Fritillaria* spp.), may be the disturbance and maintenance of open areas provided by digging and regular burning, of the type
practiced by native people. Anecdotal evidence and my own observations
suggest that such geophytes show remarkable rejuvenation after controlled
burns, natural low-intensity fire, and clearing of underbrush. Redwood lily
(Lilium rubescens), an uncommon plant along California's North Coast
whose bulbs were eaten by indigenous peoples, is a good example. According to the botanist Frederica Bowcutt, who has conducted field research in
Sinkyone Wilderness State Park, "Redwood lilies occur mainly in forest
openings created by limited logging and in berms of loose soil created by
annual grading of a dirt road in and near the park." "It is conceivable," she
adds, "that the redwood lily was at one time more abundant when large bulbs
were periodically harvested and bulblets were left to repopulate the tilled
soil. The grading may in part mimic the effects of agroecological methods
used by indigenous peoples. Grading aerates the soil, as does the traditional
digging of bulbs by native people for food."[44] Bowcutt has also observed the

rare coast lily *(Lilium maritimum)* growing elsewhere in Mendocino County in areas underneath power lines that are annually maintained to keep the brush down (pers. comm. 2004).

While the documentation of California Indian horticultural techniques has heightened over the past decade, the study of the potential ecological effects of these practices on plant population dynamics and productivity has barely begun. Studies of plant–human interactions, particularly experiments that simulate the horticultural techniques of indigenous people and measure resulting impacts, are likely to reveal horticultural practices that accommodate the basic biology of native plants.[45] Some human tending and harvesting could be an effective conservation biology approach to maintaining geophytes, particularly monocots, in the wilds. Furthermore, a fuller understanding of survival mechanisms in geophytes will aid in developing guidelines for harvesting and management regimes that allow for some use of conservation areas. These ecological field experiments will also help us to better understand the biology of rare monocotyledonous geophytes in general.

Part III

REKINDLING THE OLD WAYS

CHAPTER 11

Contemporary California Indian Harvesting and Management Practices

In spite of the difficulties so many basket weavers face, we have a strong determination to keep these traditions alive. It is a way of saying, "We are proud of our ancestors. We are proud of how they lived their lives. They are part of us and we are part of them, and we remember and honor them by continuing the traditions they began."

LINDA YAMANE, Rumsen Ohlone (1997)

Despite the history of genocide, dispossession, and assimilation described in Chapter 3, California is still home to approximately 150,000 people who trace their ancestry to the state's original indigenous inhabitants.[1] Together they represent a rich diversity of cultural groups speaking a variety of indigenous languages, organized by different social structures, and practicing different religious customs and forms of land management.

Since their world was taken over by Euro-American settlers, California Indian tribes have continued to endure major threats to their cultural survival. In the past century or so, economic and cultural pressures have caused many Indians to embrace modern Western culture, exploitative land uses have degraded their traditional lands and reduced the variety and abundance of regional floras, and restrictions on the use of public lands have limited their traditional practices. Each year, fewer elders speaking the native languages and remembering the Old Ways are left alive to transmit these essences of culture. More than half of the languages spoken in California at the beginning of Euro-American contact are now extinct, and at least a dozen are imperiled, with only one to several speakers left.[2]

Many California Indians, however, are working actively to preserve their cultural heritage and ethnic identities. Most tribes have maintained a distinct cultural identity, and many regulate their business and conduct fundraising efforts through formal tribal councils. Some tribes have Indian owned and operated cultural museums and language revitalization programs.

Intra- and intertribal gatherings—in the form of Indian days, big times, native foods festivals, and powwows—take place up and down California annually. Organizations such as the California Indian Basketweavers Association (CIBA), the California Indian Storytelling Association, the Cultural Conservancy, and Advocates for Indigenous California Language Survival are working actively to preserve traditional practices and languages.[3]

Much of the effort to preserve California Indian cultures hinges on traditional gathering and land management. Since a large part of traditional Indian culture—from basketry to spirituality—has a direct connection to the natural world, maintaining relationships to plants and animals and to the land is of critical importance. Some native people still gather materials for baskets, eat wild food, and teach the ancient lore and plant and animal knowledge of their tribe. This chapter focuses on these activities and the barriers their practitioners face. It also begins a discussion about the links between indigenous cultural survival, the revival of traditional cultural practices, and environmental conservation and restoration, a topic that is explored in detail in Chapter 12.

Native Gathering and Management Today

In nearly every California Indian tribe, one can find men and women of all ages who gather native plants, fish, and hunt. Others enjoy the foods, medicines, crafts, and ceremonial objects these weavers, craftspeople, shamans, and hunters create from native plants and animals. Black morels (*Morchella elata*) and oyster mushrooms (*Pleurotus ostreatus*) are picked and added to gravies, spaghetti, and other modern foods. Leaves from Sierra mint, pennyroyal, yerba buena, and Labrador tea plants are made into refreshing drinks, and gooseberries, blackberries, blackcap raspberries, and elderberries are used in pies, jams, and jellies. (See Figure 38.) The young leaf shoots of *Lomatium californicum* are gathered and eaten like celery, or used as seasoning in soups. Various kinds of pine nuts are kept on hand for afternoon snacks for children and visitors. Bowls of acorn mush are kept in modern refrigerators. Clovers, sour docks, watercress, and miner's lettuce are eaten as salad greens, and the bulbs of wild onions, yampahs, lilies, blue dicks, and sanicles are eaten raw, boiled, or roasted. The seeds of Nevada stickleaf (*Mentzelia dispersa*) are eaten lightly roasted, ground into a fine flour, and rolled into balls. Ailments are treated with traditional medicines such as yarrow, madrone, incense cedar, yerba mansa, yerba santa, and angelica. Basket weavers sort, debark, split, trim, soak, and dye branches, stems, roots,

Figure 38. *(From left to right)* Iliene Cape (Mono), Mandy Marine (Mono/Maidu), Edith Beecher (Mono), and Melba Beecher (Mono) gathering mountain mint *(Pycnanthemum californicum)* for tea in the Sierra National Forest. The plant benefits from periodic burning. Photograph by Kat Anderson.

and rhizomes of various shrubs, trees, ferns, and sedges and weave them into beautiful baskets. The healing aroma of smoldering white sage, sweet grass, juniper, or wormwood drifts through the air during sweats and other ceremonies. Salmon and eel are caught in northern California rivers.[4]

California Indian cultures are dynamic; accordingly, tribes have abandoned some traditions and held tight to others, and in many instances they have blended Western and traditional ways. This process can be seen in traditional management and harvesting of native species. Both former practices and traditional tools have been altered to suit present cultural needs in contemporary Western society. Metal knives and shears, for example, have replaced obsidian knives for pruning. In digging white root, the digging stick has been replaced with a crowbar or tire iron or the popular grubbing hook (also known as a three-pronged curved fork). To dig bulbs and tubers, women employ a handled pointed iron bar or a simple broom handle. Other more modern tools, including garden forks, trowels, and spading forks, are also used for digging. In California Indian kitchens, coffee grinders, electric blenders, and sometimes hand-operated gristmills are used for grinding seeds into flour.[5]

Collaboration with Public Lands Agencies

Increasingly, individual and relatively haphazard efforts to locate and harvest native species are being supplemented by more systematic initiatives to revive traditional management practices on state- and federally owned land in California. The California State Department of Parks and Recreation, the U.S. Forest Service, and the National Park Service have long acknowledged California Indians' ties to public lands by allowing Indians access for plant gathering. More recently, however, the Forest Service and the Park Service have begun to rethink their roles; they are going beyond issuing passive use permits to actively integrating native peoples' management of traditional plant resources into public lands management programs.[6] Through programs and agreements that maintain areas for populations of highly valued native plants and policies that assure the availability and preservation of cultural resources, these agencies have taken steps toward becoming advocates of maintaining, tending, and encouraging the growth of plants important to Indian people.

These programs typically begin by surveying the resources, recording their conditions and numbers, and matching this information with indigenous needs. There are at least nine areas of inquiry in this initial process: (1) Who are the Indian groups in the area who currently gather plants? (2) What plants are currently important to Indian people, and how do they fulfill Indian cultural needs? (3) What plant parts are gathered from these plants? (4) What tools are used to gather or manage these plants? (5) Where are these plants located in the field? What is their present-day abundance and distribution? (6) What are the current Indian harvesting and management techniques for the plant? What is the frequency, time of year, and intensity of harvesting or management? Is this kind of management in conflict with any of the present-day uses of land, according to public or private owners, and if so, how can such conflicts be mitigated? (7) What is the impact of Indian harvesting on plant productivity and numbers, and what is the sustainable level of utilization from field populations that would not jeopardize conservation goals? (8) What traditional collection sites exist in public areas that have a long and distinguished history of use, and how can these sites be preserved for future generations? (9) What public areas would make good new plant collection and management sites in which to tend, maintain, and encourage the growth of plants important to Indian people?

MANAGING FOR BEAR-GRASS

Contemporary northern California weavers talk about the difficulty of obtaining bear-grass as an overlay material in basket making; there is not enough of it, and the quality of the leaves is usually poor because the plants have not been burned. Pam Mendelsohn recorded among the tribes of northwestern California in 1983 that indigenous fires had traditionally been set in bear-grass patches after the first heavy rain in October or November. The new bear-grass leaves—stronger, thinner, and more flexible than older leaves—were then collected the following June or July. She concluded that controlled burning of underbrush was critical to the fitness of bear-grass.[7] In response to such recommendations, U.S. Forest Service fuels specialists, timber planners, and cultural resource managers on the Six Rivers, Shasta-Trinity, Klamath, and Plumas National Forests, together with California Indian basket weavers, have begun to burn small areas of one acre to two hundred fifty acres to favor this cultural resource. The plants are then monitored over the course of several years to see how they respond to burning, and the burning prescription is adjusted accordingly.[8]

PROTECTING AND CULTIVATING SEDGE

As noted in Chapter 6, native weavers in different tribes still gather white root for use in baskets, often from beds that have been cultivated for generations. When Henry Icho (Wukchumni Yokuts) took Frank Latta to a collecting site, he told Latta that native people "always dug the roots at this same place, had been doing so for at least four generations that he knew of."[9] Good sedge habitat, however, has dwindled drastically for wildlife and weavers alike. Basket makers face fences and No Trespassing signs in many locations. Many of the banks along streams have been channelized to reduce the threat of flooding, eliminating sedge areas. Cattle and sheep have overgrazed and trampled sedges, reducing their vitality until they eventually died out.

Sedge beds are still harvested much as they were in precontact times. Rhizomes are dug in spring, summer, or after the first rains in fall, depending on soil type and preferences of the tribe or family. According to Pomo weavers interviewed by David Peri and Scott Patterson, the best time to gather sedge growing in "sand root beds" is from late summer to the first rains in fall, after the spring growth has matured, and the best time to dig for white root in "dirt root beds" is after the early fall rains, when "the soil has loosened up."[10]

Some weavers cut back the tops of the sedge to about four to six inches before digging so that the sharp-edged leaves will not cut their hands. A weaver lays bare the rhizomes with the chosen tool, selecting those that meet her criteria for length, diameter, straightness, strength, pliability, and color.[11] She digs the soil away from either side of the rhizome until both ends are reached and severs each end from the parent stock. Weavers cut only one- and two-year-old rhizomes and are careful to leave the parent plants behind, to ensure a continuous supply. The disturbed soil is put back in place.

The collected rhizomes are washed in the creek or stream immediately. A parchmentlike brown skin that encases an inner woody core is removed and discarded. Later, the inner core is split lengthwise into two fine threadlike strands. The threads are coiled and often stored for two years. The rhizomes have lost most of their water after six months. After drying, the rhizomes are uncoiled, soaked, and then trimmed to a specific width. When soaked, dry roots regain their flexibility.

One Pomo coil can contain 80 rhizomes, each averaging 18 to 24 inches long. Those cultivated in sand root beds with loose soils typically grow to a length of 3 to 5 feet and on occasion will grow to 6 or 7 feet. Peri and colleagues recorded in the late 1970s and early 1980s that a senior Pomo weaver harvested more than 5,000 rhizomes per year for making baskets.[12]

The Sierra National Forest has been working with the Mono to identify and fence willow and sedge sites in an important high-elevation meadow that has a long history of indigenous collection of plants for basketry material.

BURNING STANDS OF DEERGRASS

Deergrass is at the crossroads of the interests of California Indians, heritage resource managers, and natural resource managers. Connected to ancient human history and still in demand by contemporary weavers in different parts of the state, it is a natural and cultural resource. Modern resource managers are finally realizing the importance of fire for maintaining deergrass and other plants of cultural importance. In the Sierra National Forest, resource managers have been working with the Mono to re-create the ancient Mono practice of burning deergrass in order to increase its flower stalk production and eliminate accumulated dead leaves. The first burn took place on a December afternoon in 1991, at a half-acre site chosen by Lorrie Planas (Choynumni Yokuts/Dunlap Mono), heritage resources manager for the High Sierra Ranger District. The meadow was accessible to both weavers and fire engines, had large stands of deergrass, and was away from public view. Assistant District Fuels Officer Jim Desmond set off a line of fire at the edge of the first deergrass cluster. The fire backed over the cluster with the help of a light westerly wind. Within minutes all of the bunchgrass clumps had caught fire, sending up green smoke. Given the time of year, it is amazing how readily the plants burned. "They look happy," said Martha Beecher (Mono) as she and her sisters, Lydia, Edith, and Melba, looked on with approval. After the fire had done its work, each plant had been reduced to a woody base where tiny buds would sprout the following spring. The burn was repeated in February 2003 by fuels specialists Dave Hill and Mark Helm, also of the High Sierra Ranger District.[13]

MAIDU STEWARDSHIP PROJECT

In 1998 the Maidu Cultural and Development Group launched the Maidu Stewardship Project—a plan to restore 2,100 acres of the Plumas and Lassen National Forests to presettlement conditions—using traditional ecological knowledge and management techniques such as burning, pruning, tilling, and selective harvesting. Near Greenville, California, the area in the Plumas National Forest has streams, meadows, oak woodlands, and mixed conifer forests. It is the ancestral home of the Mountain Maidu people and the home of the largest Native American population in Plumas County.

Specific vegetation management objectives for the area include enhancing culturally important plant species through Maidu knowledge and techniques; managing riparian areas for certain types of willow used by basket weavers; managing for increased availability of quality maple and bear-grass; enhancing food resources such as bulbs, corms, and tubers; enhancing acorn production; promoting general beautification of the forest; and reducing wildfire risk by establishing fuel modification areas.[14]

The project promotes environmental conditions in which the Maidu people can maintain their lifeways while enhancing the well-being of other species—wildlife, plants, and mushrooms. According to Farrell Cunningham, Mountain Maidu and Maidu Stewardship Project coordinator, "Better integration of Western ecological science and traditional ecological knowledge can be developed because we share the goal of ecological sustainability" (pers. comm. 2004).

TRADITIONAL RESOURCE MANAGEMENT
IN DEATH VALLEY NATIONAL PARK

The Timbisha Shoshone, a small tribe now numbering about 280 people, have lived for millennia in one of the harshest climates on the continent—

in and around what is now Death Valley National Park in the Mojave Desert. The tribe favored two plant resources as their staffs of life: the nuts of singleleaf pinyon pine *(Pinus monophylla)* and the seeds of honey mesquite *(Prosopis glandulosa* var. *torreyana).*[15]

The honey mesquite, a member of the pea family, reaches the size of a small tree (less than 35 feet high), and its roots can penetrate downward into the water table, often to great depths. Once pollinated, the greenish yellow flowers become pods containing relatively large seeds. The Timbisha Shoshone harvested the fully mature pods in early summer and pounded the hard seeds into a flour in mortars. They made this meal into a mush or formed it into dry cakes. If harvested while still green, the pods were pit roasted.[16]

The Timbisha Shoshone regularly pruned the honey mesquites, removing dead limbs and lower branches and keeping the areas around the trees clear of undergrowth.[17] Without this care, the dead branches and vegetation would have acted as a catchment for blowing sand, causing a buildup of sand that would in time have partially or fully buried the trees. The harvest of the pods may have contributed to new seedling production because the Shoshone caretakers cast aside the endocarps that encased the seeds, and these sometimes contained a few remaining seeds.

Singleleaf pinyon is an evergreen tree reaching a height of up to 40 feet. A pinyon tree usually does not start bearing cones before it is thirty-five years old and does not produce good seed crops until it has been growing for about one hundred years. (The management of singleleaf pinyon by Indian peoples in the arid eastern regions of California in general is discussed in Chapter 9.) Members of the Timbisha Shoshone traditionally gathered pinyon seeds in the fall, which were then roasted or parched, shelled, winnowed, ground into a meal, and made into nutritious pine nut soup, mush, or cakes. Traditional tools and harvesting methods were designed to enhance or maintain future pinyon cone production. A favorite gathering method was to whip the trees with poles when the cones opened, knocking the seeds out of the cones and onto the ground, where they could be easily collected. During this process, dead or dying branches were pruned back. Shoshone pine nut gatherers also enhanced future cone production by pinching back or breaking the growth tips by hand. To protect mature pinyon trees from fire, the Timbisha Shoshone removed from near the trees any live growing shrubs that might act as fuel ladders during a fire. They also pruned back low-lying branches that could catch fire and removed dead and fallen limbs. As a result, many areas in the pinyon–juniper forests in the mountains surrounding Death Valley where people habitually camped or collected pine nuts were kept open and free of underbrush.[18]

In November 2000 President Bill Clinton signed the Timbisha Shoshone Homeland Act, which established trust lands for the Timbisha Shoshone tribe within and surrounding Death Valley National Park. The law specified that the tribe's two major plant resources—honey mesquite groves immediately south of Timbisha Village at Furnace Creek and the singleleaf pinyon forests in Wildrose Canyon in the Panamint Range—be cooperatively managed with the National Park Service. According to former park superintendent Dick Martin, the joint management "will not be on a permit basis, but on a cooperative agreement basis, which we will sign as equals." This involves incorporating the knowledge and methodologies of the tribe into park management plans.

In 2001, with the assistance of consulting ethnobiologist Catherine Fowler, the tribe launched a five-year pilot project to resume traditional management within the Furnace Creek mesquite groves and the Wildrose pinyon stand. This project, now under way and scheduled to be completed in 2005, has three major objectives, as described by Fowler: "1) to explore the ecological characteristics of these two species and their habitats by gathering as much information about them as possible; 2) to establish some small study plots in different ecological settings that could be monitored to establish baseline data such as general tree/grove health, frequency of flowering, fruit set, and production; and 3) to initiate traditional care (pruning, clearing and discarding seeds in the case of the mesquites, and cleaning, clearing, whipping and pruning for the pinyons)."[19]

OTHER EXAMPLES

Many culturally important plant resources are disappearing because forest trees are encroaching into the meadows where the plants grow, shrinking their habitats. In partnership with the Shasta-Trinity National Forest, the Pit River tribe has begun to counteract this process by reestablishing the age-old tradition of maintaining the openness of riparian meadows through regular management. In the McCloud Flats area, they weeded out about five acres of small pine trees that had been advancing into the grassland and hand piled them for burning. They also removed non-native plants such as mullein. The meadow was then burned to enhance medicinal and food plants important to the Pit River culture, including cats ears, tiger lily bulbs, anise, sourdock, and blue-flowered camas.[20]

Once collection sites are identified, mapped, and set aside for Indian use, these areas may still not be safe from herbicide drift, roads, logging, grazing, or recreation trails (see below). Thus certain public lands agencies are establishing collecting areas and protecting them from conflicting land uses and management practices. The Eldorado National Forest, for example, has been working with local basket weavers and food gatherers to establish herbicide-free plant collection areas.[21]

Threats to Traditional Gathering

The continuance of traditions based on gathering is remarkable given the great obstacles California Indians have faced.[22] Culturally significant plants have declined with the loss of habitat, especially wetlands, riparian woodlands, and open lower montane forest. Former gathering sites rich in memories, the marks of human effort, and culturally significant plants are now blacktopped widened roads, rangelands, or private homes. (See Figure 39.) Sacred sites have been developed, and others are in jeopardy, including Mounts Palomar, Shasta, Pinos, and Konocti and Pincushion Mountain in the Sierra Nevada.[23] The vitally important traditional management practice of regular burning is no longer possible in many areas because of government prohibitions and the buildup of fuels from fire suppression.

LIMITED NATIVE LANDOWNERSHIP

In 1851–52 the U.S. government negotiated eighteen treaties involving about twenty-five thousand California Indians. In each of the treaties the tribes were acknowledged as sovereign nations and the Indians were promised provisions, cattle, and extensive tracts of valuable land to be set apart for reservations. In June 1852, however, the U.S. Senate, meeting in secret session, rejected the California treaties, and the vast reservations proposed were never created.[24] As a result, a state that at the beginning of Euro-American contact had one of the densest populations of indigenous people of any area in North America today has only a tiny proportion of its land under Indian control, mainly small rancherias under three hundred acres each.[25]

As Euro-American settlers multiplied in California, most tribes were forced to move into white settlements, and the best lands, stewarded by Indians for centuries or millennia, were usurped by non-Indians and used for agriculture, stock ranging, mining, or lumbering purposes. The new owners were generally not willing to allow local Indians to "trespass" for gathering purposes. Today, with an insignificant land base or none at all, many tribes are forced to gather and hunt on public or private lands to continue their traditions. And even on these lands, plant populations are often not of sufficient size or the materials of acceptable quality.

Figure 39. Florence Brocchini, a contemporary Western Shoshone weaver, who is continuing the basketry traditions of her Southern Sierra Miwok husband by collecting the young shoots of redbud in Midpines, California, one year after pruning. Many weavers are forced to gather material alongside highways in order to continue their traditions. Photograph by Kat Anderson.

TOXINS IN THE ENVIRONMENT

Pesticide spraying, contamination from toxic waste, and air and water pollution bear directly on the health of native people. There are increasing incidences of cancer, miscarriages, and other health problems associated with the gathering, processing, and eating of native plants because the plants may contain pesticide residues. California accounts for one-fourth of the pesticides used in the United States. Most pesticides are sprayed from the air over agricultural crops and forests, and these poisons drift and end up in streams, groundwater, urban parks, backyards, and protected wilderness areas. In addition, plants that are targeted for pesticide spraying are sometimes the very ones that are important to Indian tribes for teas, basketry, medicines, and foods. Plants with pesticide residues can pose hazards even if they are not used for food or medicine: often Indian weavers split and process plant material for basketry with their teeth.[26]

Toxic materials are widespread in the environment, and one never knows if the wild plants one is eating or working with are contaminated. For example, around Clear Lake, the largest natural freshwater lake entirely within California, the bulrushes and other plants absorb mercury and other industrial pollutants that enter the lake from gravel mining, causing potential health hazards to Indian weavers who gather black root (rhizomes of *Schoenoplectus fluviatilis, S. maritimus,* and *S. robustus*) for basketry.

Rosalie Bethel, a North Fork Mono weaver and spiritual leader, was poisoned by an herbicide in the early 1990s while gathering redbud along Fresno County roadsides. Ron Goode, also a North Fork Mono, described her experience:

> She was gathering redbud for making baskets as is customary. She immediately began cleaning and splitting some of the redbud shoots to check their quality. On her way home she noticed a tingling sensation around her mouth. Later, after arriving home, she continued to split her sticks. This is when she noticed her mouth had become numb and she soon became very nauseated. Two weeks later when she returned to Cold Springs Rancheria she asked the folks there why none of the beautiful redbud shoots had been gathered. She was informed that the County had sprayed with Round-up and some of the gatherers had known about the effects of chemical spraying.[27]

CONFLICTING USES OF PUBLIC LANDS

The multiple-use concept that guides management of federal lands supports many categories of land use—including livestock grazing, timber production, recreation, wilderness, and mining—that are often incompatible with traditional gathering. Where grazing is allowed, cattle trample and overgraze meadows that contain important basketry materials. In areas managed for timber by the Forest Service, the conifers that make good lumber are favored at the expense of shrub and hardwood species important to Indian cultures. In these areas, species such as hazelnut and tan oak are removed during thinning or are outcompeted by the fast-growing conifers. Where hiking and equestrian trails cut through gathering sites, Indians cannot practice traditional gathering with any privacy.[28]

DECLINE OF CULTURALLY SIGNIFICANT PLANTS AND ANIMALS

Over the past two centuries, native plants and animals have suffered declines in their populations as more and more land has been transformed from natural habitat into housing, roads, farmland, reservoirs, salt evaporation ponds, tree farms, and rangeland. And on the remaining wild and semiwild land, the by-products of human development—erosion, siltation, air pollution, water pollution, altered water flows—have taken their toll as well. In many cases, culturally significant species have been the ones most affected. The right to hunt or fish in an inanimate landscape is not a useful right; nor is the right to gather plants in areas where the plants are not of the quality or abundance necessary to sustain cultural traditions.[29]

Native people are full of stories about traditional gathering sites that no longer exist or certain formerly abundant plants that can no longer be found.

> We had our own special areas where we picked acorn. But they're developing in all of our places. We used to go up to the other side of Alder Springs on the way to Mountain Rest School and up there by Cressman's. It's all Southern California Edison Company land now. That was the last place where my grandma Ellen picked. . . . You can't get a whole bunch of straws [deergrass flower stalks] in one place. I used to go to six places to get enough for a year's supply. Even if I see a single plant along the road, I'll stop and get them. We have to stop wherever we see the straws and get them because they're so hard to find. That's the way we've had to do it for the last twenty years. (Norma Turner . Behill, Mono/Dumna, pers. comm. 1991)

> I guess they had special gathering sites a long time ago. Now they go wherever they can get it [deergrass flower stalks] because it's getting scarce. (Melba Beecher, Mono, pers. comm. 1991)

> It concerns me that I'm seeing a lot of our food sources disappear. We've removed a lot of black oaks in the Valley [Yosemite] in the last few years—nice big ones. The status of the oaks is poor today compared to when I was growing up. The impacts are from overwatering lawns, construction, sewer lines, and people. (Jay Johnson, Southern Sierra Miwok/Mono Lake Paiute, pers. comm. 1989)

Many culturally important animals, too, are now so rare that they can no longer be collected. The Pismo clam *(Tivela stultorum)*, for example, harvested judiciously by the Chumash along the central California coast for centuries and used for food and for making musical instruments, has been overharvested by commercial diggers, causing a disastrous decline in the intertidal populations in recent years.[30] The desert tortoise, which is on both state and federal threatened species lists, has suffered severe population losses as a result of habitat degradation, disease, and overcollecting.[31] Yet when the tortoise was properly harvested by desert peoples such as the Paiute and Chemehuevi for its meat, and its shells used in rituals by the Cahuilla, Cupeño, Luiseño, and Kumeyaay, it flourished.[32]

As the plants or animals become more and more difficult to find, requiring more search time and gasoline, traditions may be abandoned altogether. "We used to go to this side of Sanger to gather *monop* [deergrass]," says Hazel Hutchins. "They scraped that too and it's all gone. They scrape it with a bulldozer and it just disappears. . . . So we don't make baskets no more" (pers. comm. 1992). The inability to gather a particular plant or animal can thwart a project or end an entire tradition since substituting one material for another still leads to loss of place-based knowledge. For example, buying imported raffia at a local hobby shop is entirely different from the traditional local harvesting of real sedge rhizomes, which requires intimate knowledge of the plant's reproductive biology and growth patterns in specific soils.

Many formerly abundant culturally significant species, such as deergrass, Indian hemp, and sedge, do not appear on federal or California Native Plant Society inventories and are not considered rare and endangered from a biological standpoint. Nevertheless, they may be rare and endangered from a *cultural* standpoint.

LOSS OF HABITATS SUPPORTING CULTURALLY IMPORTANT PLANTS

Many of the ecosystems that are in jeopardy in California are the last significant habitats of important cultural resources. For example, various sedges (e.g., *Carex barbarae*), highly valued for basketry by numerous tribes, grow primarily in the riparian woodlands along central California rivers, a habitat that now occupies only a tiny fraction of its original acreage. Both the habitat and the populations of the plant have degraded in quality and dwindled in size because of dams, livestock grazing, channelization of streams, reclamation, introduction of exotics, and loss of water quality.[33]

The five million acres of wetlands that once existed in California have been reduced by 91 percent through diking, draining, and filling in for agriculture, housing, or other purposes.[34] This puts a tremendous hardship on

Figure 40. Justin Farmer, an Ipai weaver from southern California, gathering *kwanaay (Juncus textilis)* for baskets in a wetland area under sycamores near Idyllwild, California. Photograph by Kat Anderson.

weavers who depend on plants such as rushes (*Juncus* spp.), sedges, and bulrushes (*Schoenoplectus* spp.) that are gathered from wetlands. (See Figure 40.) For example, there was once a large marsh and slough at the northwestern end of Clear Lake where Pomo weavers gathered "black root" bulrush rhizomes for their baskets, but in the 1920s and 1930s this wetland area was converted to rice fields.

PROHIBITION OF TRADITIONAL MANAGEMENT

Wilderness managers have been trained to view human presence in natural areas as an impact, a factor that generally decreases diversity and may endanger plants and animals. With the exception of the moderate picking of renewable aboveground plant parts such as seeds, fruits, and leaves, they tend to perceive indigenous gathering and management (such as weeding, pruning, coppicing, and the digging of bulbs or rhizomes) as potentially destructive, or at best as a significant disturbance that might give "natural" vegetation a groomed or manicured appearance, in direct conflict with the aesthetic appeal of virgin landscapes. The contrast between this view and that of Native Americans is articulated by Fowler:

> For the Timbisha people, as well as perhaps others elsewhere in the Great Basin, it is important management should "show," thereby creating, at least in certain circumstances, habitats that appear to be tended as opposed to what they feel is unkempt. Elders explain that plants need to "feel" the presence of people—they are used to it— and that is how plant-human interrelationships are at least in part maintained. The other side is that they need to "hear" from people: hear the prayers of thanksgiving and the talk required before collecting medicinal plants or foods. Plants are not destroyed when people harvest or care for them in proper ways; they are actually enhanced.[35]

The assumption among wildland managers that human use is a negative impact is reflected in agency policy. Limits on harvesting and management are set for Native Americans, and the burden of proof is on them to show that their techniques do not deplete the resource.[36]

An elderly Yurok weaver in northern California was denied her application for a permit to pick five finger fern stems for basketry material some years ago because "there were no known areas within state park lands where the fern grew in sufficient quantity." Park officials were concerned that the amount regarded as necessary (two armfuls) would wipe out a major portion of the already limited population (Breck Parkman, pers. comm. 1989). They did not take into account the fact that collecting the five finger fern (or "black fern," as the Indians call it) is an ancient Yurok tradition practiced sustainably for centuries. Yurok basket makers say that collecting regularly from their fern patches allowed them to return to the same patch year after year because it increased rather than depleted the populations.[37]

Although some exciting projects using traditional management to enhance cultural resources are under way on public lands, these are few and affect a relatively small amount of acreage because of prevailing attitudes about traditional management. As noted previously, the general absence of traditional management across the state has had a profound influence on the landscape, not only causing a decline in culturally significant plants and animals, but also reducing biodiversity and increasing the frequency and size of catastrophic wildfires. The fire exclusion policies of various public lands agencies have facilitated the encroachment of unwanted plants on traditional gathering sites that "take over" favored plants. Yurok, Karuk, Hupa, and Tolowa people today believe that the exclusion of fire in northwestern California forests during the twentieth century has led to a decline in the populations and quality of two important basketry plants, hazelnut and bear-grass.[38]

The habitat for many of the basketry shrubs and other plants is still intact, but the quality of the material has declined. Decrepit plants are hanging on, but they are not yielding basketry material, and there are no new seedlings or saplings, because their habitat has been drastically altered by the absence of fire. When a wildfire does occur, the elders are happy because they think about all of the new sprouts that will be harvestable in less than a year—that is, if the fire does not kill all the shrubs outright because of all the accumulated fuel that makes it too hot. Because none of the basketry shrubs is rare and endangered, no federal or state agency is charged with managing critical habitat for these species.

The former ecological interactions among native people, plant populations, and plant communities that took place with traditional management greatly increased biological diversity at different scales of biological organization, from microhabitats to landscapes. Indian management, for example, created fine-grained patches of dogbane in seasonal wetlands and deergrass in open chaparral and ponderosa pine stands. These former ecological interactions between people and plants, and the resultant cultural landscapes, were extremely important for gathering because they maintained populations of culturally important plants. Today, without Indian management, these cultural landscapes and the culturally important plants they support have nearly disappeared. In spite of their crucial value to traditional cultural activities and biodiversity, they are not accounted for in current inventory systems that evaluate for elements of diversity, such as the Nature Conservancy's conservation inventory for its Heritage Program.[39]

The Importance of Traditional Gathering and Management

Where the cultural traditions of native people remain in California, they are of inestimable value. The gathering of plants and animals is a foundation of ethnic identity, provides a healthier diet, helps individuals and tribes to maintain economic independence, and can assist in efforts to conserve native plants and preserve the state's biodiversity. In addition, fostering the unique gathering and land management practices of each tribe preserves a remarkably diverse and rich cultural heritage.

ETHNIC IDENTITY

Each indigenous culture relies on flora and fauna characteristic of its tribal territory. The great variety in the plants and animals was manifest in the house types, foods, medicines, weaponry, and basketry material of each culture. Today, each tribe's use of particular plant materials marks its distinctive culture and plays a major role in shaping, defining, and maintaining its cultural identity. For example, the Karuk of northwestern California gather the tillers of bear lily *(Xerophyllum tenax)* for overlay designs in their baskets, whereas the Western Mono and Southern Sierra Miwok gather young redbud *(Cercis orbiculata)* shoots for the red designs in their baskets.[40] Thus the continuation of gathering and hunting traditions is an important component of tribal identity and preservation of culture.

Affinities for specific animals and plants grow out of a profound linkage to a place. The integral way in which a plant or an animal is interwoven into a culture gives the culture continuity with its past and grounding in a distinct bioregion that has been considered home for untold generations. It is the long-term association with a place and the plants and animals that inhabit it that translates into tangible, distinct tribal ethnicity.

HEALTH BENEFITS OF WILD FOODS

As California Indians were forced to assimilate into the dominant culture, they lost their lands, which resulted in abrupt and drastic dietary changes. Wheat flour substituted for flour made from pounded native seeds or grains; while potatoes replaced the native tubers, bulbs, and corms. Head lettuce replaced the many leafy native greens that were high in calcium, iron, phosphorus, and potassium. The ancient cooking techniques of baking, boiling, and roasting were often replaced with frying. As a result, health problems among California Indians today are frequently linked to inadequate nutrition. Particularly prevalent are heart disease and diabetes.[41]

Wild foods, in contrast, offer more variety, fiber, minerals, and vitamins than do domesticated crops, which are often grown in worn-out, mineral-depleted, and heavily fertilized soils laced with selenium and pesticide residues. Wild foods also lack the additives and high sugar, sodium, and fat of processed Western foods.

People in culturally intact hunter-gatherer groups who still eat traditional foods (e.g., the Philippine Tasaday, the Hadza of Tanzania, the Kade San Bushmen of Africa, the Ache of Paraguay) exhibit excellent health, and the diseases that plague Western culture, such as heart disease, high blood pressure, diabetes, and most types of cancer, are virtually absent.[42] In the 1870s, Stephen Powers noted of California Indians that their "uniformly sweet breath and beautiful white teeth (so long as they continue to live in the aboriginal way) are evidences of good health." Lucy Young (Lassik/Wailaki) said about her grandfather: "He had good teeth. All old people had good teeth." When Lucile Hooper interviewed Cahuilla Indians in southern California in the early 1900s, she was told by several old men that "the reason the Indians are dying so fast is that they are eating white man's food, canned goods and the like."[43]

Recent studies conducted in the southwestern United States have demonstrated that traditional foods such as acorns (*Quercus* spp.), tepary beans (*Phaseolus acutifolius*), and mesquite pods (*Prosopis velutina*) contribute to the health and longevity of indigenous peoples in arid environments. Traditional desert foods have soluble fibers and complex carbohydrates that may slow carbohydrate digestion and lower blood sugar levels. These studies and similar studies in Australia and Hawaii demonstrate the long-term involvement of native peoples with the land and suggest adaptations to their native floras.[44] Nabhan suggests that a return to traditional diets might help indigenous people of the southwestern deserts to combat diabetes.[45] The findings from these studies may be highly pertinent to native people in California, who also customarily relied on the acorns of various oaks and seed pods from mesquite for a substantial portion of their diets. Indeed, some of the healthiest and longest-lived Native Americans today eat acorns regularly.

Although edible plants and other traditional foods are still gathered today, they make up a negligible proportion of the diets of most California Indians. Historically, the shift to Western diets was not by choice, and even now, if gathering were not so difficult, many native people would eat more wild foods.

ECONOMIC INDEPENDENCE

Although many native people have been forced to assimilate into the dominant culture and take up conventional employment, some individuals still make a large portion of their incomes from the land. Among northwestern tribes such as the Karuk and Yurok, for example, the native economy is still built around the river and fishing. As they have for millennia, people from these tribes use carefully honed harvesting methods, mediated by cultural rules and rituals, to sustain fish populations over time.[46] Some California Indians make a living making and selling baskets, and this activity depends crucially on access to basketry materials and the ability to manage the plants producing them. The more opportunities that native people have for carrying out traditional gathering and harvesting, the greater their ability to maintain the economic independence of themselves and their tribes.

CONSERVATION AND RESTORATION OF ENDANGERED PLANTS

In California, as in many parts of the world, the knowledge of indigenous people has considerable modern value for managing and conserving in situ genetic resources.[47] (See Table 12.) California Indians managed the populations of the plants they gathered for food and cultural items, using a variety of techniques to sustain each species. These same techniques can inform the conservation and restoration of species today—particularly those that are endangered.

Loss of habitat, ecological degradation of streams, environmental pollution, and overexploitation of natural resources have combined to create a genuine peril for many animals and plants in California. Nearly one in three vertebrate species and one in ten of California's native plant species are in serious danger of extinction. As songbirds and salmon, amphibians and turtles decline, it is a sure warning that landscapes are less and less habitable for all of life.

Certain plants integral to traditional indigenous cultures are now on rare and endangered species lists assembled by the California Native Plant Society and the U.S. Fish and Wildlife Service or are potential candidates for such a listing. These plants include the kaweah brodiaea (*Brodiaea insignis*), harvested for its edible corms by the Wukchumni Yokuts, purple amole (*Chlorogalum purpureum* var. *purpureum*), dug and eaten by the Salinan, and the Torrey pine (*Pinus torreyana*), the nuts of which were gathered by the Kumeyaay for food.[48] Many of these plants were gathered in great quantities without contributing to their depletion. Former harvesting strategies and management practices for these species may have maintained and even expanded their populations, while removal of indigenous people from their homelands, causing the discontinuation of these practices, may have contributed to their decline. For example, showy Indian clover (*Trifolium amoenum*), gathered in parts of Marin and Sonoma Counties and eaten raw by the Coast Miwok, declined to such an extent that it was presumed extinct until its recent rediscovery by Peter Connors of Bodega Marine Laboratory. While urbanization and agriculture were undoubtedly major causes of the

TABLE 12

Possible Application of California Indian Use and Management Techniques for Enhancement of Uncommon, Rare, or Endangered Plant Populations

Uncommon, Rare, or Endangered Species	Other Species in Genus Known to Be Managed	Tribe	Part Used	Use	Management Techniques
Allium tribracteatum	Common Allium spp. (e.g., Allium validum)	Western Mono	bulb	food	tilling, burning
Clarkia australis	Clarkia purpurea subsp. purpurea	Central Sierra Miwok	seed	food	sowing, burning
Perideridia parishii subsp. latifolia	Perideridia bolanderi, P. gairdneri, P. kelloggii, P. parishii	Northern Hill Yokuts, Sierra Miwok, Western Mono	tuber	food	tilling, burning
Trifolium barbigerum var. andrewsii	Common Trifolium spp.	Northern Hill Yokuts	leaf	food	burning
Wyethia reticulata	Wyethia longicaulis, W. elata	Wiyot, Western Mono	seed	food	burning

clover's decline, the revival of traditional gathering and management may be its best hope for restoration.[49]

The Torrey pine, the nuts of which were eaten raw or roasted and used as flavoring in seed porridges and pinole, is endangered in a portion of its territory by urban development. According to the plant ecologist James Barry, pine regeneration has been sparse at Torrey Pines State Reserve, except following a 1972 wildfire that promoted the development of many seedlings. Fire is an important ecological factor in the perpetuation of the species and past human-set fires perhaps played a key role. The anthropologist Florence Shipek has documented Kumeyaay firing of areas containing Torrey pine groves and the planting of pines to enhance their populations.[50] In restoring fire cycles to these and other rare conifers, knowledge of how Indians changed the frequency and intensity of fires may be integral to the pines' successful modern management.

Humboldt County wyethia (*Wyethia longicaulis*) has uncommon status, yet formerly it was an important plant in the pharmacopoeia and food repertoire of the Wailaki, Yuki, Little Lake Pomo, and Yokia Pomo tribes. In *Plants Used by the Indians of Mendocino County, California*, Chesnut wrote that this plant "often completely covers whole acres of valley land in Round Valley, and is common everywhere in grassy openings in forests." Chesnut recorded multiple uses: the young leaves and stems were edible greens, the seeds made a pinole, and the resinous woody root was an emetic, a wash for sore eyes, and a poultice for sores and burns.[51] This plant was likely managed through burning. In 1918 the anthropologist Llewellyn Loud recorded the following about the area in which *Wyethia longicaulis* grew: "Within the forests, at all elevations from sea level to the tops of ridges, there were small open patches, known locally as 'prairies,' producing grass, ferns and various small plants. These prairies are too numerous to mention in detail. . . . Most of these patches if left to themselves would doubtless soon have produced forests, but the Indians were accustomed to burn them annually so as to gather various seeds, especially a species of sunflower, probably *Wyethia longicaulis*."[52]

Pringle's yampah (*Perideridia pringlei*) is classified as uncommon, yet at one time its tubers were dug in great quantities by the Kawaiisu, contributing starch and protein to their diet. Digging these tubers with a digging stick aerated the soil, which increased its moisture-holding capacity and prepared the seedbed, likely heightening seed germination rates. This genus is known for its tuberous roots, which are a combination of root and stem tissue. It is possible that tuberous root fragments were left behind in the tillage process and each fragment grew into an individual plant, greatly enhancing the plant's asexual reproduction. The plant's habitat may also have been burned, as suggested by the burning of more common yampah species by the Chukchansi Yokuts.[53]

The restoration and management of other rare and endangered plant species, such as the three-bracted onion *(Allium tribracteatum)* and Small's southern clarkia *(Clarkia australis),* might also benefit from applying indigenous knowledge. While we have no evidence that these species were utilized by indigenous people in California, we do know that closely related species were gathered and managed with burning.[54]

LANDSCAPE HETEROGENEITY AND BIODIVERSITY

Native peoples played a major role in the maintenance and enhancement of biological diversity by introducing disturbances that created and maintained mosaics of different vegetation types. These disturbances, caused mainly by burning, were carried out specifically to maintain populations of plants that were gathered for food, cordage, basketry, and other uses and to enhance their quality. Thus traditional gathering, practiced holistically as both gathering and management, has the potential to promote biodiversity and restore communities to their formerly more heterogeneous conditions.

Although some open areas are "naturally" created through windfalls, lightning fires, avalanches, and other disturbances, California Indians maintained additional open patches through direct cultural intervention. These patches ranged from fine-grained smaller openings or areas in chaparral, woodlands, and forests to larger, coarse-grained patches the size of whole communities (e.g., montane meadows, valley grasslands, coastal prairies). The fine-grained patches of favored plant species—yampah, deergrass, and clovers in meadows and oak woodlands; farewell-to-springs, lilies, mule ears, and sanicles in lower montane forests and chaparral—were managed at the population level with the introduction of Indian-set fires to create a density and size necessary for meeting the material demands of the particular village for food, medicine, or manufacturing of cultural items. The maintenance of many patches of different plant communities exponentially multiplied the boundary areas, or ecotones, between them, which typically support the very highest biodiversity.

The ethnobotanical uses of different tribes shaped the vegetation, dictating the patch type, frequency, and size. By documenting the plant species, the pervasiveness of the plant in the material culture, the number of plants per item, and the plant configurations on remaining traditional gathering sites, it may be possible to reconstruct former vegetation patterns and provide the data reserve managers and ecologists need to accommodate these indigenous industries.

California's Cultural Wealth

The majority of the traditional ecological knowledge concerning former plant and animal species that occupied specific habitats and former resource management practices and disturbance regimes is not housed in California museums or library archives but rests in the memories of elderly California Indians. This knowledge is passed down solely through oral tradition. With each passing generation, this knowledge base dwindles. Those who have the most knowledge are now in their sixties, seventies, eighties, and nineties. They are our strongest link to the Indian era—the twelve thousand years during which humans shaped the ecology of California's wildlands. Each year more elders die having passed along only a small portion of their rich and detailed knowledge.

The loss of this knowledge has been likened to the burning down of the great library in Alexandria sixteen hundred years ago.[55] But the loss of written knowledge is not the most apt comparison, because knowledge in books is fixed. Knowledge living within cultures is reshaped with new information, is learned by direct apprenticeship, and is multidimensional—impossible to capture completely on paper. Thus, even if ethnographers and ethnobotanists were to accurately record every bit of knowledge possessed by Indian elders, it would not preserve indigenous culture nearly as well as fostering its living practice and its transmission across generations.

We are not only losing native condors and clovers but also this invaluable cultural wealth—the vast experience of native cultures in how to use, manage, respect, and coexist with other life forms and assist their cycles of renewal. Even though this knowledge is the true gold of California, the preservation of indigenous knowledge is not supported nearly as well as the preservation of plants, animals, and ecosystems.

Our rich natural heritage is nevertheless linked with the historical use and stewardship of vegetation by native people. Time-tested horticultural techniques, such as burning, tilling, weeding, and selective harvesting practices, profoundly shaped California's ecosystems over thousands of years without appreciably diminishing its variety of plant and animal life. Moreover, there are links between the diversity and vitality of native cultures and the diversity and vitality of the biota in California. The diminishment of one adversely affects the other. Thus the battle to preserve animals, plants, and ecosystems and the struggle to save native cultures are deeply intertwined. The ancient land management systems of the California Indians can provide Western society with models for successful, long-term human interventions in the natural environment.

CHAPTER 12

Restoring Landscapes with Native Knowledge

This knowledge [traditional ecological knowledge] represents the clearest empirically based system for resource management and ecosystem protection in North America.

WINONA LADUKE, Anishinabe (1994)

In California and elsewhere in North America restorationists often assume that the ecosystems they are trying to restore are completely "natural"— that their structures and functions were formerly self-sustaining, maintained through natural processes, and that the key to restoring them is returning them to their natural condition. For example, the ecologist Donald Harker and his colleagues write about the importance of "naturalness," which they define as "the degree to which the present community of plants and animals resembles the community that existed before human intervention." Similarly, ecological restoration has been defined as "the process by which an area is returned to its original state prior to the degradation of any sort, i.e. back to a fully functioning self-sustaining ecosystem."[1]

In the context of California at least, there is clearly some conflict between this view and what has been described throughout this book. If restoration is aimed at returning ecosystems to the condition in which they existed before Western settlement degraded them, then that condition is surely not an entirely natural one. As we now know, many of the classic landscapes of California—coastal prairies, majestic valley oak groves, montane meadows, the oak–meadow mosaic of Yosemite Valley—were in fact shaped by the unremitting labor of generations of native people. Moreover, these and other communities were managed intensively and regularly by these people, and that many have disappeared or changed radically in the absence of management shows they were not self-sustaining.

What, then, should be the goal of ecological restoration? Restoring landscapes and ecosystems to a "natural" condition may be impossible if that natural condition never existed (at least not in the last ten thousand to twelve thousand years). Restorationists must at the very least acknowledge the indigenous influence in shaping the California landscape. This chapter advocates an additional step—using indigenous peoples' knowledge and methods to carry out the restoration process, to return landscapes to historical conditions and restore the place of humans in their continuing management.

Active Management and the Role of Indigenous Knowledge

Land managers are now realizing that many lands require an active management stance to maintain ecosystem integrity, rather than simply protection from degradation. The idea of active management on public lands in the United States can be traced at least as far back as the 1963 Leopold Report to the National Park Service. In recommending that each large national park restore "the biotic associations [and] conditions that prevailed when the area was first visited by the white man," the report's authors acknowledged that "most biotic communities are in a constant state of change due to natural or man-caused processes of ecological succession" and that in these successional communities "it is necessary to manage the habitat to achieve or stabilize it at a *desired* stage."[2]

The authors of the Leopold Report recognized that many habitats at earlier stages of succession were disappearing, not only in California, but in other parts of the country as well, and that they harbored unique plant and animal life. These habitats would have to be actively managed rather than passively protected to maintain their character and form. It was recognized that these habitats and communities, which included meadows, grasslands, and certain types of woodlands and forests, were created through fires, floods, or windstorms and did not receive enough natural disturbance to be perpetuated. The report stressed that maintaining and restoring these habitat types would depend on restoring the *processes* that shaped them— with fire being the most important management tool in the park manager's tool kit.

The ancient human role in creating and maintaining these landscapes was not well understood in 1963, but that is not the case today. Knowing how historic landscapes came to be and how they were maintained by indigenous peoples, we can develop restoration programs with a better chance of success and a greater level of historic authenticity.

INDIGENOUS MANAGEMENT REVISITED

In coming to understand the nature and extent of indigenous management in California, we have had to dispense with a number of erroneous preconceptions. Anthropologists have long seen the cultural environments shaped by Native Americans as limited to areas along river bottoms and adjacent to village sites that harbored domesticated plants such as corn, beans, and pumpkins. These were areas that were visually obvious to early settlers, missionaries, and explorers and, later, were discernible through the methodologies used by archaeologists and ecologists. These areas also somewhat resembled lands subject to forms of management familiar to non-Indians, including land clearing, planting in rows, and selection of one or a few favored domesticated species. The wildlands beyond the agricultural fields were viewed as "pristine." The major focus in anthropology has been on plant manipulation for *food* viewed in isolation, not in the broader context of prehistoric subsistence systems and how these systems fit within and influence dynamic and diverse forest, woodland, shrub, and grassland ecosystems.[3]

In the past two decades, however, findings from diverse disciplines—ethnobotany, pyrodendrochronology, paleoecology, archaeology, ecology—have deepened our understanding of pre-Columbian California landscapes and the significant and complex role the approximately 310,000 Indians from five to six hundred tribes played in shaping them. As detailed in the previous chapters, new evidence from interdisciplinary research has substantiated that human manipulations such as burning and tillage were regular, constant, and long-term activities that produced cumulative and possibly permanent effects in plant associations, species distributions and composition, and, perhaps, gene pools and genetic structures of the species and plant assemblages found in many parts of California. Some of the precontact ecosystems in California can be labeled anthropogenic, in that they were shaped by human activities and are not self-sustaining in the absence of these activities. This is evident in certain patchy landscapes (e.g., deergrass [*Muhlenbergia rigens*] in chaparral) and in ecosystem types such as montane meadows, coastal prairies, and desert fan palm oases.[4]

This new information about the relation between indigenous management and the land has important implications for anyone who wants to restore a piece of land. First, an analysis of former indigenous cultural practices and their influences on a landscape provide a baseline of information about the diversity, dynamics, and functioning of plant and animal communities under former indigenous disturbance regimes (i.e., a reference ecosystem). Second, such an analysis helps the restorationist to better understand the natural *and* cultural processes that created the historic ecosystem and landscape being restored.[5] Third, completing such an analysis helps to define project goals that are historically more accurate. Fourth, the study of former indigenous land uses and management practices provides restorationists with cultural models of human intervention in nature—models that combine a concern for the coexistence of many life-forms with management for diverse products. Most important, these studies open up opportunities for rekindling indigenous land relationships by applying the scientific knowledge of native peoples.

The Society for Ecological Restoration (SER) International has recognized that indigenous peoples and their land management practices should be part of any serious effort to restore and preserve ecosystems. The Indigenous Peoples' Restoration Network (IPRN) was formed at the 1995 SER International annual meeting in Seattle. Its mission statement reads as follows:

> The IPRN has two related central goals: to use the tools of ecological restoration to enhance the survival of indigenous peoples and cultures, and to incorporate the knowledge of these cultures into emerging models of ecosystem management. The IPRN aims to establish a mutually beneficial working relationship with traditional indigenous tribal and community groups needing technical and financial assistance for land restoration, and to encourage the empowerment of grassroots community activities and leaders in local efforts to implement their own vision of sovereignty and ecological restoration within their unique cultural, social, economic, and spiritual traditions.

Indigenous knowledge and ecological restoration were also addressed more recently in the SER *Primer on Ecological Restoration:*

> Many cultural ecosystems have suffered from demographic growth and external pressures of various kinds and are in need of restoration. The restoration of such ecosystems normally includes the concomitant recovery of indigenous ecological management practices, including support for the cultural survival of indigenous peoples and their languages as living libraries of traditional ecological knowledge. Ecological restoration encourages and may indeed be dependent upon long-term participation of local people.[6]

The art and science of the new field of ecological restoration, therefore, will necessarily integrate traditional indigenous ecological knowledge into its philosophies, methods, and practicums.

REESTABLISHING RELATIONSHIPS WITH THE EARTH

Restoring land to a hypothetical historic state is rapidly becoming important for enhancing and maintaining biological diversity throughout the world. Although passive preservation may have its place in conservation, the biggest challenge for the human species is to find ways to use nature without destroying it. This means moving "beyond preservation," as William Jordan and other authors advocate, by creating a real working relationship with the landscape.[7]

Native Americans had, and continue to have, a highly participatory relationship with nature, one that restorationists can experience as well. For indigenous cultures, intimate interaction with plants and animals comes from the rituals of prayer, offerings, songs, and ceremonies and through keen, long-term observation, communication, stewarding, and judicious use. Native elders repeatedly remind non-Indians that we will not learn to live compatibly with nature simply by locking up lands in large tracts of restored wilderness.

Spending time with a Yurok fisherman on the Northwest Coast, a Pomo weaver in Ukiah, or a Yokuts saltgrass collector in Visalia, one learns that indigenous people achieve deeper intimacy with nature by *using* it. The character of each plant takes on new dimensions when transformed by human hands through scraping, skinning, soaking, peeling, boiling, mashing, grinding, fire hardening, splitting, and decorating. An elk horn is used to make a spoon; the wolf moss on an old oak serves as a dye; fibers of yucca leaves make paintbrushes; the sharp end of a nutmeg leaf serves as a needle. Each time a person transforms a plant, animal, or mineral into a useful item it is an acknowledgment that he or she belongs to a place.

Although nature and culture are most profoundly united through the harvesting of plants and the hunting of animals, use is a double-edged sword. Once humans harvest nature there is the possibility of overuse. Removing key elements from nature means the possibility of ecological degradation, but it can also lead to deeper understanding and awareness of nature's processes and limits. Removing elements from natural systems with thoughtfulness and respect, one finds oneself asking questions that address the complex interplay between resource production and the conservation of biological diversity.

Judiciously harvesting, crafting, and using products from nature continue to be the three cornerstones that keep Indian relationships with nature alive, rich, and sustainable. In traditional cultures, there are connections between the plant growing in its environment, its creative transformation into useful items, and the use itself. Today we have taken these three connected practices and made them into the distinct realms of science, industry, and private life.

The postmodern world's disconnection from nature causes us to tend to think about human–nature interactions in terms of the two diametrically opposed extremes discussed earlier—leaving nature alone or destroying it.[8] Clear-cutting, urban development, overgrazing of livestock, and mining, all of which cause profound alterations in the natural environment, fall at one end of the spectrum of interaction; protected wilderness falls at the other end. But these are only the extremes. In between there is a continuum of resource-utilization systems involving a wide variety of techniques and varied ideas about nature and about human relationships with it. It is somewhere in the middle of this continuum that human interaction with nature is most complex and sophisticated. This characteristically involves use and disturbance but rarely pushes the habitat beyond its capacity for natural regeneration and often increases both biodiversity and the abundance of plants and animals. Indigenous cultures provide examples of this throughout California, and it is this middle ground that the restorationist also inhabits, both in the sense that he or she is trying to re-create complex systems that arise from intimate interaction between nature and culture and in the sense that in doing this he or she is beginning to create a culture of place—a culture that is capable of interacting with nature without destroying it.

Applying Indigenous Knowledge to Restoration Projects

Eric Higgs, former chair of SER International, writes in *Nature by Design*, "Ecological restoration is the best hope we have not only for repairing damage but also for creating new ways of being with wild processes."[9] The templates for judicious use of the wild can come from the rich human history in California. Native plant cultivation techniques practiced for millennia by California Indians can become important methods for rejuvenating plant communities. Using ethnobiological research to reconstruct California Indians' interactions in specific areas will make a significant and lasting contribution to our understanding of indigenous land uses and their ecological influences while yielding a set of management techniques that can be used in the restoration and ecosystem management of these lands. This work, in its rediscovery of the linkages among restoration, maintenance of field populations of specific species, and judicious use of nature, will create exciting new models of human interaction with nature different from the restore-and-leave-nature-alone models that currently exist. These models can be called indigenous management models.[10]

Restoring landscapes that existed in the past requires knowing four things about them: the species that composed them, the techniques indigenous people used to harvest their useful plants, the ways in which indigenous people used fire and other means to shape them, and the basic economies and rituals of the people who lived there. Discovering this information is complicated and, to achieve historical fidelity, requires multiple lines of evidence. Fortunately, the disciplines of archaeology, anthropology, ethnobiology, paleoecology, pyrodendrochronology, and history can assist restorationists to more accurately reconstruct the former interactions between indigenous people and the natural environment. For example, pollen studies from meadows or bogs near archaeological sites can register not only the local floristic composition but also episodes of disturbance or replacement of vegetation such as one might expect with human intervention. Sediment columns from meadows can record disturbances, including fire. Archaeological sites reflect past dynamic relationships between people and their environs. Studies of historic vegetation changes on abandoned late prehistoric sites might elucidate precontact environmental manipulation.[11]

RECONSTRUCTING SPECIES COMPOSITION

Thomas Fleischner, professor of environmental studies, reminds restorationists that "it is hard to get somewhere if you don't know your starting point."[12] Restoration of a historic landscape starts with learning about its biotic components: the species of plants, animals, and fungi that lived in it and influenced its structure and functioning as an ecosystem.

If areas to be restored have been subjected to modern land uses and management, the probability is high that many of the plant and animal species that once resided there have vanished or occur in drastically reduced numbers and densities. Thus, gaining an accurate picture of a historic biological environment means inventorying the species that are present now, as well as determining which species are missing. Some of these species may lie dormant in seed and bulb banks in the soil and may require only the reintroduction of light surface fires or some other missing human or natural disturbance. In other cases, the propagule bank is gone, and it is necessary to learn the identity of the species so that they can be reintroduced.[13]

Sources of Information

There are various sources of information that can be explored to determine indigenous land management practices. In some cases, the species composition of historic landscapes were recorded or collected and can be found in herbaria as actual specimens and in museum collections as components of cultural objects. They can also be discovered by reading and analyzing historical public land surveys, historical literature describing the flora and landscapes of early California, and ethnographies (i.e., the early descriptive monographs on Native American cultures), and by gleaning the memories of Indian elders. In addition, examples of historic landscapes or communities may still exist relatively unchanged and used as reference sites, and, if we study them carefully, we may be able to infer that the species that exist there now could have occurred in the area to be restored. Using these resources, restorationists can often reconstruct relatively complete animal, fungal, and plant species lists for historic landscapes.[14]

Ethnographies and Ethnobiologies Harrington did extensive interviews with elders in many tribes in California between 1907 and 1957 and left more than a million pages of field notes. These notes are rich in descriptions of native plant and animal life in particular regions, as well as the various indigenous gathering and hunting practices, preparation methods, and uses for these species. He made large herbarium collections that are deposited at the Smithsonian Institution. Many of these collections have not been analyzed or correlated with Harrington's ethnobotanical field notes.[15]

In addition to published descriptive ethnographies, there is also an enormous corpus of published and unpublished tribal ethnobiologies. These often contain detailed information about plant, fungi, algae, mammal, reptile, bird, amphibian, fish, and insect species, their uses, and in some instances the locations where they were once gathered.[16]

Historical Surveys A variety of old geologic, zoological, and botanical surveys of California contain much important information about the former abundance and diversity of species in different regions. One example is the Pacific Railroad Surveys, whose findings appear in reports bound in multiple volumes that are housed in large libraries with special California collections. The volumes contain beautiful botanical and zoological plates. The value of this information for generating more complete species lists for habitat types cannot be underestimated.

For example, the surveys contain information about twining brodiaea *(Dichelostemma volubile)* that could be valuable for restoration projects in the Sacramento Valley, where twining brodiaea was once common. (See Figure 41.) The distinguished botanists John Torrey and Asa Gray wrote the following passage in their botanical report of 1856 about the twining brodiaea specimen that was collected:

> In rocky places, Knight's Ferry, Stanislaus River, May, (in flower and fruit); also at Sonora, Mokelumne Hill; Valley of the Sacramento,

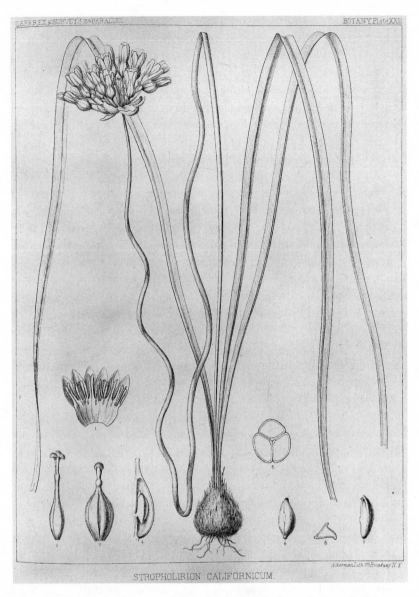

STROPHOLIRION CALIFORNICUM.

A.kerman Lith 379 Broadway N.Y.

Figure 41. Twining brodiaea *(Dichelostemma volubile,* formerly *Stropholirion californicum)* drawn by either Sprague or Riocreux from a botanical specimen collected by Dr. J. M. Bigelow during the Pacific Railroad Survey of 1853–54 under the charge of Captain Whipple.

Colonel Fremont, Mr. Rich, and *Dr. Stillman.* It is No. 1992 of Hartweg's Californian collection. A remarkable plant, of which we have had specimens for many years. It seems to be common in the Valley of the Sacramento. The tall stem, which is not larger than a crow-quill, and often more than 4 feet (Dr. Kellogg, of San Francisco, found it even 12 feet) in length, twines around other plants. In Dr. Bigelow's specimens they were on Calliprora [now *Triteleia*]. Not unfrequently several stalks are twined together.[17]

Elders' Memories Often, Indian elders, if taken to areas to be restored that lie within their traditional territories and asked questions about what the areas looked like when they were growing up, can provide valuable information. They may be able to tell the restorationist about the former diversity of species in the ecosystem and their arrangements on the land, including their associations, densities, numbers, and extensiveness in acreage.[18] Elders of the Western Mono tribe, for instance, taken to blue oak *(Quercus douglasii)* woodlands in the central Sierra Nevada foothills, have informed ethnobiologists that several kinds of edible mushrooms, now absent or in reduced numbers, grew under the blue oaks, that many different large patches of edible wildflowers, including chia *(Salvia columbariae),* blue dicks *(Dichelostemma capitatum),* Hall's wyethia *(Wyethia elata),* and farewell-to-spring *(Clarkia* spp.), grew in association with the oaks, and that these wildflower areas were periodically burned by the tribe.[19] This information may be very useful in restoring the blue oak woodland and helping blue oaks to regenerate. The formerly rich wildflower and grass layer, for example, was perhaps important in nutrient flow and blue oak tree regeneration. The diverse herb layer that existed may also have been important in energetic and trophic interactions, supporting a rich and varied fauna of butterflies, birds, and small mammals that is now largely absent. The mushrooms may have represented beneficial mycorrhizae important for the health of the oaks.

Present-day Remnant Vegetation Scattered through the state are areas that contain remnant native vegetation. These may occur along railroad rights-of-way, in areas such as parks in which most adverse human impacts have been excluded for many years, and in isolated patches that have survived by chance or intention. For example, it is still possible to find examples of perennial valley grassland, coastal prairie, and riparian woodland that bear varying degrees of resemblance to the communities that existed two hundred years ago. Species inventories of these remnant patches can be helpful in reconstructing the historical species composition of similar habitats elsewhere. Restorationists may also observe and record the structure of remnant trees, hydroperiods of wetlands, riparian areas, and vernal pools, as well as the way in which species are assembled in groups and/or edaphic continuums.

Remnant vegetation can also be a good source of seeds and plants for restoration projects. Disturbed land along roadsides that is not overrun by exotic species may support populations of early-successional-stage native species from which seeds or plants may be collected. (To the extent possible, seeds or plants should be collected from sites near the restoration site and where environmental conditions are similar. Moreover, collecting in such an area should be limited; overharvesting can harm existing populations.)[20]

Understanding Species' Requirements

To reconstruct the species composition of former communities, it is important to understand the autecology, or environmental requirements, of each species that may have been present. These include shade tolerance, moisture needs, and prerequisites for reproduction. Combined with knowledge about the likely environmental conditions in various parts of the former ecosystem, this information can help the restorationist to make more accurate determinations of each species' structural role or location in the former community.[21]

For example, in the mid-elevation montane forests of the central Sierra Nevada, various microhabitats occurred much more commonly in historic forests when Indians were frequently burning the understory. These human-maintained habitats were more open, with higher light levels, less duff, and less competition from brush, shrubs, and young trees. Knowing that such plants as *Bromus carinatus, Muhlenbergia rigens, Sanicula tuberosa, Madia elegans,* and *Wyethia elata* are relatively intolerant of shade, and that *Juncus effusus, Lupinus latifolius* var. *columbianus,* and *Pycnanthemum californicum* prefer mesic sites where the forest duff is thin, allows the restorationist to match these plants (all once used by the Western Mono) with the former environmental conditions of the mid-elevation montane forests. Examining these same forest areas today, one would find that much of the forest floor is deep in duff and litter and sparsely populated with shade-tolerant species such as *Goodyera oblongifolia, Sarcodes sanguinea,* and *Pyrola picta.*

Assessing the Former Importance of Species

It is particularly important that restorationists try to identify the key economic species of former ecosystems—those plants and animals that were gathered or hunted in large quantities and thereby shaped both the management of the ecosystem and its ecology. Artifacts in museums, such as arrows, nets, and baskets, could be analyzed for both species utilized and the number of plant parts needed to complete each individual item. From discussions with elders and non-Indian craftspeople, estimates could be made of the numbers of shoots, rhizomes, or flower stalks needed for each item, and from these the number of shrubs, ferns, and bunchgrasses needed. (See Figure 42.) The restorationist could then calculate the number of plants needed to meet cultural needs annually for an average-sized Indian village. This figure would give restorationists an estimate of the sheer quantity of plant material harvested from ecosystems and insights into the horticultural methods required to achieve sustained yields of this magnitude and the desirable characteristics of the various plants used in manufacture.[22]

Often, a clue to the importance of a species is the pervasiveness with which it was used. There are some species that were used by a number of tribes over wide geographic regions. This speaks to their long history of use, as it takes time for such information to diffuse between tribes. A plant species used over a large geographic region and across ethnographic-linguistic boundaries forms a "regional complex."[23] Use of chia for its highly desirable seed is an example of a "food complex." (See Map 3.) The widespread use of a species also frequently points to its reproductive capabilities: large quantities can be gathered, repeatedly, without hindering future productivity. It is important to identify these species because they had a critical role in the sense that elimination of the harvesting and management techniques used to perpetuate them would have had a major impact on community structure and composition, biological processes, and indigenous subsistence economies. For example, large milkweed *(Asclepias speciosa)* stands and wildflower *(Perideridia* spp., *Clarkia* spp.) patches managed by California Indians supported vast migratory populations of butterfly and other insect species that no longer exist in those numbers. These patches have been reduced to scattered, sparse populations. Only by reconstructing and restoring the populations of the plant species composing these patches to some threshold size can we then restore the migratory populations of butterflies and other species that utilized them.

Understanding species requirements and assessing their former importance must be tied to the actual planning and implementing of a restoration project. For example, knowing a species' environmental requirements helps the restorationist to plant it in the best location for success or to plant it at the best time in the life history of the ecosystem. Meanwhile, assessing the importance of a species (ubiquitous, common, conservative) could give the restorationist valuable information for developing seed mixes and planting materials lists, both in terms of the species to include and in terms of their density in the planting. Finally, a plant or an animal might be "used" by

A Woodwardia fern stems   **D** Hazelnut sticks

B Maidenhair fern stems   **E** Red willow roots

C Bear-grass

Figure 42. Fanny Smoker, a Yurok, with a year's supply of hazelnut sticks, woodwardia fern stems, maidenhair fern stems, coils of bear-grass, and coils of red willow roots for basketry. These materials came from burned or pruned plants. Reconstructing indigenous influences on historic landscapes involves estimating the numbers of plant parts and plants required for various cultural purposes of a village. These estimates make it possible to develop more accurate models of the kinds of management systems that would have had to exist in the past to meet those requirements. Courtesy of the Phoebe Apperson Hearst Museum of Anthropology and the Regents of the University of California. Photograph by B. F. White, 1930, #15–9020.

Map 3. California Indian tribes
that utilized chia *(Salvia columbariae)*
for their edible seeds. This is an
example of a regional food complex.

Within the map:

Tolowa
Chilula Yurok Karuk
Chimariko
Wiyot Hupa
Whilkut
Mattole Wintu
Redding
Nongatl Lassik
Sinkyone Wailaki
Cahto Yuki
Yuki

Shasta
Klamath
Modoc
Alturas
Pit River
(Achumawi)
Yana Atsugewi
Northern
Paiute
Nomlaki
Yahi
Mountain
(Northeastern)
Maidu
Chico
Concow

Pomo
Lake
Miwok
Wappo
Coast
Miwok
Patwin
Nisenan
Sacramento

San Francisco
Plains
Miwok
Sierra
Miwok
Washoe
Mono Lake
Paiute

Ohlone
(Costanoan)
Stockton
Sonora
Yosemite

Valley
Yokuts
Bishop
Owens Valley
Paiute
Mono
(Western Mono)
Esselen
Fresno

Timbisha Shoshone
(Panamint)
Salinan

Foothill
Yokuts
Tubatulabal

Kawaiisu
Chumash
Kitanemuk
Santa Barbara
Tataviam
Barstow
Southern
Paiute
Serrano
Island
Chumash
Tongva
(Gabrielino)
Los Angeles
Chemehuevi
Mojave
Tongva
(Gabrielino)
Ajachmem
(Juaneño)
Luiseño
Palm Springs
Cahuilla
Cupeño
Halchidhoma
San Diego
Kumeyaay
(Diegueño,
Ipai, Tipai)
Quechan
(Yuma)

Legend:
☐ California Indian tribes
that utilized chia *(Salvia
columbariae)* seeds for food

people living outside the plant's or animal's natural range via trade or commerce. Therefore, it is important to study a tribe's consumption of a plant or animal not only for its own use but also for trade with other tribes.

RECONSTRUCTING AND REINTRODUCING HARVESTING STRATEGIES

Continual human intervention in the form of judicious harvesting may be a necessary part of maintaining ancient, species-rich ecosystems that were shaped by Native Americans over millennia. Plucking tules from wetlands, picking savory fruits from shrublands, or digging bulbs and tubers from meadows and riverbanks—if done judiciously—may be beneficial to the long-term health of particular ecosystems and species. For this reason, an additional step in reconstituting a historic landscape or community is to reconstruct and reintroduce the indigenous harvesting techniques known to have existed in the area.

Restorationists and ethnobiologists, in collaboration with native people, can uncover examples of different kinds of plants and animals that were sustainably harvested in an area that is the target of restoration efforts. For each of these species, they can rediscover the cultural rules and prescriptions that regulated harvest and the on-the-ground harvesting variables—seasonal timing, frequency, intensity, and long-term patterning—associated with that species and its use. These harvesting variables, discussed in Chapter 4, together form a "harvesting regime."

Restorationists need to know how successful a particular harvesting regime was in maintaining a plant population and how quickly densities returned to an equilibrium after harvesting. For example, were there rest periods that allowed for population recovery? Was there an attempt to keep the population density above a certain threshold? Were the technologies chosen appropriate? Are there new technologies or tools that could be used to appropriately harvest the crop? Harvesting strategies that demonstrate compatibility between use and conservation should be favored by the restorationist.

For the most part, there is limited information on harvesting practices in the historical and ethnographic literature. While there are some native people gathering plants and animals in various regions today, gaining access to their knowledge will take long-term concerted effort and trust building on the part of restorationists. In addition, the techniques they use may differ from those employed in precontact times, when conditions were different, more people participated in harvesting, and larger amounts needed to be harvested.

To reintroduce harvesting, special gathering arrangements can be made between restorationists, native peoples, and the managers or owners of the land being restored that allow native people to continue their gathering traditions and co-manage the land.[24] In some cases, ecosystems will require long-term harvesting, which demands the restoration of ecological associations between people and wildland environments.

Volunteers who live in the vicinity of the restored area could assist with periodic harvests of different plants. There is a danger that harvesting the land could quickly become a mere mechanical exercise for non-Indians, devoid of any lasting meaning, unless individuals and communities close to restored areas develop their own connections and contexts with the land beyond the mere borrowing of techniques from native cultures. One way to do this is to integrate the harvested plant material into the daily lives of the local residents through activities that fulfill their artistic, culinary, and spiritual needs. Examples include making a pie with wild fruits gathered locally or making a soup with locally harvested mushrooms.[25]

It is even possible that limited, controlled collecting for commercial purposes can have a place in restoration projects. Some commercial collectors take all the plants of an area without sparing any individuals, and they leave no seeds, rhizome fragments, bulblets, or cormlets for future production. But it is possible for collectors to follow indigenous guidelines and thus benefit the population being collected. A historical example is Carl Purdy, who started harvesting geophytes for the horticultural trade in the 1870s and replanted the bulblets and cormlets of the bulbs and corms he harvested, a strategy he probably learned from the Indians. Purdy seemed to be attuned to the reproductive biology of geophytes and had this to say about sustainable harvesting: "When we dug the bulbs [Lilium pardalinum], we selected salable ones, and then carefully replanted the small bulbs. In three years or so we could return for another crop. There were beds which we had cropped for fully thirty years before this resource for the bulbs failed." The bulbs failed, according to Purdy, not from harvesting pressures but rather from the "changed conditions" brought on by the suspension of burning: "In the earlier years brush fires were periodic and these fires opened up the too shady patches and fertilized them. Brush fires became much less frequent, and many beds were shaded out and died."[26]

In general, before Westerners can begin to harvest native plants judiciously in wildland environments, more knowledge has to be acquired about the life cycles and biology of the native species of interest—to develop the principles and methods that are the basis for long-term sustainable harvesting and that also allow for the coexistence of many other species. This will require the study of measurable parameters and indicators connected with sustainable harvesting and management. This could be accomplished through conducting well-designed scientific experiments that rigorously explore the ecological effects (at different scales of biological organization) of various kinds of harvesting regimes.

Collecting local seeds and other propagules for use in restoration projects is an important activity in terms of making the restoration both more economically affordable and genetically authentic. It is also an important way to make connections and establish social bonds with like-minded people.

RECONSTRUCTING AND REINTRODUCING DISTURBANCE REGIMES

Disturbance regimes—both anthropogenic and nonanthropogenic—typically interact with ecosystems and affect ecosystem processes, structure, and composition at various scales. Ecologists have studied disturbance processes caused by nonanthropogenic sources, such as lightning, droughts, hurricanes and other wind events, insect outbreaks, floods, and volcanoes, and their effects on ecosystems and natural areas.[27] Some ecologists have also explored the effects of Indian-set fires on ecosystems, although there remains a lively debate over the magnitude of those effects.[28] Almost no one has studied the effects of other Indian-caused disturbances on ecosystems, namely, cutting and pruning, harvesting and digging, hunting, moving of species, and developing preferred strains of species. This lack of studies is indicative of an inability by even those ecologists who claim to be interested in whole-system studies to consider human disturbances that aided ecosystems, such as burning under oaks to destroy insects that might damage the acorn crop, in their experiments. The tendency is to study only those natural disturbances or disturbances, such as clear-cutting, that destroy segments of or entire ecosystems.

Ecologists' definitions of disturbance also indicate a temporal dimension. As noted in Chapter 1, in *The Ecology of Natural Disturbance and Patch Dynamics*, White and Pickett define disturbance as "any relatively discrete event in time that disrupts ecosystem, community, or population structure and changes resources, substrate availability, or the physical environment." In a more inclusive manner, the Forest Service has defined disturbance as "a discrete event, either natural or human induced, that causes a change in the existing condition of an ecological system."[29] The authors of both of these definitions visualize disturbances as only relatively large scale, forceful, onetime events that cause significant change within an ecosystem.

I would like to suggest that this definition should be expanded to include

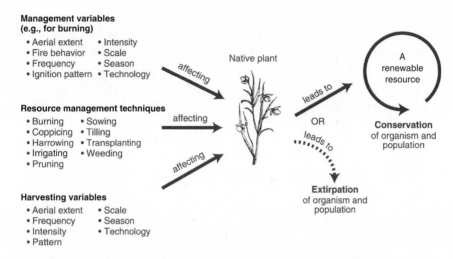

Management variables (e.g., for burning)
- Aerial extent
- Fire behavior
- Frequency
- Ignition pattern
- Intensity
- Scale
- Season
- Technology

Resource management techniques
- Burning
- Coppicing
- Harrowing
- Irrigating
- Pruning
- Sowing
- Tilling
- Transplanting
- Weeding

Harvesting variables
- Aerial extent
- Frequency
- Intensity
- Pattern
- Scale
- Season
- Technology

affecting

Native plant

leads to

A renewable resource

OR

leads to

Conservation of organism and population

Extirpation of organism and population

Figure 43. A template for documenting California Indian disturbance regimes.

disturbances, especially those created to support indigenous economies, that tend to be smaller scale, seemingly low-impact activities that occur with enough frequency to cause permanent changes in species populations, ecosystem structure and function, and landscape appearance.

The key assumption here is that native species in many ecosystems evolved under conditions of frequent and regular natural and human disturbance and require such disturbance for optimal growth, reproduction, and survival in a particular area over the long term. Thus bringing back the kinds of disturbance that likely occurred under indigenous management may be a prerequisite for restoring the former biodiversity of an area.[30]

Restorationists will need to rediscover not only the array of resource management practices (e.g., burning, pruning, weeding) utilized in the type of ecosystem that is being restored but also the disturbance variables associated with the application of each of these techniques (e.g., season, frequency, intensity, and extent). Together, these variables form a disturbance regime for a specified species of plant or animal in the context of its community. (See Figure 43.)

Understanding the Goals of Indigenous Disturbance

In some cases, it will be possible to determine details of the goals native people strove for in managing the landscape. These cultural objectives— producing nonwormy acorns, increasing the numbers of young sprouts for basketry, encouraging young, palatable forage for wildlife—can then form the basis for determining the "desired future conditions" of the landscape being restored. Important information about former management goals can be obtained by asking native people who are living near the area to be restored how their elders managed the natural environment. In their investigations, restorationists will find that frequently an ecosystem was managed for multiple objectives because typically its plants and animals had myriad cultural uses. This complex management regime was reflected in the diverse landscape patterns of aboriginal California.[31]

Closely tied to the former management objectives for an ecosystem were the food and material culture of the tribe that had stewardship of the area. Knowledge of the tribe's food and material culture, which can be a critical aspect of the reconstruction of disturbance regimes, can be acquired from ethnographic materials and from a study of the manufacturing techniques of cultural artifacts housed in museums. The important idea is that indigenous people manipulated landscapes to enhance not only the quantity of desirable plants but also the *quality* of their parts, as manifested in their size, shape, color, length, diameter, taste, nutritional value, and other parameters.

Reintroducing Disturbance

The reintroduction of former disturbance regimes—particularly those involving fire—can present a chicken-or-egg dilemma. On the one hand, a fire or other major disturbance may be the only practical means of re-creating the former structure and composition of the community. On the other hand, at the beginning of the restoration process fire would likely behave very differently than desired and have many adverse ecological effects, because the structure of the vegetation, the fuel types and amounts, and the plant and animal assemblages are often very different from what was once present. The solution to this problem is to initiate the restoration with a series of intermediate steps, especially thinning to reduce fuel loads and change the vegetation structure and reseeding to restore the former populations of herbaceous species.

The potential ecological effects of reconstructed indigenous disturbance regimes at different scales of biological organization could be initially investigated without setting up experiments. Let us take a real-life example. The Western Mono historically burned under California black oaks every one to three years, usually in October, November, or December, to reduce an insect pest that feeds on acorns. Existing knowledge of the insect's life cycle and its habits could be compared with details of the reconstructed indigenous fire regime to assess the probable effectiveness of the reintroduced burning in breaking the pest's life cycle.

Figure 44. The experimental approach to testing the ecological effects of indigenous harvesting and management practices. *(Top)* An experimental harvest of blue dicks *(Dichelostemma capitatum)* corms with a digging stick at the University of California Albany field station, which simulates indigenous harvesting regimes designed to determine the growth response of blue dicks to these regimes. *(Bottom)* An experimental burn in the Sierra National Forest that simulates indigenous burning to increase flower stalk production of deergrass *(Muhlenbergia rigens)* for basketry. Photographs by Kat Anderson.

Because the information for reconstructing disturbance regimes is highly fragmented, ecological field experiments could also be conducted to investigate the probable environmental impacts of various kinds of practices. The restorationist, with assistance from statisticians and quantitative ecologists, could design treatments to test the ability of various practices to produce desirable effects in particular plant species—"desirable" in the sense that they produce plant materials exhibiting the qualities and quantities useful to native people.[32] These experiments could reveal how production systems might be differentially intensified so as to maintain biological diversity in other parts of the system and how increased exploitation of certain species affects other species and general ecosystem properties. (See Figure 44.) Restorationists could use some of these tested techniques in the restoration of areas set aside for indigenous gathering, for demonstration purposes, or for the general purpose of restoring and maintaining biological diversity.

Restoration is a long-term process. It may require years of thinning, controlled burning, reseeding, removal of exotics, soil stabilization, and other work before an area is returned to a condition in which natural processes and indigenous-style disturbance and harvesting can sustain it. Over the long term, a restored ecosystem's integrity will depend on reestablishing the mutually beneficial ecological associations between it and people.

Restoration and the Conservation of Indian Cultures

Western culture's indebtedness to native cultures is immense. Many plant-derived pharmaceuticals were first discovered by traditional healers in indigenous societies, and many of our food crops were domesticated or partly domesticated through indigenous horticultural activities. Modern genetics has often tapped indigenous peoples' intricate knowledge of minute and often hidden differences in species. In California specifically, Indians often acted as guides for early overland parties of Euro-Americans, and many modern highways follow ancient Indian trails. There are many instances of Indians sharing food with European newcomers and teaching them how to harvest and use the flora and implement agricultural techniques. Furthermore, early Euro-American ranchers, trappers, farmers, and loggers built their fortunes on the vast biological wealth that was stewarded by California Indian tribes.[33]

Native knowledge will further contribute to modern society by helping us to restore our "wild" landscapes. However, if restorationists use traditional ecological knowledge without concern for the well-being of the cultures in which the knowledge was fostered, it may be a hollow effort. The use of native knowledge detached from its source and rich context (termed "cultural appropriation" by anthropologists) can be meaningless and exploitative, like taking a sacred symbol and placing it on a T-shirt.[34] One way to avoid this kind of exploitation is to make sure that restoration and related activities proceed with a concerted commitment to the conservation of indigenous cultures. Below are examples of what can be done.

- Many botanic gardens and arboreta are already involved in the conservation of rare taxa through their propagation and cultivation. These organizations could begin to establish relationships with California Indian groups of the region and grow culturally important plants that are currently dwindling on reservations, rancherias, and public lands. These plants could be reestablished on traditional gathering sites, on both public and tribal lands. As an example from outside California, Navajo Nation community members have joined together, with the help of the Desert Botanic Garden in Phoenix, to restore the Navajo sedge *(Carex specuicola)* on tribal lands.

- Natural history museums can reorient their exhibits to portray "wild" California as an inhabited land with a lengthy and rich Indian land management history. Cultural museums can develop exhibits that more accurately represent the continuity of California Indian cultures as vibrant and living, instead of focusing on the past. They can begin to invite guest Native American curators to plan and design exhibits and thereby have a major voice in interpreting their own history. Some museums, such as the Grace Hudson Museum in Ukiah and the Oakland Museum, have already begun to initiate symposia that bring together California Indians, government officials, and the general public to show that native cultures are not extinct or static.

- Restorationists could be involved in the development of teaching aids for environmental education for Indian children. These aids might include native plant gardens, plant collections with detailed ethnobotanical information, or photographs that connect ethnobotanical knowledge with plants. Financial donations could be given to native language programs and other cultural revitalization efforts.

- Restorationists and environmentalists could work with tribal leaders to establish programs for the reacquisition of lands within reservation boundaries that have passed out of Indian trust status;

or to develop prospects for Indian collaboration in land restoration and management arrangements on public and private lands.[35]

- Public agencies can become advocates of maintaining, tending, and encouraging growth of native plants important to Indian people. This would involve surveying the resources, recording their conditions and numbers, matching these data with indigenous needs, and reintroducing indigenous horticultural techniques such as fire. Thanks to education by members of the California Indian Basketweavers Association and others, many agencies have already begun to take on this more active role.

- Public agencies can develop programs for management of traditional plant resources for Native Americans as a component of public lands management programs. Under such programs, Indian gathering sites would have land use status equal to that of other land use categories. In other words, this category would have its own research and management funding within each of the agencies that administer our public lands.

- In silvicultural certifications, prescriptions could be written that would consider cultural values of indigenous people.[36]

Rekindling Human–Nature Relationships

Ecological restoration, by definition, has an interest in reconstituting landscapes from earlier times. By and large, these landscapes were richer in biodiversity than the remnant vegetation existing today. Restoring plants, animals, and the ancient cultural practices of native peoples to landscapes can be a way to honor our human predecessors and to make history highly relevant to present and future generations. Restoration is also a discipline of the present: it fulfills people's inherent needs to experience and heal nature, regardless of race, ethnic background, class, or gender. And last, it is a visionary field, offering a hopeful future in which it is possible to mend relationships with native people and create a world where humans can coexist with all of nature's denizens. It will accomplish these goals by offering a wider arena of choices for human–nature interactions than just destructive land uses *or* preservation of nature without humans.[37]

Landscapes are not just made up of assemblages of species; rather, they are expressions of species evolution and species behavior.[38] The adaptations of plants and animals that exist today are responses to past sequences of environmental conditions. In many cases, environmental conditions have been altered by the history of the land's use by humans. Therefore, past and present use of the earth must be understood in order to frame effective environmental policies, management regimes, and restoration strategies for the future. We must integrate both environmental and cultural information at a variety of temporal and spatial scales.

Restoration of historic landscapes will depend on the inventory of existing species and populations, the reintroduction of plant and animal species that are known to occur there but are now gone, and the reintroduction of harvesting and disturbance regimes that originally shaped the landscape. The rich indigenous history with the land still remains largely untold and is waiting to be rediscovered, acknowledged, preserved, and reenacted in landscapes across California.

Indigenous Wisdom in the Modern World

We are living today only because the generations before us—our ancestors—provided for us by the manner of their responsible living.

SIMON ORTIZ, Acoma (1998)

The word *tending,* as in the title *Tending the Wild,* is meant to encapsulate the essence of the relationship that the indigenous people of California had with the natural world in pre-Columbian times. It also suggests the timeless wisdom inherent in this relationship, wisdom that we sorely need today. *Tend* means "to have the care of; watch over; look after." Thus the word connotes a relationship of stewardship, involvement, and caring very different from the dualistic, exploit-it-or-leave-it-alone relationship with nature characteristic of Western society. The role of nature's steward was (and still is) one shared by many indigenous peoples. Calvin Martin notes that the Navajo, for example, "regard the world as utterly beautiful" but know that "beauty—*hózhǫ́,* they call it—unravels, succumbs to entropy. We were created, they say, to restore the beautiful."[1]

The word *tend* is derived from the root *ten,* meaning "to stretch." *Ten* is related to the ancient Sanskrit, Persian, and Greek words for "string," all of which imply tension—such as that of a properly tuned string on a musical instrument.[2] This sense of the word suggests that it will not be easy to find ways to interact with nature that allow both humans and nature to flourish. But it also suggests that the key to this process lies in achieving a creative, even tension between nature and culture, a tension that our human antecedents in California understood well.

There are many ways in which the wisdom of the tending relationship can have relevance in our modern society. Chapter 12 is devoted to one specific application of this wisdom—ecosystem restoration and wildland management—but there are other elements of indigenous wisdom with more general application. These are grouped below into five essential principles, or focal points.

1. *The ecological history of the land matters for management today, and indigenous practices are at the roots of this history.* As detailed in Chapter 5 and elsewhere, there are many areas in California where indigenous interactions with nature are an important part of the land's ecological history. The interruption of indigenous management in these areas has been in part responsible for decreased biodiversity, the loss of open habitats due to natural successional processes, and increasing wildfire severity. In the words of Nels Johnson, "The lesson here is that sudden departures from long-established human–ecological interactions can lead to yet another ecological transformation. This new ecological condition may bear little resemblance to a primeval state that disappeared when prehistoric people arrived."[3] Thus reintroducing indigenous management practices, or, more generally, practices based on the traditional ecological knowledge of the people indigenous to the area (as argued in Chapter 12), can be an effective way of restoring landscapes, conserving species, and enhancing biodiversity.

Using indigenous management practices may mean embracing paradoxes. Wendell Berry, farmer, poet, and essayist, reminds us of the essential paradox that Native Americans have known for millennia: "The natural forces that so threaten us are the same forces that preserve and renew us."[4] Fire, in particular, both destroys and renews vegetation. For much of Euro-American history on the North American continent, fire has been viewed solely as an enemy to life, yet today it is accepted among ecologists that fire plays a major role in the maintenance of many plant communities and that fire suppression creates its opposite extreme: catastrophic fire. The way to prevent catastrophic fires is to burn areas regularly, and to set fires during times when fires are more controllable, as the California Indians did. When used appropriately, then, fire becomes the instrument for its own subduing.[5]

2. *Humans can use natural resources to meet their needs without destroying those resources.* This idea is both a summarization of prehistoric indigenous management practices in California and a concept with relevance for wildland management today. With respect to basketry materials (Chapter 6), woody materials for arrows and other cultural items (Chapter 7), and food (Chapters 8, 9, 10), indigenous people in California developed ways of using natural resources to meet human needs without degrading the ecological basis for their renewal. Indeed, California Indians found it possible to be both users and benefactors of native plant and animal populations. As dis-

cussed in Chapter 4 and elsewhere, the keys to this tempered use of nature lay in a number of strategies and management guidelines:

- Through daily, firsthand observation, become intimately familiar with the needs, characteristics, growth, and reproduction of the plants and animals being used. (See Chapter 2.)

- Pass on and accumulate this knowledge from generation to generation. (Create and maintain a rich storehouse of traditional ecological knowledge.)

- Harvest plant materials in a way that does not interfere with the plant's inherent capacity for renewal. (Adjust the season, frequency, aerial extent, and quantity of harvest to the biology of the harvested species.) (See Chapter 4.)

- Use fire regularly because it clears out dead material and helps plants come back stronger. Dig bulbs, corms, and tubers as gophers do, leaving behind parts that will come back and make new plants. (In managing habitats and populations of useful plant species, mimic the processes of natural disturbance to which the plants are already adapted.) (See Chapters 4, 5, 6, 7, 9, 10.)

- Rely on a broad base of plants and animals from each habitat.

- Trust in the resiliency of nature. Know that life is a cycle of destruction and rebirth.

Native people's judicious use of nature illustrates an important, self-reinforcing dynamic: Tempered, sustainable use of natural resources demands responsibility, responsibility generates respect for nature, and respect for nature is a basis for sustainable use.

3. *Human prehistory is more complex than the simple dichotomy between hunter-gatherer and agriculturist would indicate.* The management practices of the indigenous people of California expand our understanding of the foundation on which agriculture rests. Practices such as burning for straighter, longer dogbane stems, replanting the bulblets of edible bulbs, and pruning hazelnut to create long sprouts for basketry, though not agriculture per se, were nonetheless much more purposeful than the opportunistic, haphazard gathering of hunter-gatherers. The tending, harvesting, and management practices of many California tribes constituted a form of protoagriculture (see Chapter 8) that helps us to understand how the very long transition between hunting and gathering and plant domestication and agri-

culture might have occurred in various parts of the world. It also helps us to draw a much richer and more complex picture of possible human–nature interactions. In North America, where plant domestication extends back only about five thousand years,[6] our growing knowledge of protoagriculture allows us to speculate about the ways in which the first humans on the continent made a living off the land before the advent of agriculture.

4. *Achieving sustainable use of the earth's resources will involve cultural changes as much as advances in knowledge and transformation of economies.* The lifeways of California Indians past and present show us that sustainable resource use and human coexistence with plants and animals are based not only on factual knowledge but also on cultural tradition, cultural values, and societal organization. Tempered gathering and caring for native plants is not solely a matter of following nuts-and-bolts prescriptions for when, where, and how to harvest and manage a plant or population; it comes from human motivations. Motivation is fostered within the culture itself—through art, legend, kinship systems, ceremonies, and its overarching worldview. The beliefs and values reinforced by and expressed through these elements of culture define human ways of being with and in nature, and they go hand in hand with the actual individual behavior that is acted out on the land.

The ways in which the indigenous cultures of California fostered beliefs and values consistent with nondestructive use of nature suggest principles that we can apply in effecting the changes in values, beliefs, and behavior that will be necessary for modern society to move toward sustainable resource use:

- Knowledge of the finiteness of natural resources must be accompanied by mores and values that limit and constrain resource use. As noted in Chapters 2 and 4, cultural rules such as "Do not needlessly kill life," "Do not waste resources," and "Do not take everything— leave some for tomorrow; leave some for the other animals" reflected an awareness that resources could be overexploited and were what ultimately regulated resource use in many areas.

- Nondestructive relationships with the natural world need to be reinforced through a variety of expressive and spiritual mediums. Throughout this book (especially Chapter 2) it has been stressed that just as important as the actual harvesting and management practices were the often unseen social and religious meanings and contexts, encoded in prayer, song, dances, offerings, and taboos, that motivated conservation of natural resources. These fed the human spirit and instructed people about right and wrong behavior and the position and obligations of each person toward nature.

- Social structure has a strong bearing on how societies relate to the natural world. As discussed in Chapter 8, the wealth-spreading devices operating in many California Indian societies, along with the lack of strong economic hierarchies, discouraged hoarding of resources and encouraged cooperation. The chiefs divided important harvests among families so that no one went hungry. In many tribes, when a hunter hunted and brought home game, it was distributed among all the village members. These formal cultural rules acted to ensure broad community access to the staples of the indigenous diet and prevented food sources from being raided and exhausted by a few families or powerful individuals. Moreover, norms of sharing and cooperation between many groups, and mutual respect for territorial limits, meant there was little fear of theft of food stores and, therefore, minimal competition for resources and less motivation to overexploit them.

- A society's view of humans' place in nature shapes collective behavior toward other species. The kincentric worldview of California Indians (see Chapter 2), wherein plants and animals are seen as blood relatives, fostered responsible treatment of other species because it established that they had an equal standing in the world and in fact had much to teach humans.

5. *Establishing more intimate relationships with nature and rooting ourselves in a place foster self-fulfillment and a more responsible stance toward the natural world.* Indigenous people show us, through example, the value of intimate relationships with the natural world. Their rich ties to the land, expressed in their culture and experienced directly through their senses, show us the foundations of a true culture of place. Underlining the importance of a strong connection to the natural world even today, the conservation biologist Edward Grumbine says, "Biological diversity will not be sustained if new ways of managing nature do not also transform how we experience our place in nature, how we manage ourselves."[7] Many aspects of California Indians' embeddedness in nature provide us with useful principles:

- Awareness of natural cycles—seasonal changes, migrations of animals, reproductive cycles—connects one with the pulse of the natural world. (See Chapter 2.)

- Creative expression linked to the natural world can bring nature more fully back into our lives. Drawing and painting natural scenes, singing and writing about nature—all inspire the heart, feed the intellect, satisfy the curiosity, and give spiritual sustenance to humanity.

- Working directly with materials one has collected from the natural world creates a direct, sensual link to nature and place. Making a musical instrument from a native hardwood, fashioning a fish weir from young shoots, or weaving a basket with sedge roots and deergrass stalks satisfied California Indians' inner urge for self-expression, reenacted hand movements familiar to their ancestors, and provided a means for intimately knowing the places of gathering. (See Chapters 6, 7.)

- A kin-based view of humans' place in nature (see Chapter 2) allows no room for alienation from nature. California Indians have always told non-Indians, "All of life is related—you and I and plants and wildlife are kin." Although some Westerners have adopted a bio-centric view of nature, in which all life "has a right to exist for its own sake," we still operate in a largely anthropocentric world wherein nature is viewed as "a stockpile of resources, lifeless matter-in-motion, a standing reserve for human appropriation."[8]

- By using nature, we begin to know our place in it. In other words, our investment in a place is deepened when that place feeds our bodies as well as our spirits and minds. For California Indians, nature was not an abstract concept relegated to the remote fringes of human communities but was intimately intertwined with daily living. Indigenous peoples' lifeways show us that intimacy comes from interacting with plants and animals *where they live* and establishing relationships with them. Through direct contact and harvest for the satisfaction of human needs, the separation of culture from nature that modern society creates can be mended.

- Living in the same place as one's ancestors establishes a multidimensional bond to that place. It has been noted throughout this book that it is tremendously important for California Indians' connection to a place to have the depth of time, for them to be able to point to a particular harvesting patch, shrub, tree, living site, or sacred spot and know that many generations before them used the same plants, walked the same paths, tended the same land. To many

California Indians, a gathering site is consecrated by gathering carried on over long periods. Each generation honors the ancestors, honors the customs, and honors the plants by using and tending them. The tremendous pool of intergenerational knowledge from which to draw is a blessing and a gift. In general, each of us is deeply dependent on the trials and accomplishments of many generations of humans before us.[9]

Aboriginal California is long gone, and we cannot bring it back. But the ancestors of present-day Indians have left us a multifaceted legacy: landscapes that still bear the unmistakable imprint of indigenous management, a priceless body of knowledge about native plants and animals and their uses and their responses to tending and harvest, techniques for managing the land to maximize biodiversity, a philosophy that sees humans' role in the environment as one of stewardship, and descendants who preserve many elements of their ancestors' ancient traditions. All these legacies have relevance today, despite and even because of the changes that have occurred since pre-Columbian times. The extent to which we listen to what they teach us may very well determine the chances for the long-term survival of our species on this planet.

Notes

PREFACE

1. The discipline of ethnobiology encompasses the totality of the place of nature in a culture—including the role of plants and animals as tribal characters in legends, songs, or rituals designed to regulate resource use; the innovative ways in which humans have used plants and animals and their constituent parts; and the manners in which human activities have altered the natural environment and augmented or decreased plant and animal populations. For more information, see the Society of Ethnobiology website at www.ethnobiology.org.

INTRODUCTION

1. Cronon 1995:79. For articles that address wilderness as a cultural myth, see Gómez-Pompa and Kaus 1992; Denevan 1992.

2. The ecologist Jon Keeley (2002:314) makes the case that "it is very likely that through the use of fire, native Californians markedly altered vegetation patterns over much of the region" in the coast ranges of California. However, he does not argue for a ubiquitous human presence: "Indeed, considering the rugged terrain and pattern of Indian impacts hypothesized by Cooper (1922), it is very likely there were significant portions of the central and southern California landscape that were remote and inaccessible, particularly as tribes did not generally occupy lands near the border of their territory (Kroeber 1962)" (p. 314). And he points out: "Indirect evidence of wilderness areas is suggested by the extraordinary diversity of languages in the state, which may have their origins in the isolation and long-term stability of tribelets induced by wilderness areas between groups. Pre-contact California is surpassed in the density of languages only by the island of New Guinea, where rugged topography and village isolation are believed to have played prominent roles in the evolution of language diversity (Diamond 1992)" (p. 315).

3. For a discussion of how much of the land was used by indigenous people in California, see Keeley 2002:307. For an opposing view that much of Califor-

nia and the West was a wilderness, with little or no indigenous influence, see the essays in Vale 2002.

4. For the absence of the word *wilderness* in tribal dictionaries, see Callaghan 1984; Bethel et al. 1984; Pitkin 1985; Zigmond, Booth, and Munro 1991.

5. Oelschlaeger 1991:3.

6. Lawlor 1991:307; Neihardt 1972:9. For indigenous views of wilderness, see Forbes 2001; Martinez 2003; Standing Bear 2001.

7. For a definition of traditional ecological knowledge, see Inglis 1993. For books and articles on the subject of traditional ecological knowledge and indigenous resource management, see Berkes 1999; Ford and Martinez 2000; Nazarea 1999; Margolin 1997–98.

8. For gathering peppernuts, see Smith 1990. For gathering bear root (yerba mansa), see Franco 1993. For gathering deergrass, see Ortiz and Staff 1991.

9. The prehistoric record in many parts of the world consists of examples of native peoples bringing in exotic species and overharvesting nature, processes that led to soil erosion, extinction of native species, and deforestation. Familiar to most Americans are such examples as Easter Island, the Hawaiian Islands, and Mesa Verde in Colorado. For Pleistocene extinction of the megafauna, see Martin and Klein 1984; Pielou 1991. For extinction or reduction of California's fauna, see Gutherie 1993; Raab and Bradford 1994; Hildebrandt and Jones 2002.

10. Preston 1997:266, 270.

11. Stephen Krech's *The Ecological Indian* (1999) is a case in point. He summarizes the research of other scientists, showing that Native Americans across the continent had "impacts." But he carefully selects most examples that show negative impacts on the natural environment leading to overharvest, extinction, and deforestation. The title of the book is misleading—because "ecological" connotes concern for the welfare of nature and *relationship* with other beings.

12. Hughes 1983:4.

13. Biodiversity can be defined at several levels of biological organization: "Biodiversity is the variety of life and its processes. It includes the variety of living organisms, the genetic differences among them, the communities and ecosystems in which they occur, and the ecological and evolutionary processes that keep them functioning, yet ever changing and adapting" (Noss and Cooperrider 1994:5). For data on the vanishing ecosystem types of California, see Barbour et al. 1993; Noss, LaRoe, and Scott 1995. For a look at the positive effects native peoples can have on biodiversity, see Minnis and Elisens 2000; Dasmann 1991.

14. On vanishing plants, see California Native Plant Society 2001. On vanishing animals, see Peterson 1993. On vanishing plants and animals, see California Resources Agency and Department of Fish and Game 1992. For rates of extinction, see LaRoe et al. 1995:6.

15. Diverse, fully functional ecosystems are important to humans for a variety of reasons. For the ecological services that ecosystems provide, see Jones and Stokes Associates 1987. For the human benefits that unaltered nature provides, see Jensen, Torn, and Harte 1993. For causes of dwindling biodiversity, see Ehrlich and Ehrlich 1981.

16. For a discussion of the values of setting aside wilderness, see Noss 1991; essays in Callicott and Nelson 1998. For the pros and cons of reintroducing indigenous management in national parks, see Egan 2003.

17. For the difficulties of gathering cultural resources, see back issues of *News from Native California* and *Roots and Shoots* (newsletter of the California Indian Basketweavers Association). For co-management and restoration projects that involve native people, see chapter 11.

18. Worster 1994:60. For a definition of *resource,* see *American Heritage Dictionary* 1992:1536.

19. These meanings are taken from Snyder 1990:179 and *American Heritage Dictionary* 1992:873.

20. Leopold 1949:258. For an enlightening overview of how disturbances of different types and scales can benefit or harm the functioning and structure of ecosystems, see Botkin 1990. For a discussion of the impacts and unsustainability of industrial agriculture, see Kimbrell 2002. For a review of overgrazing of livestock, see Burcham 1982. For an overview of overharvesting of timber in national forests and destruction of forest ecosystems, see Hirt 1994.

CHAPTER 1. WILDLIFE, PLANTS, AND PEOPLE

1. On the diversity of rivers in California, see Mount 1995.

2. On number of endemic plants in California, see California Native Plant Society 2001. On species of oaks in California, see Tucker 1993. On number of oaks in the United States, John Tucker pers. comm. 2004. On species of manzanitas, see Wells 1993. On species of California lilacs, see Schmidt 1993.

3. For an overview of California Indian languages, see Hinton 1994a. She says that the state is the most linguistically diverse place on the continent (Hinton 1994a:13). For the number of languages in the United States, see Goddard 1996:3. Culture areas in California can be found in Kroeber 1939a. Kroeber assigned the southernmost groups to the Southwest culture area; groups of southeastern California to the lower Colorado River culture area, the northeastern part of the state to the Northeast culture area, the east side of the Sierra Nevada and parts of interior southern California to the Great Basin, the northwestern corner to the Pacific Northwest coast, and the area between the Pacific coast and the west side of the Sierra Nevada to central California. For the Moratto quote, see Moratto 1984:3.

4. Margolin 1989:67. For numbers of tule elk, see Holing 1988.

5. On bandtail pigeons, see Mayfield 1993.

6. Cook 1960:248–49. Jerry Powell, an entomologist at the University of California, Berkeley, confirms the butterfly identification, based on the journal description.

7. Carson 1950:110.

8. Taylor [1850] 1968:xiii–xiv.

9. Wolfe [1938] 1979:30.

10. Greene 1892:57; Whitson et al. 1996:519.

11. Rice 1920a:17; Stadtman 1967.

12. National Geographic Society 1924:198.

13. Vogl 1967, in Hanes 1988:420.

14. Purdy 1976:35.

15. On the definition of disturbance, see White and Pickett 1985:7. Fire ecologists are rethinking the definition of disturbance as it relates to fire. Neil Sugihara and Michael Barbour state in *Fire in California Ecosystems* (in press) that "fire is so fundamental and its biological influence so pervasive that it is appropriate to view fire as an ecological process and not as a disturbance. [Fire is] an ecological process that is a part of the environment, as are the patterns of temperature, moisture, wind, flooding, soil development, erosion, predation, herbivory, carbon cycling, and nutrient cycling."

16. On small mammal digging, see Lovegrove and Jarvis 1986. On wave action contribution to intertidal rock biodiversity, see Sousa 1985. On fires maintaining species-rich grasslands, see Risser 1988:177; on flooding and wildlife diversity, see Mitsch and Gosselink 1993. On the link between pyrodiversity and biodiversity, see Martin and Sapsis 1992.

17. On disturbance as a recurrent feature, see Christensen 1988. On the Connell hypothesis, see Connell 1978.

18. Frenkel 1970; Spurr and Barnes 1980; Sauer [1952] 1969:125.

19. On the ecological degradation of Europe, see Ponting 1991; Darby 1956; Dawson and Brechin 1999:43. On deforestation in different parts of the world, see Perlin 1989.

20. On the former range, diet, and habits of grizzly bears, see Storer and Tevis 1955. On Bidwell's account, see Bidwell 1928:73, 75–76. For sightings of grizzlies feasting on beached whales, see Wagner 1929:247; Brewer 1966:130.

21. For an excellent reference book on the range, diet, and habits of tule elk, see McCullough 1969. For early sightings of tule elk, see Evermann 1916:1; Lynch 1954:25; Bosqui 1904:62; Taylor [1850] 1968:77. On their whistles, see Audubon 1989:196.

22. On the former range of pronghorn, see Jameson and Peeters 1988:21. For early sightings of pronghorn by Fages, see Bolton 1911:9. For Muir's description, see Wolfe [1938] 1979:235.

23. On black jaguars, see Jameson and Peeters 1988. On mountain lions, see Young and Goldman 1946; on their screams, see Carson 1950:16 and Dasmann 1985 in Leopold and Blake 1985:98. On wolves, see Jameson and Peeters 1988. On bighorn sheep, see Muir 1961:234–35, 240; Monson and Sumner 1980.

24. Duhaut-Cilly 1929a:239. On condors, see Johnstone 1888.

25. For sparrow migration, see Weidensaul 1999:8. On sooty shearwater migration, see Cogswell and Christman 1977:28.

26. Weidensaul 1999:22.

27. Dawson 1923:1586.

28. Welch 1931 quoted in Leopold 1977:26.

29. For the Martínez quote, see Nathan and Simpson 1962:46. On La Pérouse's account of the California quail, see Margolin 1989:65. For Nordhoff's

passage, see Nordhoff 1873:81. On the diet and nesting habits of the California quail, see Spaulding 1949:14.

30. Mount 1995:80.

31. For eyewitness accounts of geese abundance, see Brewer 1966:216; Hutchings 1990:178. On the comment that presenting geese to a hostess was considered stingy, see Hutton 1942 in Garner 1970:103.

32. Dawson 1923:1846.

33. Nordhoff 1873:81; Mayfield 1993:75.

34. Miller 1987:91.

35. Loud 1918:237.

36. Taylor [1850] 1968:222.

37. Van Dyke 1886:114.

38. On river otters, see Grinnell, Dixon, and Linsdale 1937:275–85. On the ease of shooting otters from a boat, see Nordhoff 1873:81.

39. Loud 1918:238.

40. Brewer 1966:130.

41. On abalone, see Kofoid 1915:128; Howorth 1978:15. On the Emory passage, see Calvin 1951:178.

42. Wilken 1993:55; California Native Plant Society 2001.

43. For the quote, see Oosting 1948 cited in Munz and Keck 1973:11. For general information on plant communities, see Munz and Keck 1973.

44. For a discussion of vegetation types, see Anderson, Barbour, and Whitworth 1997. For the Bakker quote, see Bakker 1994:6.

45. On coastal salt marshes, see Barbour et al. 1993; Macdonald 1988.

46. Hutton 1942 in Garner 1970:103.

47. Barbour et al. 1993:44.

48. Mason 1957.

49. For the size of Tulare Lake, see Margolin 1993:9. On the Narváez map, see Thelander and Crabtree 1994:170. For the Carson quote, see Carson 1950:92. On white-fronted geese, see Dawson 1923:1856.

50. On otters and beavers, see Nordhoff 1873:81. On American white pelicans, see Dawson 1923:1967–68.

51. On the canvasback, see Nordhoff 1873:81. On American egrets, see Dawson 1923:1899. On cormorants, see Dawson 1923:1947.

52. Dasmann 1965:63.

53. Heady et al. 1988; Stromberg, Kephart, and Yadon 2001.

54. Heady 1988; Beetle 1947.

55. On vernal pools, see Barbour et al. 2003. On the Muir observation, see Wolfe [1938] 1979:21.

56. On chaparral, see Wieslander and Gleason 1954; Hanes 1988:418.

57. See Hanes 1988; Ornduff 1974; Keeley and Keeley 1989.

58. Barbour et al. 1993.

59. Belcher 1843.

60. Dawson 1923:1893.

61. Brewer 1966:278.
62. Pavlik et al. 1991; Jepson 1923:88.
63. Nordhoff 1873:112. Vancouver 1999:90.
64. Dawson 1923:1901.
65. Barbour et al. 1993; Nordhoff 1873:122.
66. Rundel, Parsons, and Gordon 1988; Reynolds 1959; Anderson 1993b.
67. Hutchings 1990:326.
68. Barbour et al. 1993; Zinke 1988.
69. Barbour et al. 2001.
70. Munz and Keck 1973; Sawyer, Thornburgh, and Griffin 1988.
71. For population density estimates, see Merriam 1905:594; Moratto 1984.
72. On population estimates, see Cook 1971:66, 72; Cook 1978; Kroeber [1925] 1976:883; Powers [1877] 1976:416. On Old World diseases and reduction in New World Indian populations, see Dobyns 1983; Ramenofsky 1987. On the passage from Cabrillo's journal, see Kelsey 1986:145. Fray Narcisco Durán noted in his diary on May 20, 1817, "Once the pass in the Sierra is discovered, which the said end seems to offer, we would be able to ascertain the truth of what the Indians have told us for some years past, that on the other side of the Sierra Nevada there are people like our soldiers. We have never been able to clear up the matter and know whether they are Spanish from New Mexico, or English from the Columbia, or Russians from La Bodega" (in Chapman 1911:15).
73. Kroeber 1962.
74. Duhaut-Cilly 1929b:313.
75. Boscana [1933] 1978:24. Henley 1854, quoted in *History of Stanislaus County* 1881:87.
76. For Tubatulabal beliefs on the length of their tenure in their territory, see Voegelin 1938:1. On the length of time indigenous people have been in California from a Western science perspective, see Fiedel 1999. On tribal perspectives on time in their territories, see Anderson 1991b; Potts 1977:8. On Modoc beliefs regarding the center of the world, see Faulk and Faulk 1988:16. On the absence of migration stories, see Lee 1998:6–7; Voegelin 1938.
77. See Timbrook 1990; Loud 1918:225.
78. Nunis 1968:35–36.
79. Warner n.d.
80. Van Etten 1994:8.
81. For Wiyot place-names, see Loud 1918:234. For Tolowa place-names, see Waterman 1925:536. For Wintu place-names, see Knudtson 1977:81. For Nomlaki place-names, see Goldschmidt 1951:312. For Cahuilla place-names, see Barrows 1967:33.
82. Merriam n.d.; Brown 1876:157.
83. For the Jack Stewart quote, see Steward 1934:430. Foster 1944:157.
84. Powers [1877] 1976:109–10.
85. Simmons 1997:68.
86. For Sierra Miwok trading patterns, see Barrett and Gifford 1933:251, 254. For trading patterns of different tribes, see Davis 1961; Sample 1950.

87. Davis 1961.

88. For the Dixon passage, see Dixon 1905:201. For the highway route over Pacheco Pass as a former Indian trail, see Dillon 1965:135.

CHAPTER 2. GATHERING, HUNTING, AND FISHING

1. For division of labor by gender, see Wallace 1978:683. There is some evidence that a clear division of labor between the sexes with regard to gathering and hunting was not always the case throughout prehistory. We know, for example, that in historic times, entire families participated in gathering essential plant resources. See McGuire and Hildebrandt 1994.

2. California Native Plant Society 2001.

3. Harris 1891.

4. For uses of the California poppy and blue dicks, see Strike 1994; Bocek 1984; Russell 1927.

5. For Pomo uses of clamshell beads strung with Indian hemp, Sherrie Smith-Ferri pers. comm. 2004; and see Barrett 1952. For Sierra Miwok uses of plants, see Anderson 1988.

6. For the use of moss for bedding, see Chesnut [1902] 1974:302. For the use of Indian paint fungus, see Murphey 1959. On the use of lace lichen for baby diapers, see Goodrich, Lawson, and Lawson 1980:123. On Kawaiisu beliefs, see Barras 1984:32. On the use of brown, red, and green algae for foods, see Goodrich, Lawson, and Lawson 1980:124–25, 127.

7. On numbers of different kinds of animals in California, see Bowman 1990.

8. Driver 1936:191.

9. On traps for catching kangaroo mice, see Hedges and Beresford 1986:46. On capturing large mammals with pit traps, see Potts 1977:43. On the origin of the Pit River name, see Dixon 1908. On use of hunting blinds, see Mayfield 1993:100.

10. For the use of mudhen skins for gloves, see Dixon 1908. For the use of bighorn sheep horns for glue, see Coville 1892. On the use of condor feathers for regalia, see Bates 1982b:17. For the use of whale baleen for combs, see Miller 1991:85. On the use of abalone shells for jewelry, see Mason 1912:129. For the use of grizzly bear femurs for daggers, see Storer and Tevis 1955.

11. On Choynumni rabbit drives, see Mayfield 1993:99–100. On the use of eggs of various kinds, see Driver 1936:185. On the use of gull eggs, see Le Conte 1994.

12. Baker 1930:222.

13. On Menzies's observations, see Menzies 1924:320. For capturing surf fish, see Dillon 1965:146. On smelt fishing, see Kroeber and Barrett 1960. On salmon fishing, see Lufkin 1991.

14. On insects for food, see Bean and Saubel 1972; Underhill 1941:19; Sutton 1988. On the Tubatulabal harvest of honeydew from aphids, see Voegelin 1938. On the use of creosote bush gum, see Coville 1892; Barrows 1967. On the collection of army worms, see Peri and Patterson 1979:23.

15. On the use of the whitelined sphinx moth, see Fenenga and Fisher 1978.

On the pandora moth, see Sutton 1988. On the use of the California tortoise-shell butterfly, see Anderson 1993b.

16. G. Simpson 1961:3 cited in Berlin 1992:5.

17. Berlin 1992:8.

18. Conklin 1962 cited in Berlin 1992:13.

19. For one of the few folk taxonomy studies done in California, see Whistler 1976.

20. Hunn 1990:115.

21. Powers [1877] 1976.

22. Merriam told this story to Frank Latta; see Latta 1977:617–19. For the dates and the localities of the manzanita collected as voucher specimens, see Merriam 1918.

23. On the number of manzanita species, see Wells 1993. On Karuk identification of manzanitas, see Schenck and Gifford 1952.

24. Hudson n.d.b: G.H.M. Acc. No. 20,211.

25. On Luiseño uses of wild cucumber, see Beemer 1980:60. On Native American uses of cascara sagrada, see Strike 1994. On Ohlone uses of willow for aspirin, see Bocek 1984:249. For modern uses of milkweed, see Knudsen and Sayler 1992.

26. Barrows 1967:66.

27. On harvesting seeds within a week, see Underhill 1941:17. On Karuk harvesting of ponderosa pine roots, see Schenck and Gifford 1952:378.

28. On Kashaya Pomo harvest of wild oats, see Goodrich, Lawson, and Lawson 1980:85. On the toxicity of shellfish, see Ross 1997:24. On Tubatulabal cues for when to harvest pinyon pines, see Voegelin 1938:16.

29. On harvesting plants in relation to environmental conditions, see Heffner 1984.

30. On Nisenan selection of oaks for acorns, see Wilson 1972:37. On Pomo harvest of *Angelica,* see Chesnut [1902] 1974:371. On Hupa harvest of *Iris,* see Goddard 1903–4:35. On Paiute harvest of *Yucca,* see Hudson 1904: G.H.M. Acc. No. 20,017:28.

31. For Kashaya Pomo timing of the dogwood harvest, see Goodrich, Lawson, and Lawson 1980:42. For knocking the acorns of California black oak, see McCarthy 1993. There are dozens of references to knocking acorns from trees in many parts of California.

32. On Cahuilla acorn storage, see Barrows 1967:52. On Sierra Miwok placement of granaries, see Anderson 1993b.

33. On California Indian spiritual beliefs about nature, see Swezey 1975; Nomland 1935:169; Deloria 1994. On gathering restrictions, see Abel-Vidor, Brovarney, and Billy 1996:20; Ortiz 1993a:196–99.

34. On the rule for not taking everything, see, e.g., Potts 1977; Beard 1979:52; Murphey n.d.

35. Peri, Patterson, and Goodrich 1982:30.

36. On Gary Snyder's culture of place, see Snyder 1990. Paul Ehrlich (2000:

309) believes that the human capacity for ethics has "its primary roots in biological evolution but that ethical standards themselves are based in cultural evolution."

37. On the Concow Maidu goose dance, see Dixon 1905. For clover dances among the Maidu, see Powell 1877:286. On Sierra Miwok acorn feasts, see Curtis 1924b. On Gashowu Yokuts crop-prophesying dances, see Gayton 1948b:173.

38. On Tubatulabal digging of jimson weed, see Voegelin 1938:5. On the Pomo myth, see de Angulo 1935:240, 247.

39. On the Miwok myth, see LaPena and Bates 1981. On the Pomo myth, see Loeb 1926a:489. On Yurok views of greed and catastrophic fire, see Thompson [1916] 1990:174.

40. For the Native American view that all life is kin, see Salmón 2000; Gayton 1946:262; Cajete 2000.

41. Peri and Patterson 1979.

42. On Hupa relationships with Douglas-fir, see Davis 1988:336–37.

43. For Yuki names after plants, see Foster 1944:182. For Wappo names, see Driver 1936:205. For Miwok names after plants, see Gifford 1916:146–61. For Ohlone naming of a child under a redwood, see Mason 1912:160.

44. Callahan 1979:240.

45. Strong [1972] 1987:116.

46. For Paiute views of killing wildlife, see Underhill 1941:23. On Ishi's views about killing wildlife and luck, see Pope [1918] 1979:197.

47. E. Anderson 1996:64–65.

48. Shepard 1996:xi–xii.

49. Perlot 1985:230.

50. Foster 1944:203.

51. On the Wappo calendar, see Driver 1936:195. On the Northern Maidu calendar, see Kroeber [1925] 1976:438. On the Central Pomo calendar, see Kroeber [1925] 1976:209. On the Cupeño calendar, see Strong [1972] 1987:253. On the Klamath calendar, see Kroeber [1925] 1976:323.

52. Krupp 1987.

53. Patencio 1943:113.

54. Swezey and Heizer 1993.

55. On world renewal ceremonies that balance and replenish the earth, see LaPena 1997–98; Nomland 1935:167; H. H. Roberts 1932; Lang 1991, 1994; Kroeber 1971; Keeling 1992; Hudson and Underhay 1978:62–63.

CHAPTER 3. THE COLLISION OF WORLDS

The epigraph to this chapter is drawn from Young 1992:51.

1. Polk 1991.

2. Nash [1967] 2001.

3. "Frontier" means a region just beyond or at the edge of a settled area (American Heritage Dictionary 1992:729). It is the edge between the known and unknown.

4. Costo and Costo 1987:1.

5. For resistance tactics, see Davis 1971; Phillips 1993; Fuster 2001; Toypurina 2001; Asisara 2001; González 2001.

6. Thornton 1987; see also American Friends Service Committee 1960. For accounts of California Indian deaths due to disease and starvation, see Cook 1976. For accounts of California Indian wars, massacres, and murders, see Bledsoe 1956; Quinn 1997; Gray 1993; Crampton 1957; Secrest 2003; Norton 1979.

7. Haas 1995.

8. See Hurtado 1988 for Indian labor provided during the period 1840–60. For the Chumash hunting of otter for the missionaries, see Nathan and Simpson 1962:64. For Indian sowing of barley and peas, see Marryat 1855:58. For Kitanemuk herding of sheep, see Heizer [1974] 1993:116. For Wappo cutting and hauling of timber, see Yount 1923. For Cupeño plowing and sowing of wheat, see Morrison 1962:45. For Indians shearing sheep, see Barras 1984:70. For Pit River Indians building Peter Lassen's house and cultivating his farm, see Hutchings 1962c:387. On the Tubatulabal as vaqueros, see Walker 1970.

9. Wagner 1928b:391.

10. Engstrand 1997:84–85.

11. Wagner 1928a:46.

12. Wagner 1928a:46–47.

13. Kelsey 1986:153.

14. Kelsey 1986:158; Engstrand 1997:86.

15. Father Andrés de Urdaneta is credited with finding the most direct passage from the Philippines to New Spain. His log book recorded the route followed as "northeast till the northwest winds were encountered near the California coast, and then southeast, parallel to the coast, to Acapulco" (Wagner 1923:140).

16. Wagner 1923:140; Engstrand 1997:87; Bicknell and Mackey 1998:9.

17. Engstrand 1997:87–88.

18. Heizer 1947:264, 279.

19. Penzer 1926, quoted in Allen 1971:3, 16.

20. Heizer 1947:267.

21. Engstrand 1997:90; Wagner 1923.

22. Wagner 1923:154.

23. Wagner 1923.

24. Wagner 1924:13–14.

25. Wagner 1924:15–16.

26. Wagner 1924:17, 18.

27. Engstrand 1997:91.

28. Wagner 1928b:348.

29. Engstrand 1997:91; Wagner 1928b:347.

30. Wagner 1928b:356.

31. Wagner 1928b:361, 366.

32. Engstrand 1997:92.

33. Bicknell and Mackey 1998. Bicknell and Mackey point out that Manila

galleons that were built in Spain would have had to sail around the Cape of Good Hope to reach the Philippines.

34. Heady 1988:497.

35. See Lightfoot and Simmons 1998 for a discussion of possible introduction of Euro-Asiatic pathogens to California Indians during this period.

36. Castillo 1978:100. During the Mission period, the Russians also influenced the coastal environment and changed the Indians' ways of life, though to a much lesser extent than the Spanish. Mostly to the north of Spanish settlements, the Russians expanded their sea otter fur trade and set up permanent colonies. In 1811 the Russian-American Fur Company set up a colony eighteen miles north of Bodega Bay and called it Rus (later mispronounced "Ross" by Americans). Russians, their Aleut assistants, and the Pomo felled trees to build a large log fort, agricultural crops were planted, and livestock raised. From this base they began hunting sea otters from December through March as far south as Monterey. The Pomo worked as agricultural laborers, vaqueros, and servants for the Russians and were generally treated much better than were the missionized Indians (Sandos 1997).

37. Rawls 1984:13.

38. Castillo 1978:100.

39. Chapman 1939:385.

40. Castillo 1978:101; Forbes 1982, quoted in Caldwell 1995:3.

41. Cook 1943:73–80, cited in Castillo 1978:102.

42. Simpson 1962:6.

43. For a discussion of how Native Americans were viewed by non-Indians, see Pearce 1988. On how Native Americans were viewed by missionaries, see Costo and Costo 1987.

44. Boscana [1933] 1978:55.

45. Chapman 1939:387.

46. Scanland 1894:636.

47. Simpson 1962:67.

48. Cook 1976:97.

49. Engelhardt 1929:8.

50. Hickman 1993.

51. Heinsen 1972.

52. The bark of cascara sagrada (*Rhamnus purshianus*) was used as a laxative for centuries by the Modoc, Yurok, and other tribes.

53. Consult Capt. Alejandro Malaspina's account in Cutter 1960 for a glowing report of the missions' accomplishments with the Indians.

54. For the Vancouver quote, see Vancouver 1999:92; Langsdorff 1927:171, cited in Heizer and Almquist 1971:7.

55. For a discussion of diseases at the missions see Cook 1976; Walker and Johnson 1994; Hurtado 1988.

56. Engelhardt 1929:35.

57. Cook 1976:16.

58. Kroeber [1925] 1976:888.

59. Cook 1943:91–135, cited in Castillo 1978:101.

60. Rawls 1984:14, 16.

61. Cook 1976:70.

62. Cook 1976:79. For other forms of Indian resistance, both passive and active, see Castillo 1978:103–4; Haas 1995:13–34.

63. Cook 1976:39.

64. Tarakanoff 1953:15–18, quoted in Rawls 1984:38.

65. Castillo 1978:102.

66. Hafleigh 2000; Lummis 1929:332.

67. Dasmann 1999:106.

68. Cook 1976:39; Lummis 1929:332.

69. For killing off birds and mammals that might consume crops, see Simpson 1962:163; Nathan and Simpson 1962:28.

70. Randall 2000:193–98; Pitcairn 2000:321–26.

71. Lummis 1929:332.

72. Hendry 1931.

73. Dasmann 1999:109.

74. Holliday 1999.

75. Cowan 1977:10.

76. Chapman 1939:455.

77. Castillo 1978:104–5.

78. Cook 1976:211, 213.

79. Castillo 1978:106.

80. Cook 1978:92.

81. Goetzmann 2000; Nunis 1968:34.

82. For accounts of beaver trapping and products from furs, see Dillon 1967:125; Leopold and Blake 1985:66.

83. Nunis 1968:34.

84. Walker 1970:4; Mason 1881:64.

85. Holliday 1999:16.

86. For the worth of sea otter skins and numbers captured, see Davis 1929:4, 22. For the fur trade to the Orient, see Ogden 1941:2. For an estimate of otters' historical population, see Benz et al. 1994:503.

87. Palazzo 1994a:466.

88. Webb 1952:288–89.

89. Cook 1976:63.

90. Webb 1952:291.

91. Davis 1929:6. The Spanish term for these Indians was *Indios manzos* (tame Indians); see Wierzbicki [1849] 1970:18.

92. Castillo 1978:105.

93. Colton 1948:7.

94. Robinson 1948.

95. Colton 1949:222–23. Although the Californio life was admired by Englishmen such as Colton, other newcomers condemned the Mexicans' lack of ambition and failure to improve the land. Kevin Starr (1973:5) sums up the French,

Russian, English, and American visitors' response to Mexican rule: "[They] confessed themselves astonished at the frailty of the Spanish and Mexican hold: a society without schools, without manufactures, without defenses, administered by a quasi-feudal mission system and inhabited by a population that barely exceeded 1500. Travelers complained of difficulties in obtaining supplies, lack of transportation, an absence of skilled workmen, poor houses and furniture, sour wine, indifferent food, and persistent fleas. It was a society so backward that its plow and ox-cart were those of ancient times, so disorganized that in spite of the fact that countless cattle roamed its hills, it had to secure dairy products from the Russian colony at Fort Ross and have leather shoes shipped around the Horn from Boston."

96. Garner 1970:128.

97. Robinson 1948:61.

98. Cook 1976:304.

99. Cook 1976:302.

100. Castillo 1978:106. For Indian resistance during this era, see also Sanchez and Piña 2001; Machado 2001; Espinosa and Garcia 2001.

101. Castillo 1978:106.

102. For reduction in beaver numbers, see Leopold and Blake 1985:66; Grinnell et al. 1937. For reduction in wolverine, fisher, marten, mink, and river otter, see Dasmann 1965:45.

103. Colton 1948:15, 17.

104. For descriptions of large numbers of cattle and sheep during the Mexican era, see Brewer 1966:93; Davis 1929.

105. Holliday 1999:75; American Friends Service Committee 1960:6.

106. Cook 1978:93.

107. Mason 1881. Note that Indian burning of the vegetation interfered with the gold mining, lumbering, and grazing plans of Sutter, Marshall, and other new settlers, and thus agreements were drawn up to prohibit Indian burning.

108. Kemble 1971:123.

109. Holliday 1999:59.

110. Holliday 1981; Carson 1950:19; Holliday 1999:46.

111. Paddison 1999:ix.

112. Perlot 1985:110.

113. Livingston [1974] 1993:90.

114. Mayfield 1993:122.

115. Brewer 1966:295.

116. Miller 1987:90.

117. Kirkpatrick 1962:52.

118. Cook 1976:37.

119. Heizer [1974] 1993:xvi–xvii.

120. LaPena 1978:324–25.

121. Castillo 1978:107–8.

122. Cook 1976:262–63.

123. Perlot 1985:132.

124. Heizer and Almquist 1971:202. For the source of few whites answering for killing Indians, see Heizer and Almquist 1971:200. The Bancroft Library at the University of California, Berkeley, contains two thick scrapbooks called the "Bancroft Scraps" that contain dozens of articles from California newspapers documenting hate crimes against California Indians.

125. Heizer and Almquist 1971:67–68. For U.S. treaties and agreements, see Kvasnicka 1988.

126. Heizer 1972; Rawls 1984:146.

127. Castillo 1998:8–9.

128. Castillo 1978:110. For accounts of Indian slavery and massacres during this period, see Beale's report of 1853 (Beale 1957).

129. Rawls 1984:152.

130. Wooman and Johnson [1974] 1993:28.

131. Castillo 1978:110.

132. Burgess 1860.

133. Heizer and Almquist 1971:42.

134. Castillo 1978:111.

135. Rawls 1984:152–53.

136. For Native American labor in California, see Hurtado 1988, Davis 1929:13, 43, Barrows 1967:71.

137. Holliday 1999:63, 68; Rawls 1976, cited in Rawls 1999:5.

138. Carson 1950:60–61.

139. Hill 1999:70.

140. Teggart 1923:199.

141. Heizer [1974] 1993:xvi.

142. American Friends Service Committee 1960:6.

143. Bell 1927:35.

144. See Heizer and Almquist 1971 on voting rights (96), testifying against a white person (47), all-white juries (61), and the enslavement and forced yielding of land (48, and Heizer [1974] 1993:32). On carrying a gun, see Heizer [1974] 1993:11. On child slavery, see Heizer [1974] 1993:219. On carrying a gun see Heizer [1974] 1993:11. On forced prostitution and rape of California Indian women, see Hurtado 1988:179–80; Heizer [1974] 1993:271.

145. Rawls 1999:1.

146. For accounts reflecting these assumptions, see Brewer 1966:276; Bryant 1985:272; History of Stanislaus County 1881:16; Carson 1950:100.

147. See Hutchings 1962d:250.

148. For commercialization of hot springs, see Calvin 1951:165. For the use of asphaltum, see Brewer 1966:59. For the harvest of salt, see Barrows 1967:30.

149. Borthwick 1948:40.

150. For the eating of sandhill crane for Thanksgiving or Christmas dinner, see Steinhart 1990. For grizzly bear meat served at restaurants, see Marryat 1855:121; and for pressure on the resource, see Grinnell, Dixon, and Linsdale 1937:74. For a description of the taste of rabbit meat, see Van Dyke 1886:32, 133. For a description of the meat one might be served on a train, see Nordhoff 1873:26.

151. For accounts of the plentiful supplies of wildlife, hunting method, price, and preferred species for eating, see Borthwick 1948:38–39; Brewer 1966:220; Nordhoff 1873:81; Dawson 1923:1856, 1868.

152. For accounts of the harvest of murre eggs, see Hutchings 1962b:43; Dawson 1923:1501, 1503.

153. See Kirk 1994; Braham et al. 1994; Calambokidis, Feldman, and Hanni 1994.

154. Carson 1950:95–97.

155. Borthwick 1948:92.

156. Taylor 1951:75.

157. Dasmann 1965:86.

158. Evans 1990, cited in Arno and Allison-Bunnell 2002.

159. Grinnell, Bryant, and Storer 1918.

160. Dasmann 1965:45.

161. For the fate of snow and Ross's geese, see Dawson 1923:1847; Cogswell and Christman 1977:120. For the fate of greater white-fronted geese, see Dawson 1923:1856. For the hunting and decline of American egrets and Western grebes, see Dawson 1923:1898, 1899, 2045, 2046.

162. For the trapping of quail see Hutchings 1962a:258; Leopold 1977:29.

163. For reduction in quail numbers see Spaulding 1949; Dawson 1923:1586; Leopold 1977:29.

164. Dawson 1923:1502.

165. For the fate of bighorn sheep, see Dasmann 1965:45; Monson and Sumner 1980:290–91.

166. Bleich et al. 1994.

167. Brewer 1966:185. For numbers of elk killed and non-Indian products from elk, see also Davis 1929:31; Merriam [1921] 1994:106. For information on the decline of pronghorn antelope, see Barrett and Weiss 1994:53; Brewer 1966.

168. Howorth 1978:15.

169. For the population of northern and Guadaloupe fur seals, see Palazzo 1994b:507. For numbers of sea lions killed, see Daugherty 1979:36. For the population of sea otters, see Davis 1929:22; Steinhart 1990:82.

170. For use of strychnine to poison wildlife, see Muir [1911] 1944:208 and Henshaw 1876:308. For reduction in grizzly bears and their uses, see Storer and Tevis 1955; Grinnell, Dixon, and Linsdale 1937:93.

171. Holliday 1999:280.

172. Preston 1981; Kemble 1971:315; Eigenheer 1976.

173. For a thorough history of the "Europeanization" of the flora and fauna of the New World, see Crosby 1972, 1986. For acres in cultivation in California, see U.S. Dept. of Commerce 1975.

174. For an account of wheat production in early California, see Rothstein 1999:33; Baxley 1865:413. For information on steam-powered harvesters, see Jelinek 1999:240.

175. Van Dyke 1886:162.

176. For Peter Shield's observations, see Rothstein 1999:38. For information

on early wheat farmers' ties to global markets, see Rothstein 1999:36. For the impact of industrial farming, see Kimbrell 2002. For the benefits of native grasses to land health, see back issues of the newsletter of the California Grassland Association.

177. For an estimation of the size of the original grasslands, see Dasmann 1965:63. For number of cattle and sheep in 1862, see Dasmann 1965:67–68.

178. U.S. Forest Service 1936, cited in Burcham 1982:100.

179. For the spread and uses of exotic plants, see Crosby 1986; Frenkel 1970; Gerlach 1998; Bossard et al. 2000. The ecologists John Randall and Marc Hoshovsky (2000:322–23) state that today, more than two hundred years after the first invasive introductions, alien plant problems are widespread and severe, posing threats to biodiversity second only to direct habitat loss and fragmentation.

180. Donahue 1999.

181. Rothrock 1876:210–11.

182. Burcham 1982:196. For negative impacts of historical overgrazing, see also Perkins 1863.

183. Bidwell 1866:202–13.

184. Quoted in Fletcher 1987:87.

185. Holliday 1999.

186. Alverson, Kuhlmann, and Waller 1994:121.

187. For uses of tan oak, see Tappeiner, McDonald, and Roy 1990. For non-Indian uses of coast redwoods, see Stanger 1967.

188. For non-Indian uses of oaks, see Barbour et al. 1993:106–8; Brown 1949:15. For non-Indian uses of sycamores, see Garner 1970:139.

189. For non-Indian uses and logging of giant sequoias, see Johnston 1983; Otter 1963:16. For non-Indian uses of sugar pine, see Cermak 1996:16–17.

190. For an overview of Indian boarding schools in the United States, see Archuleta et al. 2000. For a regional view, see Ross 2000.

191. Rawls 1984:213.

192. American Friends Service Committee 1960; Castillo 1978.

193. Hurtado 1982:269, 260.

194. On the difficulty of gathering, see Redding 1880:364.

195. Browne 1877:20.

196. Heizer [1974] 1993:224–25.

197. *San Francisco Bulletin* 1861, cited in Heizer [1974] 1993: 240–41.

198. Heizer [1974] 1993:229.

199. For accounts of Indians laboring against their will, see Baxley 1865:466; Garner 1970:91. For the article on kidnapping of Indian children, see Carpenter 1893:391.

200. Castillo 1978.

201. Prucha 1990, quoted in Goodyear 2000:10.

202. Lummis 1929:306.

203. Archuleta et al. 2000:19.

204. Castillo 1998:10.

205. See Crum 2001.

206. Faulk and Faulk 1988.

207. Fletcher 1888:217.

208. Rawls 1984:166–67.

209. Goldberg-Ambrose and Steward 1997.

210. Fletcher 1888.

211. Shipek 1987, cited in Haas 1995:61.

212. ARCIA 1897:117, quoted in Castillo 1978:116. For a report on the conditions and villages of the Mission Indians leading up to the establishment of reservations, see Jackson and Kinney 1883.

213. American Friends Service Committee 1960.

214. Rawls 1984:171.

215. Heizer [1974] 1993:xiii.

216. *Bodie Weekly Standard* 1877, cited in Fletcher 1987:73.

217. Heizer [1974] 1993:253.

218. Heizer [1974] 1993:259–60.

219. Johnston-Dodds 2002:21.

220. Brown 1876:163.

221. ARCIA 1895:131, cited in Castillo 1978:115–16; Castillo 1978:117.

222. Castillo 1978; Bledsoe 1956.

223. Kasch 1947, cited in Castillo 1978:118.

224. Nash [1967] 2001.

225. Muir [1911] 1944:58–59.

226. Anderson and Nabhan 1991; Muir [1911] 1944:78.

227. Muir [1911] 1944:176, 79. For Muir's view of Native Americans, see Cohen 1984:185.

228. Cook 1978:93.

229. Castillo 1998:14.

230. Indian Board of Co-operation 1919–20:3.

231. Kickingbird and Rhoades 2000.

232. Raup 1959:12, quoted in Castillo 1978:118.

233. ARCIA 1903:14–15, cited in Castillo 1978:121.

234. O'Brien 1989:80.

235. O'Brien 1989:81.

236. Spott 1926.

237. Advisory Council on California Indian Policy 1997.

238. Castillo 1978:118. For an overview of American Indian policy in the twentieth century, see Deloria 1985.

239. Castillo 1978.

240. Indian Board of Co-operation 1919–20:8.

241. Castillo 1978:119.

242. Johnson, Haslam, and Dawson 1993.

243. Kelly 1932.

244. For information on the trapping of predatory animals in California, see McIntyre 1995:174; Dasmann 1965; Grinnell et al. 1937.

245. Fox 1985:116–17.

246. Fox 1985:351.
247. Fox 1985; Dasmann 1965.
248. Clary 1986:2–3, 6; Hirt 1994:xviii.
249. Robinson 1988:16–17.
250. Dasmann 1965.
251. Barnes 1908.
252. Donahue 1999:36.
253. Grinnell et al. 1937, vol. 1.
254. Leopold 1977:29; Elliot and Littlefield 1994.
255. Pyne 1982:262–63, cited in Arno and Allison-Bunnell 2002:18.
256. U.S. Dept. of Interior 1903:6.
257. Arno and Allison-Bunnell 2002.
258. Pyne 1982, cited in Arno and Allison-Bunnell 2002:20.
259. Medvitz 1999:13.
260. Crosby 1972:66.
261. Callahan 1979:226.

CHAPTER 4. METHODS OF CARING FOR THE LAND

1. Posey 1984, 1985; Lewis 1989, 1991a, 1991b, 1994.
2. For an example of sucker fishing by the Pit River people in ways that conserve the resource for future generations, see Foster 1999. "Keep it living," or *q'waq'walatowkw*, is a Kwak'wala term shared with Turner by the hereditary chief Kwaksistala Adam Dick of the Pacific Northwest. Deur and Turner in press.
3. Heizer and Elsasser 1980.
4. For the rule of not taking too much, see R. K. Roberts 1932; Heizer 1978. For Alaska and Canada rules, see E. N. Anderson 1996:56; Nelson 1983; N. J. Turner in press.
5. Potts 1977:36.
6. Barrett and Gifford 1933:212.
7. Driver and Massey 1957.
8. Latta 1977:551.
9. Schilling 1939:8; Nabhan 2001.
10. On Chemehuevi, Washoe, and Paiute timing of the willow harvest, see Fulkerson 1995; Kelly 1932:120. On Pomo timing of the hazel harvest, see Goodrich, Lawson, and Lawson 1980. On experimental results of sprouting after pruning at different seasons, see Buell 1940.
11. Savelle 1977.
12. On Pomo frequency of the harvest of sedge, see Peri, Patterson, and Goodrich 1982. On Klikitat frequency of the harvest of western red cedar roots, see Schlick 1994:97. On Navajo frequency of the harvest of medicinal roots, see Anderson and Nabhan 1991:28.
13. On the harvest of seaweeds in California, see Lewallen and Lewallen 1995:76. On Hawaiian harvest of seaweeds, see Fortner 1978. On mushroom harvest and management, see Anderson 1993d; Templin Richards 1997.

14. On Paiute harvest of juniper trunks for bow wood, see Wilke 1993.

15. Mooney 1890:259.

16. On the mobility of agriculturists versus hunter-gatherers, see Lee and DeVore 1968; Sauer 1967.

17. For Pomo marking of trees and shrubs for use, see Barrett 1952:50; Kniffen 1939:379.

18. Beemer 1980:33–34.

19. On repeat use and "ownership" of specific gathering and fishing sites, see Aginsky 1943; Voegelin 1938; Anderson 1993b; Latta 1977; Driver 1936; Drucker 1937:243; Garth 1953.

20. On Pomo repeated use of willow and sedge tracts, see Peri, Patterson, and Goodrich 1982. On Cahuilla ownership of gathering areas, see Bean and Saubel 1972:32.

21. Gayton 1948b:181, 175.

22. Barrett 1952:50.

23. Anderson 1993a.

24. Kroeber and Heizer 1968:24.

25. Driver and Massey 1957.

26. On slow matches, see Olmstead and Stewart 1978:229; Kroeber [1925] 1976:332; Barrett 1910:257; Nomland 1935:153. On fire fans, see Hudson n.d.b:G.H.M. Acc. No. 20,207a; Gayton 1948b:220.

27. Marryat 1855:135; Martin and Sapsis 1992.

28. On accounts of California Indian burning, see Timbrook, Johnson, and Earle 1993; Collings 1979; Stewart 2002; Lewis [1973] 1993.

29. On Pomo burning of fern patches, see Peri and Patterson 1979:20. On Wiyot burning for feed for deer, see Driver 1939. For Muir quote, see Muir 1961:154.

30. Lawton et al. 1993; Steward 1934.

31. McMillin 1956:29.

32. Clark 1894:15.

33. On Timbisha Shoshone pruning, see Fowler 1996. On Miwok gathering of firewood, see Barrett and Gifford 1933. On Yuki and Pomo pruning of redbud, see Chesnut [1902] 1974:357.

34. Peri and Patterson 1979:39.

35. On Tubatulabal pruning, see Smith 1978:444.

36. For quote, see Peri and Patterson 1979:39. For a description of knocking pinyon trees, see Dutcher 1893. For a description of knocking oak trees for different tribes, see Anderson 1993b.

37. Coville 1892; Havard 1896:43.

38. Pimienta-Barrios 1993:186.

39. On Paiute sowing of seeds, see Steward 1938a:104. On Quechan sowing of seeds, see Drucker 1939:11. On planting the seeds of the desert fan palm, see Nabhan 1985:26; Cornett 1987b:17.

40. Peri, Patterson, and Goodrich 1982; Anderson 1997.

41. Latta 1934.

42. For the Martínez quote, see L. B. Simpson 1961:45. For the uses of chuchupate by the Chumash, see Timbrook 1990.

43. For Maidu cultivation of bulbs, corms, and tubers, see Potts 1977. For Pomo cultivation of underground swollen stems, see Peri, Patterson, and Goodrich 1982:119, 121, 125.

44. Ornduff 1974.

45. Peri, Patterson, and Goodrich 1982:21.

46. For information on nutrient cycling, see Barbour, Burk, and Pitt 1980:292. For fire as a mineralizing agent, see St. John and Rundel 1976.

47. Franco 1993:18.

48. See Anderson 1996a; Shipek 1989.

49. For burning in oak woodlands, see Anderson 1993c; Heffner 1984; Schenck and Gifford 1952. For Tolowa burning under oaks, see Driver 1939:381.

50. Jack 1916:195.

51. On insects that feed on acorns, see Gibson 1969:259. On studies of insects and their damage to acorns of specific California oak species, see Keen 1958; Brown 1979:190; Lewis 1991; Swiecki, Bernhardt, and Arnold 1990.

52. For information on fungi and bacteria that colonize acorns, see Swiecki, Bernhardt, and Arnold 1991:10, 35. For Lassik and Pitch Wailaki burning under oaks, see Keter 1987:4. On Karuk burning in forests, see Schenck and Gifford 1952. On Luiseño burning to destroy insect pests and diseases, see Shipek 1977:118. For discussions of a new paradigm for control and/or eradication of forest insect outbreaks and disease using prescribed burning, see Mutch 1994; Brennan and Hermann 1994.

53. Kroeber [1925] 1976:396.

54. Bolton 1930:417; Muir 1961:154; Drucker 1937:233. Drucker recorded the following: "[L]ate spring, when the old fern was quite dry and the new growth just starting, is said to have been the time for burning off the hillsides to improve the hunting grounds" (p. 232).

55. For studies that document the effects of burning on improving forage for wildlife, see Reynolds and Sampson 1943; Carlson et al. 1993; Klinger, Kutilek, and Shellhammer 1989.

56. Kroeber [1925] 1976:259.

57. On Paiute hunting with fire, see Steward 1935:9; 1938b:188. On Miwok hunting, see Barrett and Gifford 1933:179. On Hupa hunting, see Driver 1939:374.

58. Voegelin 1938:11,13; Fletcher 1987:4.

59. On Tubatulabal smoking for ground squirrels, see Powers 1981:31. On Foothill Yokuts techniques, see Gayton 1948a:75. On smoking for ground squirrels by other tribes, see Du Bois 1935:14; Gayton 1948a:75; Bright 1978:181.

60. Gayton 1948a:77; Marryat 1855:136.

61. For use of fire to capture grasshoppers, see Foster 1944:167; Merriam 1955:39; Olmstead and Stewart 1978:228; Powers [1877] 1976:379.

62. Clark 1973:212.

63. Du Bois 1935:14.

64. Mooney 1890:260.

65. On Karuk burning to reduce brush, see Harrington 1932:64. On Maidu burning to eliminate underbrush, see Dixon 1905:201. On Lassik burning to clear underbrush, see Essene 1942:55. On Chukchansi burning, see Merriam 1902b: 239. On Tolowa burning to reduce brush, see Drucker 1937:233.

66. Stromberg, Kephart, and Yadon (2001) found that the species richness in coastal prairies was nearly twice that of relatively diverse serpentine California grasslands and other North American grasslands. They conclude, "If protection of biodiversity is a management goal in land use plans, coastal grasslands should be protected as biodiversity 'hotspots' " (p. 236).

CHAPTER 5. LANDSCAPES OF STEWARDSHIP

1. Yount 1923:52–53. Using fire scar data and fire behavior modeling, Greenlee and Langenheim (1990) estimate the mean fire interval for lightning fires to be 135 years in the Monterey Bay area in coastal redwoods. They estimate the mean fire interval for aboriginal fires to be 17 to 82 years. They conclude that "although fire scars cannot be used to estimate early fire periodicity in the prairies, this vegetation type could not persist without burning at an interval of 1 to 15 years" (p. 245).

2. For books and articles that document landscapes that have been cultivated by indigenous peoples, see Doolittle 2000; Bonnickson 2000; Blackburn and Anderson 1993b; McCann 1999a; Stewart 2002; Mann 2002.

3. Blackburn and Anderson 1993a.

4. Taylor 1932.

5. Gordon-Cumming 1883:415.

6. Clark 1894:14–15.

7. Perlot 1985:63.

8. For a description of and case studies in historical ecology, see Merchant 1993; Crumley 1994; Egan and Howell 2001. For an article that combines pollen studies with archaeology to reveal vegetation change with indigenous settlement, see Anderson and Carpenter 1991.

9. Nabhan and Rea 1987; Nabhan et al. 1981. For cultivation of devil's claw by the Paiute in the Bishop and Big Pine areas, see Hudson 1904.

10. Anderson, Barbour, and Whitworth 1997.

11. On the introduction of plants through human agency, see Gilmore 1977. On assessing potential archaeological sites by plant associations, see Jones 1942. For phytoarchaeological studies, see Minnis and Plog 1976; Gilmore 1931.

12. For the association of *Agave parryi* with archaeological sites, see Minnis and Plog 1976. For the association of plants with Pueblo ruins, see Yarnell 1965. For the association of nettles with Salish camps, see White 1975.

13. Jimson weed association with village sites from Ken Hedges, pers. comm. 1990. Jepson quote from Jepson 1943:457.

14. Jepson 1943:457; Hester 1978:502.

15. On walnut in association with archaeological sites in California, see Ornduff 1974.

16. Jepson 1910:166.

17. Timbrook 1993:57; Schumacher 1875.

18. Collins 1991:414; Guthrie 1993.

19. Cornett 1987b, 1989a, 1995.

20. Fiedler and Ahouse 1992.

21. Cornett 1985a, 1986, 1989b.

22. McClenaghan and Beauchamp 1986.

23. Cornett 1989b. For quote, see Cornett 1987b.

24. On indigenous burning of desert fan palms, see Cornett 1987a, 1989a; Nabhan 1985; Parish 1907; Shipek 1989:163. For Jepson quote, see Jepson 1910:173.

25. On animal dispersal of palm seeds, see Cornett 1985b; Bullock 1980.

26. See Fagan 2003:76 for the length of time humans have inhabited the northern coast. Grass grains were an important food source during the earliest habitation.

27. Collier and Thalman 1991.

28. Nordhoff 1873:119.

29. For tule elk in coastal prairies, see Mason 1970. For a description of coastal prairie, see Heady et al. 1988; Stromberg, Kephart, and Yadon 2001. For Indian burning in coastal prairies, see Greenlee and Langenheim 1990.

30. Chamisso 1999:147.

31. Lutke 1989:257.

32. Menzies 1924:302.

33. Gibbs 1853:134.

34. Schenck and Gifford 1952.

35. For Pomo burning of California fescue, see Peri, Patterson, and Goodrich 1982:120. For the use of iris, see Nomland 1935. For Karuk burning of iris, see Gifford 1939.

36. On burning of grassy areas to drive and roast grasshoppers, see Nomland 1935; Loeb 1926b; Foster 1944. On Pomo burning of soaproot, see Peri, Patterson, and Goodrich 1982. On the uses of soaproot by coastal tribes, see Nomland 1935; Bocek 1982; Collier and Thalman 1991. For burning of areas for tarweed, see Gifford [1939] 1965:17.

37. Bolton [1927] 1971:216.

38. Kroeber 1939b.

39. Thompson [1916] 1991:26, 33.

40. de Massey 1926:154.

41. Other studies that document succession in coastal grasslands are Elliott and Wehausen 1974; McBride and Heady 1968.

42. Bicknell 1989, 1990, 1992; Bicknell and Austin 1991; Bicknell, Austin et al. 1993; Bicknell, Bigg, and Austin 1993; Bicknell, Bigg et al. 1993; Bicknell, Godar et al. 1993; Bicknell, Hansen, and Mackey 1987, 1988.

43. Bicknell 1989, 1992; Heusser 1983; Adam and West 1983; West 1990; West 1993; Popenoe et al. 1992:1951.

44. Great gray owls' acute sense of hearing, facilitated by huge, hemispher-

ical facial disks, allows them to prey by ear on gophers and meadow mice concealed beneath dense meadow vegetation.

45. For evidence of Indian burning in meadows, see Reynolds 1959. For an eye-opening account from a rancher of the loss of dozens of meadows in the central Sierra Nevada due to absence of Indian and, later, rancher burning, see O'Neal 1953. For Native American weeding of young hardwoods after burning, see Clark 1894. For the view that the meadows are pristine, see Ratliff 1985.

46. For articles on tree invasion of meadows, see Taylor 1990; Bradley 1911; Helms and Ratliff 1987; Ernst 1949.

47. Wood 1975.

48. On Wintu use of hazelnuts, see Du Bois 1935:20. On the storage of hazelnuts in the shell, see Chesnut [1902] 1974:333; Peri, Patterson, and Goodrich 1982:121. On Yurok uses of hazelnuts, see Thompson [1916] 1991:31.

49. For the use of hazelnut for arrows, see Goodrich, Lawson, and Lawson 1980; Bocek 1984. On hazelnut for looped sticks, see Barrett and Gifford 1933:220.

50. For Karuk and Wiyot burning of hazel, see Driver 1939:334. For Yurok burning of hazelnut, see Thompson [1916] 1991:31.

51. On Karuk burning, see Baker 1981:26. On the quote by Georgia Orcutt, see Schenck and Gifford 1952:382.

52. Turner et al. 1990:190–91.

53. For contemporary pruning of hazelnut, see Heffner 1984.

54. On tobacco uses, see Lang 1996.

55. Gayton 1948b:269.

56. Zigmond 1981:43–44.

57. Driver 1937:84, 125.

58. Stewart 1941.

59. Barrett and Gifford 1933.

60. Thompson [1916] 1991:249.

61. Drucker 1937:239.

62. Silver 1978:222.

63. Harrington 1932:64.

64. On Concow Maidu uses of valley oak, see Curtis 1924a. On Yuki uses of valley oak, see Foster 1944. On Kawaiisu uses of valley oak, see Zigmond 1981.

65. On former extent of riparian forests, see Holing 1988. On Derby's comments, see Derby 1932.

66. For the Durán quote, see Chapman 1911:9.

67. Leonard 1978:140.

68. Bryant 1985:266.

69. Burcham 1982:79; Hilgard, Jones, and Furnas 1882.

70. Jepson 1923:167.

71. Cook 1960:248.

72. Moerenhout 1934a:74.

73. Moerenhout 1934a:73.

74. Stewart 1935:59–60.

75. Peri and Patterson 1979:25–26.

76. For acorn periodicity of valley oak, see Tietje 1990:80.

77. Priestly 1937:77–79.

78. Bryant 1985:267.

79. Farnham 1846:330–31.

80. For detritus accumulation in grasslands, see Knapp and Seastedt 1986. For light requirements of herbaceous plants, see Canham and Marks 1985.

81. See Arnold 1950; Biswell 1972:82–83.

82. Pavlik et al. 1991:11.

83. For uses of the resin from sugar pine, see Alta California, April 23, 1851, in Mitchell 1957:57. For other uses of sugar pine, see Farris 1982; Barrett and Gifford 1933:151; Hudson 1901b:G.H.M. Acc. No. 20,002; Powers [1877] 1976: 380; Zigmond 1981.

84. Barrett and Gifford 1933:150.

85. Merriam 1967:358–59.

86. For retrieval of pinecones, see Barrett and Gifford 1933; Goodrich, Lawson, and Lawson 1980.

87. Kilgore 1973:347; Chang 1996.

88. Sugihara and McBride 1996:44.

89. Mason 1955:42.

90. Native people also remember burning in areas of the giant sequoia, which grows in association with sugar pine:

> My husband's family talked about burning all the time. . . . They burned in areas where the giant sequoias grew. The trees were sacred. Burning helped the giant sequoia have a longer lasting life. It would also help the small sequoias. If there is an overgrowth of trees the sequoias don't have root space. Fires burn off a lot of duff so it helps the trees to breathe. (Clara Charlie, Chukchansi/Choynumni, pers. comm. 1991)

> They'd burn every year in some places for certain plants. In other areas they'd burn once every three years or once every two years in the giant sequoias. Burning cleaned the old brush out and helped the animals also. There would be more animals that would come—deer, bears, birds, squirrels, and they would find more food. Also the old people knew what trees were healthy and which trees were unhealthy. They knew that fires were going to help get rid of this tree because if too many bugs or woodpeckers mess with this tree it's going to cause harm to other trees. Fire was a cleanser. (Lalo Franco, Wukchumni Yokuts, pers. comm. 1991)

91. Ron Goode learned about burning from Lena Walker (who is three-fourths North Fork Mono) and Lena's father, Joe Kinsman.

92. On fire scar studies in montane forest with sugar pine, see Caprio and Swetnam 1995; Swetnam 1993; Kilgore and Taylor 1979. For the Christensen statement, see Christensen et al. 1987:iii.

93. On Native American burning augmenting lightning fires in montane

forests, see Reynolds 1959:60. The ecologist Bruce Kilgore (1973:347) says: "Growing evidence suggests the Indian people living on the western slope of the Sierra Nevada had developed a cultural pattern which had profound influence on the vegetation of the entire area, particularly on the foothill brush, woodland, and mid-elevation mixed conifer zones. By burning periodically, the Indians kept successional stages at pre-climax levels, thus insuring their supply of plants that provided suitable foods and materials."

94. See Otter 1963; Hitchcock 1950.

95. On burning of areas by sheepherders and cattlemen to improve the annual browse, see Sudworth 1900:572; Peattie 1953:13; Johnston 1970:100. For Muir quote, see Muir 1962:154.

96. Betty Jamison's father, John O'Neal, wrote an article in 1953 that described the rich pasturage for cattle and sheep in the open understories of the montane forests of the central Sierra Nevada in the late 1800s and early 1900s. These were kept open by Indian burning and later by burning by ranchers.

97. Bade 1905.

98. Sudworth 1900:554–55.

99. Lanner 1996:15.

CHAPTER 6. BASKETRY

1. On wastebaskets, see McCawley 1996:30. For baskets as funerary urns, see McCawley 1996:163.

2. For a description of baskets for removing the brown skin of ipo, see Howe 1979:110. For a description of baskets for serving trays, dishes, and cups, see Foster 1944:167; Bibby 1996:91; Clarke Memorial Museum 1985:72; Barrett 1910:268–69.

3. Brandes 1970:91.

4. Merrill 1923.

5. Forbes [1839] 1937:120.

6. For a discussion of the basket styles of weavers, see Abel-Vidor, Brovarney, and Billy 1996:16; Fulkerson 1995:5. For the quote concerning the Cahuilla, see Barrows 1967:43.

7. For Karuk basketry and specifically Karuk weavers commanding a high bride-price, see Clark Memorial Museum 1985:34. For baskets of foods being given as peace offerings, see Bolton [1927] 1971:141, 149, 152, 153.

8. For the Modoc legend, see Faulk and Faulk 1988:13.

9. See Yamane 1991:7.

10. For a description of sedge root harvesting, see Peri and Patterson 1993. On the meaning of *temalpakh*, see Bean and Saubel 1972.

11. Kroeber [1925] 1976:819.

12. O'Neale 1995:130.

13. Merrill 1923:217.

14. For the number of deergrass stalks in various baskets, see Anderson 1996a. For the number of rhizomes of sedge in various baskets, see Stevens 1999:43.

15. For the edibility of bear-grass rhizomes, see Weeden 1996; Jepson 1911:107.

16. For a description of bear-grass used in hats and pendants, see Driver 1939:392. For a description of the use of bear-grass in dance dresses, see Mason 1889:212–13; Nomland 1938:107.

17. Schlick 1994:99.

18. Jepson 1911:107.

19. For a description of harvesting and drying the leaves of bear-grass, see Nelson 1988:15. For a description of soaking bear-grass leaves, see Baker 1981:63. For a description of dyes, see Brodo, Sharnoff, and Sharnoff 2001; Goddard 1903–4.

20. For Hupa burning for bear-grass, see Goddard 1903–4; Sweet 1918:15.

21. Gibbs 1853:133.

22. Harrington 1932:63–64.

23. Gifford 1939.

24. Kroeber 1939b.

25. Although *Carex barbarae* and *C. obnupta* are most often mentioned in the literature, it is likely that most sedges with creeping rhizomes that formed beds were used, including *C. nebrascensis, C. utriculata, C. vesicaria* (mentioned in Mathewson 1998:165), *and C. lanuginosa* (mentioned in Allen 1972).

26. For the quote by Chesnut, see Chesnut [1902] 1974:314. For numbers of rhizomes in different basket types, see Stevens 1999:43.

27. For the importance of basketry to the economy of Indian households, see Bates and Lee 1990; Patterson 1998:13–14; Smith-Ferri 1998. For the price of baskets, see Chesnut [1902] 1974:317.

28. Gayton 1948b:159.

29. For raking around sedges, see Peri 1985.

30. On density of Pomo sedge beds, see Peri, Patterson, and Goodrich 1982:31.

31. Peri 1985.

32. For the quote concerning the ecological effects of pruning sedge root by Pomo weavers, see Allen 1972:16.

33. Allen 1972:19.

34. Latta 1977:539.

35. Beetle 1947.

36. For the number of tribes that used deergrass, see Anderson 1996a. On use of deergrass in Sierra Miwok coiled basketry, see Bates 1982a. On Prince Paul's observations, see Clark 1959. For Barrows quote, see Barrows 1967:42.

37. Although she is of Shoshonean ancestry, Florence Brocchini follows the basket weaving traditions of her husband's tribe, the Southern Sierra Miwok.

38. Patencio 1971.

39. Farmer 1993:145.

40. Anderson 1993b.

41. Shipek 1989; Lee 1989; Anderson 1996a; Lathrop and Martin 1982.

42. Lathrop and Martin 1982:9.

43. Cain 1961:98; Mathewson 1998.

44. See Franco 1993:18. For Kumeyaay burning of tules and cattails, see Shipek 1993:383. Driver (1937:374) recorded of the Mattole: "Annual burning in September. One side of river burned one year, other side following year." For Mojave burning of tules along the Colorado River, see Möllhausen 1858.

45. For burning of tule areas for diseases, see Mathewson 1998.

46. Sutter 1939:97, 113. Taylor [1850] 1968:73.

47. On the importance of bulrushes to wildlife, see Martin, Zim, and Nelson 1951. On the enhancement of wetlands with controlled burning, see Schlichtemeier 1967. For succession in wetlands, see Odum 1969; Noble and Slayter 1980.

48. For burning tule ponds for beaver capture, see Barrett and Gifford 1933. For Tachi Yokuts burning of tules, see Driver 1937:110.

49. Mason 1912:155.

CHAPTER 7. FROM ARROWS TO WEIRS

1. For a description of this archaeological find, see Coles 1989.

2. Rackham 1967. Coppice woodlands are cultural artifacts formed by silvicultural systems using the technique of pruning or burning to shape the plant architecture of the shrubs or trees. They are a worldwide phenomenon. See Buckley 1992 for coppice woodland management in Great Britain.

3. For a comparison of the anatomical structure of basketry materials before and after fire and the effects on the quality of the material, see Rentz 2003. See also Anderson 1991a, 1999; Mathewson 1998.

4. Anderson 1993b.

5. Anderson 1993b.

6. Kozlowski 1971.

7. Anderson 1993b.

8. Panshin and de Zeeuw 1980:292.

9. McMinn 1939; Panshin and de Zeeuw 1980:292.

10. Cornett 1995:17.

11. Zimmermann and Brown 1971:30, 35.

12. Not all shrubs regenerate vegetatively after fires. Some, after being burned, release seeds.

13. See Keeley 1986; Naveh 1975; Wells 1969.

14. Anderson 1993b.

15. Bates and Lee 1990.

16. For the use of willow by the Yosemite Miwok, see Clark [1904] 1987:38. For the use of young black oak by the Southern Sierra Miwok, see Merriam 1955:111.

17. Nicholson 1923.

18. Beemer 1980:23.

19. For a list of historical references to burning for basketry, see Anderson 1999.

20. For a more exhaustive list of elders and their quotes and general references on burning for material culture, see Anderson 1999, 1993b; Ortiz 1993a; Heffner 1984; Bates 1984.

21. Bohrer 1983.

22. For a description of basketry fragments found on the edge of Lake Winnemucca, see Orr 1956. For a description of the diffusion of basketry into Oregon and California, see Adovasio 1974:121. For a description of the archaeological evidence for baskets being utilized in California, see Dawson 1973. For a description of the archeological findings on San Miguel Island, see Connolly, Erlandson, and Norris 1995.

23. Anderson 1993b.

24. Dixon 1905:146.

25. Thompson [1916] 1991:31.

26. Merriam 1955:11.

27. Gayton 1948b:237; Clark [1904] 1987:67.

28. Dawson 1973.

29. For eaglet cages made by the Tubatulabal, see Voegelin 1938:21. For pigeon decoy cages, see Kroeber [1925] 1976:529. For pigeon snares made by the Yokuts, see Latta 1977:493. For rattlesnake cages, see Gayton 1948b:255. For Pomo traps for catching quail, see Barrett 1908.

30. For a description of baskets used in conjunction with fish weirs, see Hoover 1973:7; Fowler 1986:88; Spier 1978:473. For a description of basket traps for catching fish, see Kroeber [1925] 1976:529; Gayton 1948b:263; Fowler 1986:88; Kniffen 1939:376; Downs 1966:15.

31. For descriptions of skirts of native hardwoods, see Font 1971; Merriam 1955:27. For dance regalia made of maple, see Merriam 1955:27. For headdresses, see Bates 1982b:14–15; Merriam 1955:27.

32. For descriptions of rod armor, see Dixon 1905:205; Hudson 1903:G.H.M. Acc. No. 4455; Johnson 1978:353; LaPena 1978:329; Mason 1889; Wallace 1978:169; Silver 1978:218; McCorkle 1978:696; Hough 1893:641.

33. See Potts 1977:17; d'Azevedo 1986:477.

34. On the definition of cordage, see Mathewson 1985. On the antiquity of cordage as a craft, see Adovasio 1974. For maple lashings for ladders, see Barrett and Gifford 1933:195. For tule boat construction, see Bryant 1985:360.

35. On the range and distribution of Indian hemp, see Dempster 1993.

36. For the variety of items made from Indian hemp, see Anderson 1993b; Peri, Patterson, and Goodrich 1982; Mathewson 1985. For contemporary gathering of milkweed and Indian hemp, see McKinney and Castro 1999–2000; Ortiz 1999–2000.

37. On the extent of hemp fields, see Merriam 1955. On the numbers of stalks required for different cultural items, see Anderson 1993b. On the number of stalks for Washoe and Northern Paiute gill nets, see Lindstrom 1992.

38. For the season of harvest, see Barrett and Gifford 1933; Mathewson 1985. On details of burning Indian hemp by different tribes, consult Kniffen 1939; Anderson 1993b; Peri, Patterson, and Goodrich 1982:121, 122.

39. For games with ground lances, see Barrett and Gifford 1933:264; Culin [1907] 1975:498. For counter stick descriptions, see Barrett and Gifford 1933:266; Culin [1907] 1975:140; Fowler and Liljeblad 1986:453.

40. For descriptions of the hoop-and-pole game, see Voegelin 1938:48; Chalfant 1933:87; Culin [1907] 1975:482, 523; Gayton 1948b:148; Barrett and Gifford 1933:263. For a description of basketball play by the Sierra Miwok, see Barrett and Gifford 1933:259. For a description of a ball game played by the Yokuts, see Culin [1907] 1975:596. For information on popguns, see Culin [1907] 1975:759.

41. Wallace 1978:647. On contemporary gathering and preparing of elderberry and the making and playing of elderberry flutes, see Cunningham-Summerfield 1998.

42. For descriptions of hemispherical houses, see Gayton 1948b:220; d'Azevedo n.d.; Voegelin 1938:24; Nabokov and Easton 1989:56; Latta 1933a; Latta 1977:349. For the Northern Paiute house description, see Wheat 1967:104.

43. For descriptions of conical houses, see Barrett and Gifford 1933:198; Gayton 1948b:260; Gayton 1948a:63.

44. For a description of fish weirs, see Drucker 1937:232; Gayton 1948b:263; Voegelin 1938:14; Goddard 1903–4:24; Hudson 1899. For a description of a Wukchumni Yokuts fish weir, see Gayton 1948a:75.

45. For a description of rodent sticks, see Wheat 1967:116; Fowler 1986:83; Gayton 1948b:262. For a description of looped stirring sticks, see Clark [1904] 1987:43; Gayton 1948:79, 266; Voegelin 1938:18; Inter-Tribal Council of Nevada 1976:95. For the memory of harvest of young growth after fires or pruning for the manufacture of household utensils by contemporary Native American elders, see Anderson 1993b.

46. For a description of plants and their uses for arrows, see Merriam 1955:22; Chesnut [1902] 1974:35, 332; T. Kroeber 1961:191; Goddard 1903–4:34; Barrett and Gifford 1933:217; Anderson 1993b.

47. For the pruning of spicebush and snowberry by the Pomo, see Peri and Patterson 1979:21, 28. For a description of burning shrubs for arrows and bows by the Luiseño, see Shipek 1977:118. For the quote regarding Paiute and Shoshone pruning of snowberry, see Murphey 1959:52.

48. Anderson 1993b.

49. Anderson 1993b.

50. For a description of harpoons, see Barrett and Gifford 1933:189; Voegelin 1938:14; Gayton 1948a:75; Potts 1977:44; Inter-Tribal Council of Nevada 1976:9; Wheat 1967:65.

51. Allen 1972:14.

52. For the necessity of pruning mature shrubs for dead wood, insects, and diseases, see Brown 1977. For a discussion of root-to-shoot ratios of shrubs and trees, the changes in photosynthesis, and ecological effects on the plant, see Barbour, Burk, and Pitt 1980.

53. Wilson 1970.

54. For information on parasites in plants used by the Pomo for basketry,

see Goodrich, Lawson, and Lawson 1980:42; Hudson n.d.:G.H.M. Acc. No. 21,170. For insect problems experienced by contemporary weavers, see Ortiz 1998:25; Mathewson 1998. For use of indigenous fire to reduce insects, see Mathewson 1998. For studies confirming that fire eliminates or reduces diseases and insects, see Hardison 1976; Abrahamson and Hartnett 1990; Mutch 1994.

55. Martinez 1993.

CHAPTER 8. CALIFORNIA'S CORNUCOPIA

1. Latta 1977:703.

2. On descriptions of abundance in Cahuilla territory, see Barrows 1967:70. On the abundance of yampah, see Luther Burbank Society 1914.

3. Carson 1950:94.

4. Fowler 1996.

5. For the portion of the California Indian diet devoted to plant foods, see Heizer and Elsasser 1980; Powers [1877] 1976:406; Kroeber [1925] 1976:814. For quote, see Kroeber [1925] 1976:523.

6. For the harvesting of salt from saltgrass, see Essene 1942:9. For the preparation of coltsfoot for salt by the Yuki, see Murphey 1959:26. For the harvest of pitch for sugar, see Barrett and Gifford 1933. For the harvest of sugar from common reed, see Sutton 1988. For the harvest of a sweet substance on willows by the Kawaiisu, see Zigmond 1981.

7. For the number of edible wild and cultivated plants, see Prescott-Allen and Prescott-Allen 1990.

8. For mention of salal berries, see Powers [1877] 1976:51. For the use of deer brush seeds by the Concow Maidu, see Duncan 1964:12. For the Sierra Miwok way of sipping manzanita cider, see Barrett and Gifford 1933:162.

9. For the taste of balsam-root seeds, see Murphey 1959:26. For the taste of tarweed, see Powers [1877] 1976:425. For the taste of buttercup seeds, see Powers [1877] 1976:426. For the taste of purple nutsedge, see Murphey 1959:16. For the taste of the buds of barrel cactus and Mohave yucca, see Bean and Saubel 1972:68. For contemporary uses of plants for foods by Sierran tribes, see Anderson 1988, 1993b.

10. The first Muir quotation is from Fleck 1985:89; the second is from Muir [1911] 1944:79.

11. For use of lemonadeberry by the Kumeyaay, see Hedges and Beresford 1986:37. For use of the barrel cactus by the Cahuilla, see Bean and Saubel 1972:68. For the use of chia, see Hedges and Beresford 1986:41; Murphey 1959:28.

12. For the use of buds, inner bark, and young pine cones, see Powers [1877] 1976:196, 235. For a description of the diet of the lone woman on San Nicolas Island, see Nidever 1976:12.

13. Vizcaíno quote in Bolton 1916:85. For a description of the giving of brine flies by the Mono Lake Paiute to Brewer, see Brewer 1966:417. For the giving of pine nuts to Lynch and the regiment, see Lynch 1954:31. For offering food to a member of the Donner party, see Reid 1983.

14. On sharing among the Wiyot, see Driver 1939:374. On sharing among the Desert Cahuilla, see Strong 1972:77–78. On sharing among the Mattole see Driver 1939:383. On sharing among the Tolowa, see Driver 1939:380. For wealth-spreading devices operating among other tribes, see Curtis 1924a; Bee 1983.

15. Garth 1945:557.

16. For an assessment of the honesty of Native Americans, see Powers [1877] 1976; Kneeland 1872; Gayton 1948a:65.

17. For the Barrows comment, see Barrows 1967:53. For legends that reinforce sharing and shun greed, see LaPena and Bates 1981; Blackburn 1975:75–76; Margolin 1981:78–79.

18. For respecting the resources of other villages or tribes, see Frémont 1887:446–47; Goldschmidt 1951:333; Barrett 1952:50. For resource relationships among the Achumawi, Northeastern Maidu, and Northern Yana, see Garth 1945:359. For the Tolowa sharing of gathering and hunting tracts with outsiders, see Drucker 1937:243. Throughout California's prehistory there was some intergroup conflict for both personal and social reasons. Some of these conflicts involved trespassing on a group's territory or contests over group rights to food and other economic resources. See McCorkle 1978; Nelson 1997.

19. For Wappo names, see Driver 1936. For Yokuts names, see Latta 1977:695.

20. For Sinkyone names, see Nomland 1938. For Miwok names, see Gifford 1916.

21. For the ethnobotany and myths of desert cultures, see Nabhan 1985. For the Cahuilla belief that all food was once human, see Hooper 1920:356. For Washoe legends, see Dangberg 1968:32.

22. For gathering prayers of the Atsugewi, see Schulz [1954] 1988:60. For Modoc adornments worn during root digging, see Ray 1963:26. For examples of first food rites and harvest season ceremonies, see Barrett and Gifford 1933:157–58; Swezey 1975.

23. For example, archaeological models based on notions of political economy that ignore intensification of food resources as possible causes and consequences of cultural evolution have their roots in the thinking and writings of Morgan, Spencer, and other social evolutionists of the nineteenth century (Raab 1996:67).

24. Because most California Indian tribes did not practice agriculture, they were tagged as hunter-gatherers by anthropologists. This term connotes a nomadic people living in small groups, using a home base or camp, sharing collected food resources, and having males doing the hunting and females the gathering. Although some of these characteristics can be accurately applied to California Indians, others (particularly the idea of being nomadic) distort the ways most groups actually lived. Even worse, in the popular mind *hunter-gatherer* is equated with people who live at the whim and fancy of nature because they are incapable of directing or managing nature for food.

25. For descriptions of the intellectual superiority of civilization over savagery, see Lubbock 1882; Morgan 1877; Pearce 1988. For an explanation of the

evolutionary stages of humankind, see Morgan 1877. For the Powell quote, see *Star* 1889.

26. Lubbock 1882:2. See Brace 1869.

27. Hittel 1897:732. For a discussion of the term "digger" and the Yount quote, see Lonneberg 1980:5.

28. Powers [1877] 1976:419.

29. Kroeber [1925] 1976:524–25.

30. Harlan and de Wet 1973.

31. For an example of a human ecology textbook that is organized according to the notion of human progress, see Richerson, Borgerhoff Mulder, and Vila 1996. For a discussion of the evolutionary progression from hunter-gatherer to agriculturist, see Trigger 1980; Lomax and Arensberg 1977.

32. In recent years archaeological evidence for intensification of food resources, particularly abundance of small seeds, has mounted, leading to hypotheses that suggest deliberate indigenous management such as burning to increase production of small-seeded plants. See Wohlgemuth 1996; Raab 1996; Fagan 2003; Basgall and Bouey 1991.

33. Harlan 1995:15. There is strong evidence for multiple domestications of the same species in different parts of its range. This is true for chile peppers (*Capsicum* spp.), cotton (*Gossypium* spp.), and maize (*Zea* spp.) (Pickersgill 1989:436). "Domestication is thus no longer seen as a unique or particularly difficult event" (Pickersgill 1989:437).

34. See Bean and Lawton 1993.

35. See Jackson and Jackson 2002; Buchmann and Nabhan 1996.

CHAPTER 9. PLANT FOODS ABOVEGROUND

1. The importance of seeds and grains in the indigenous diet is represented in the archaeological record. See Fitzgerald and Jones 1999; Wohlgemuth 2004b. For a recent description of locations with abundant and diverse wildflower displays in California, see Griggs 2000; Faber 1997. For the use of the seeds of *Plagiobothrys canescens* among the Tulare County Indians, see Eastwood n.d. For the use of the seeds of *Plagiobothrys nothofulvus*, see Anderson 1993b. For the gathering of seeds of *Balsamorhiza sagittata* by the Sierra Miwok and the Atsugewi, see Barrett and Gifford 1933:152; Garth 1953:139. For the gathering of seeds for food of *Castilleja exserta* (formerly *Orthocarpus attenuata*) by the Sierra Miwok, see Barrett and Gifford 1933. For the harvest of seeds of *Layia platyglossa* by the Ohlone, see Bocek 1984.

2. For gathering the seeds of *Blennosperma nanum* in the vernal pools of the lower foothills of Placer County by the Nisenan (Southern Maidu), see Powers [1877] 1976:425. For gathering of *Lasthenia glabrata* in vernal pools by the Cahuilla, see Barrows 1967:65. For gathering *Wyethia mollis* seeds for food, see Miller 1928. For gathering *Clarkia purpurea* seeds by the Yuki and Sierra Miwok, see Chesnut [1902] 1974; Barrett and Gifford 1933. For gathering *Epilobium densiflorum* (formerly *Boisduvalia densiflorum*) seeds by the Pomo and Sierra Miwok, see Barrett 1952; Barrett and Gifford 1933. For harvesting of *Am-*

sinckia menziesii seeds by the Atsugewi, see Garth 1953:139. For gathering of *Madia elegans* seeds in grasslands by the Indians on Tule River Reservation, see Eastwood n.d. For gathering *Madia elegans* seeds in coastal prairies, see Schenck and Gifford 1952. For the Will Green quote, see Green 1895:282.

3. For use of grass grains, see Bocek 1984; Steward 1933. For the Latta reference, see Latta 1977:551. Wohlgemuth has found increased caryopsis size through time for native barley (*Hordeum* sp.) and maygrass (*Phalaris* sp.) in Sacramento Valley samples. While these data are preliminary, it may mean that native people were selecting for larger grain size, one of the crucial steps in grain domestication. See Wohlgemuth 2004b.

4. The record of small seed and grain use is not static through time, or the same in different regions. In interior central California (the Central Valley, the southern North Coast Ranges, and the Bay Area south to Gilroy), small seed use increases dramatically at about 1,200 to 1,000 years ago. This intensive use continued into the historic period. Before that, small seeds and grains are much less common in the record, and one or two taxa dominate (maygrass and goosefoot). After about 1,200 years ago, not only are seeds and grains much more common in the record, they are much more diverse, with no taxon dominant, as in earlier sites. Outside of central California the sequence of small seed and grain use is much different. The data are sparse, but we have no evidence of intensive small seed and grain use throughout prehistory (Eric Wohlgemuth pers. comm. 2004).

5. Gayton 1948a:115.

6. Shirley 1937:7.

7. For gathering of red maids seed, see, e.g., Goldschmidt 1978; Barrett and Gifford 1933; Bocek 1984. For red maids found in a Santa Rosa Island grave, see Orr 1968:200. A radiocarbon date of 600 ± 70 B.P. was obtained for the seeds. For Nomlaki and Yuki quantities of seed harvested, see Chesnut [1902] 1974: 346–47; Rusby 1906:69. According to Wohlgemuth, "A late prehistoric cremation from Pleasanton had tens of thousands of charred *Clarkia* seeds, and achenes of an unidentified plant in the aster family, and hundreds of cleaned acorn kernels as offerings. Also, a historic Patwin burial between Grimes and Knights Landing had tens of thousands of cleaned *Atriplex patula* seeds" (pers. comm. 2004).

8. For quantities of chia gathered by the Paiute, see Embody 1907. Chia was (and still is) important to many tribes. For use of chia among the Chumash, see Timbrook 1986. For Southern Maidu quantities of seeds, see Duncan 1964. For Modoc and Klamath quantities of seeds, see Kroeber [1925] 1976:325.

9. On Muir's observations, see Muir [1911] 1944:226. On grains of *Achnatherum speciosum* gathered by the Mono Lake Paiute, see Fletcher 1987:4. On gathering of *Elymus condensatus* and *Achnatherum speciosum* grains by the Owens Valley Paiute, see Steward 1934. On grain gathering by the Washoe, see DeQuille 1963:83.

10. For storing grass seeds in granaries, see Barrett and Gifford 1933:207. For storing different seed varieties separately, see Foster 1944:166; Barrett 1952:86. For Washoe storage methods, see Curtis 1924d:93–94.

11. Bunnell 1980:71.

12. Barker 1955.

13. Brandes 1970:92.

14. Latta 1933b:5.

15. On eating seeds dry with the fingers, see Gifford [1939] 1965:26. For a discussion of size of mortar cups for different seeds, see Hudson 1901b:G.H.M. Acc. No. 20,002; Barrett and Gifford 1933:208; Melba Beecher, pers. comm. 1997. For a discussion of the earliest seed processing equipment, see Moratto 1984. For the Curtis quote, see Curtis 1924c:62.

16. For moistening of seed meal by the Southern Maidu, see Hudson n.d.a:G.H.M. Acc. No. 20,011. On yampah seeds (also called anise seeds) used by the Pomo, see Barrett 1952:62.

17. For a discussion of pinole, see Barrett 1952:85.

18. On seeds gathered by tribes of Mendocino County, see Chesnut [1902] 1974. On seeds gathered by the Kitanemuk, see Harrington 1986.

19. On a description of balls of seeds eaten by the Pomo, see Barrett 1952:52. For the process of forming seed balls described by Palóu, see Dane 1935:109.

20. Doebley 1984:53. See Hillman and Davies (1999), who suggest that sickle harvesting and subsequent sowing of wild wheat and barley can lead to genetic isolation of strains and promote, in a matter of as few as two decades, cultigen traits such as reduced grain dormancy, shatter-resistant inflorescences, and perhaps increased grain size.

21. For the broadcasting of seeds by seven tribes, see Driver and Massey 1957:225, 228. On the Wappo sowing of seeds, see Beard 1979:52. On the sowing of seeds by the Chumash, see Miller 1988:87. On the sowing of seeds by the Southern Paiute, see Bye 1972:91.

22. For sowing of seed by the Paiute, see Steward 1938a:104.

23. Kroeber [1925] 1976:736.

24. Kroeber [1925] 1976:467.

25. Moncada 1774 quoted in Clar 1959:7.

26. Driver and Massey 1957:225.

27. Gayton 1948b:176.

28. On burning on the Eel River, see Gibbs 1853. For burning for better wild seed crops by tribes of northwestern California, see Driver 1939:314, 381. For burning wild rice plants by the Karuk, see Harrington 1932:64.

29. On Pomo setting of fires, see Peri and Patterson 1979:43. On Voegelin's documentation of tribes burning to improve wild seeds, see Voegelin 1942:57. On Washoe setting of fires, see Van Etten 1994:24. On Kumeyaay burning of fields and broadcasting of seed, see Shipek 1989:3. On burning for chia by the Cupeño and Cahuilla, see Drucker 1939:9; Bean and Saubel 1972:137.

30. Stewart 1941:376.

31. Among the California Indians who remember burning for seed crops are Clara Charlie, Rosalie Bethel, Francys Sherman, and Lalo Franco.

32. Harrington 1932:85–86.

33. For Central Sierra Miwok cultivation of seeds, see Hudson 1901a:

G.H.M. Acc. No. 20,004. For the *nu'wa si* identification, see Eastwood n.d. For the *noco'pai yu* identification, see Barrett and Gifford 1933:154.

34. Smith 1995:21.

35. On grasslands and detritus, see Knapp and Seastedt 1986.

36. Cronise 1868:525.

37. Mayfield 1993:34.

38. Lowry (pers. comm. 2004).

39. Barrett and Gifford 1933.

40. Muir [1911] 1944:58.

41. Perlot 1985:194.

42. Szczawinski and Turner 1980:13; Kuhnlein and Turner 1991.

43. On the leaves of a *Phacelia* saving a tribe from starvation, see Eastwood n.d. For greens being mentioned in indigenous calendars, see Strong [1972] 1987:253.

44. See Barrett and Gifford 1933; Bocek 1984.

45. On the harvest of yampah for greens by the Wiyot, see Loud 1918. On the use of cow parsnip for food, see Goodrich, Lawson, and Lawson 1980; Collier and Thalman 1991. For the use of cattails and tules, see Knap 1975:99.

46. Barrett and Gifford 1933; Sapir and Spier 1943:251.

47. Goodrich, Lawson, and Lawson 1980:108.

48. For the use of *Yucca whipplei*, see Bean and Saubel 1972; Sparkman 1908.

49. Barrett and Gifford 1933:160–61.

50. On ways to prepare clover, see Schulz [1954] 1988:62. For a description of eating clover among the Choynumni, see Mayfield 1993:68.

51. On a description of the Wintun (Wintu) harvest of clover, see Powers [1877] 1976:235. On the Nisenan harvest, see Duncan 1964:34.

52. Curtis 1924a:107.

53. Dixon 1905:183.

54. Mooney 1890:259.

55. On Coast Miwok use of checker mallow, see Collier and Thalman 1991. On Nisenan use of dandelion greens, see Duncan 1964:12. On Kawaiisu use of phacelias, see Zigmond 1981.

56. Harlan 1992:233; Barrett and Gifford 1933:159–60.

57. On cooking milkweed, see Lucas 1945. On use of California poppy, see Powers [1877] 1976:425. On gathering of *Stanleya pinnata*, see Murphey 1959:23. On preparation of *Stanleya* and *Caulanthus* by the Panamint, see Coville 1892:354.

58. Paige 1992. For a definition of compensation, see Ricklefs and Miller 2000:725. For a definition of overcompensation, see Ricklefs and Miller 2000:737.

59. Knap 1975:43.

60. Chalfant 1933:76.

61. Fowler 1996:96.

62. Harrington 1986.

63. Duncan 1964:75.

64. On Wappo burning for greens, see Beard 1979:52. On Maidu burning

for greens, see Duncan 1964:9. On Yokuts and Sierra Miwok burning for greens, see Aginsky 1943:403.

65. On burning for clovers and bracken fern among the Pomo, see Peri, Patterson, and Goodrich 1982:119. On the Stewart quote, see Stewart 1935. On burning by the Yana, see Sapir and Spier 1943:246.

66. Simpson 1938:51.

67. On the ecology of lupines, see Haq 1993:114. On uses of lupines for edible greens, see Timbrook 1990; Clark [1904] 1987:48; Anderson 1993b; Barrett and Gifford 1933.

68. For harvest of wild fruits by California tribes, see Barrett and Gifford 1933; Bean and Saubel 1972; Chesnut [1902] 1974.

69. On wild strawberries, see Heizer and Kroeber 1979:37. On harvest of high elevation elderberry, see Gayton 1948b:175. On the nutritional value of wild fruits, see Turner and Szczawinski 1979; Kuhnlein and Turner 1991.

70. For harvesting methods of the Sierra Miwok and Chukchansi Yokuts, see Neely 1971; Thrall and Gayton 1938; Barrett and Gifford 1933. On the harvesting methods of the Cahuilla, see Bean and Saubel 1972. On the harvesting methods of the Yokuts, see Latta 1977:555. On harvesting and cleaning methods of the Yurok, see Warburton and Endert 1966:107–10.

71. Latta 1977:555.

72. On wild plum abundance, see Wistar 1937:262. For a description of holly-leaved cherry abundance, see Barrows 1967:70.

73. Powers 1856:187.

74. On Washoe travel to gather chokecherries, see Downs 1966:19. On distance to gather huckleberries, see Chesnut [1902] 1974:378. On Cahuilla labor shifted to domesticated fruits, see Barrows 1967:70.

75. On Yokuts blackberry gathering, see Gayton 1948a:115. On Karuk harvest of huckleberries, see Schenck and Gifford 1952.

76. On Kumeyaay harvesting of prickly pear, see Hedges and Beresford 1986. On gooseberry harvesting and preparation by the Pomo, see Goodrich, Lawson, and Lawson 1980:51.

77. On processing and use of holly-leaved cherry by tribe, see Timbrook 1982.

78. On use of fan palm fruits by the Cahuilla, see Bean and Saubel 1972:146. On fruit drying by the Kumeyaay, see Hedges and Beresford 1986. On fruit drying near Cahuilla homes, see Barrows 1967:52. On the use of California bay nuts by the Kashaya Pomo, see Goodrich, Lawson, and Lawson 1980: 91.

79. On fruit cooking methods, see Brubaker 1926.

80. For the use of sourberries, see Anderson 1988.

81. Goldschmidt 1951:410. For palm fruit preparation, see Bean and Saubel 1972:146.

82. Merriam 1902a:210–11. One of the most common fruits in archaeological sites is manzanita. At one site in Redding there are about 10 million charred manzanita pits (E. Wohlgemuth pers. comm. 2004).

83. Quick 1962; Keeley and Keeley 1989.

84. For Wiyot burning for berries, see Driver 1939:381. On pruning and

burning of berry bushes by the Pomo, see Peri, Patterson, and Goodrich 1982:122, 126. On Karuk burning, see Harrington 1932:63.

85. Jack 1916:194–95.

86. Jack 1916:194–95. On Concow Maidu burning, see Jewell 1971.

87. On Yurok burning, see Warburton and Endert 1966:104–5, 110.

88. For how pine nuts were prepared, see Goodrich, Lawson, and Lawson 1980; Farris 1982:39. For use of pine nuts in northern California, see Farris 1993. For the meaning of Atsuge, see Garth 1978:236.

89. For uses of pines, see Barrett and Gifford 1933; Gifford 1932; Lowie 1939:328; Price 1980:49, 50; Mathewson 1998.

90. On Sierra Miwok roasting of young gray pine cones, see Neely 1971.

91. For the Muir quote, see Muir 1962:72. On putting gray pine cones in the fire, see Merriam 1902a:206–7.

92. Jepson 1923:45.

93. For the Merriam quote, see Merriam 1905:595; see also Clyman 1926b: 277. For fire return intervals, see McClaran and Bartolome 1989, cited in Stephens 1997. For Mensing, see Mensing 1988, cited in Skinner and Chang 1996.

94. Stewart 1935:40–41.

95. Hutchings 1990:189.

96. Gayton 1948b:176.

97. Ledig 1999.

98. Jackson and Spence 1970:615.

99. On pinyon evidence at Gatecliff Shelter, see Lanner 1981:56. On Owens Valley archaeological evidence, see Bettinger 1976. On archaeobotanical research, see Basgall and McGuire 1988.

100. Fowler 1996.

101. Burcham 1940:5.

102. On fire frequency and tree encroachment, see Wright, Neunschwander, and Britton 1979; Young and Evans 1981. On increasing herbaceous production through controlled burning, see Barney and Frischknecht 1974.

103. On historic overgrazing of pinyon woodlands, see Everett 1987:152.

104. Burcham 1940:8.

105. On contemporary pinenut gathering by the Washoe, see Fillmore 1995–96. On contemporary pinenut gathering by the Southern Sierra Miwok, see Anderson 1988.

106. For the Gifford quote, see Gifford 1971:301. On species of oaks exploited for acorns generally, see Driver and Massey 1957:209. On black oak acorn use, see Latta 1977:393; Chalfant 1933:80; Bean and Saubel 1961:183; Merriam 1955:89. On tan oak acorn use, see Levy 1978:491; Bright 1978:182; Kelly 1978:416. On coast live oak acorn use, see Grant 1978:516; Mason 1912:118.

107. See Heizer 1978:690; Pavlik et al. 1991:97. For Paiute travel to the west side for California black oak acorns, see Muir [1911] 1944:228. On contemporary use of acorns, see Ortiz 1991.

108. Colton 1860:119–20.

109. On estimates of the beginning of intensive exploitation of acorns in central California and the Sierra Nevada, see Wohlgemuth 2004a.

110. On the use of fire as a management tool for oaks by California Indians generally, see McCarthy 1993. On records of fire use by specific tribes, see Peri and Patterson 1979:25–26 for the Dry Creek and Cloverdale Pomo; Kniffen 1939 for the Kashaya Pomo; Beard 1979:52 for the Wappo; Driver 1939:314, 381, for the Yurok and Tolowa; Shipek 1977:117 for the Luiseño; Duncan 1964 for the Maidu; Kroeber [1925] 1976 for the Ohlone.

111. Duncan 1964:9.

112. For the Kniffen citation, see Kniffen 1939:378. On Yurok burning to improve visibility of acorns, see Driver 1939:314.

113. On burning by the Tolowa to kill parasites, see Driver 1939:381.

114. Kauffman and Martin 1987:124.

115. Roos-Collins 1990:1–2.

CHAPTER 10. PLANT FOODS BELOWGROUND

1. Bolton [1927] 1971:291.

2. Latta 1933b.

3. Hargreaves 1996:88. For a review of the botany and physiology of flower bulbs, see De Hertogh and Le Nard 1993. For an overview of California's geophytes, see Rundel 1996. The dictionary definition of geophyte is a land plant that survives an unfavorable period by means of underground food storage organs. Buds arise from these to produce new aerial shoots when favorable conditions return (Allaby 1985).

4. For information on edible bulbs, see, e.g., Barrett and Gifford 1933; Bean and Saubel 1972; Chesnut [1902] 1974; Goodrich, Lawson, and Lawson 1980. For use of tubers of *Sanicula tuberosa*, see Eastwood n.d.

5. "Root" foods is not technically correct except for tap-rooted species such as balsam-root (*Balsamorhiza* spp.) and young rootstocks of plants such as cattails.

6. Using a digging stick was universal among tribes. See, e.g., Kelly 1932; Nomland 1935. For reference to the fire hardening of the digging stick, see Powers [1877] 1976.

7. For management of geophytes, see Peri 1985; Anderson 1997; Anderson and Rowney 1998, 1999.

8. While it is true that root foods are underrepresented, we do find charred *Brodiaea/Dichelostemma/Triteleia* corms in archaeological sites that date as early as 5500 B.P. (E. Wohlgemuth, pers. comm. 2004).

9. For prominence of tubers, bulbs, and corms in the Modoc diet, see Spier 1930:164. For prominence of corms and roots in Great Basin diets, see Fowler 1986. For prominence of *ipos* (*Perideridia* spp.) and camas in the eastern Apwaruge (division of the Atsugewi) diet, see Garth 1978:243. Root foods were important enough in the Cahto diet to be assigned as the name for the winter season called "root time" (Myers 1978:246). For the importance of geophytes in the Pacific Northwest, see Hunn and French 1981:87.

10. Harshberger 1928:1.

11. Latta 1977:144.

12. For the Hupa generic name for "bulbs," see Goddard 1903–4. For the Central Sierra Miwok names of plants with edible roots, see Barrett and Gifford 1933. For how the Indians of Mendocino County distinguished a plant with an edible bulb or tuber, see Chesnut [1902] 1974. For the Sinkyone myth, see Nomland 1935:172.

13. For an analysis of the protein content of camas bulbs, see Driver and Massey 1957:209. (According to Powers [1877] 1976, camas was used as an emergency food if the acorn crop failed.) For an analysis of the protein content of acorns, see Baumhoff 1963:162.

14. For a Wukchumni Yokuts woman's testimony to the benefits of eating grassnuts, see Latta 1962. For the foods eaten and avoided by the Pomo after childbirth, see Hudson n.d.b:G.H.M. Acc. No. 21,029:6.

15. Timbrook 1993.

16. Inulin is a carbohydrate that becomes sweet upon cooking, due to a partial conversion to the sugar fructose (see Kuhnlein and Turner 1991:9–10).

17. For accounts of cooking and storing bulbs, see Spier 1930; Chesnut [1902] 1974; Barrett and Gifford 1933. For Breen's diary entries, see Reid 1983.

18. For length of time to store camas, see Spier 1930:164; Miller 1987:22.

19. For reference to *koo-nuk*, see Latta 1934:26. For a description of the density of *Triteleia laxa*, see Chesnut [1902] 1974. For arrowhead gathering by the Klamath, see Spier 1930. For brodiaea gathering by the Yahi, see Pope [1916] 1979:229. For gathering of wild onions by the Washoe, see Powers 1856:189–90.

20. For yampah gathering, see Coville 1897:101. For a description of grassnuts in Paiute territory, see Hudson 1904:G.H.M. Acc. No. 20,017:44.

21. Faulk and Faulk 1988:18.

22. Latta 1933b.

23. Purdy 1976:114.

24. Nomland 1938.

25. For potential benefits of small and large mammal digging of geophyte propagules, see Roof 1981; Lovegrove and Jarvis 1986; Work 1995. The relationship of the decline in geophytes to the absence of Indian management is unknown and much research is needed.

26. Harrington 1932:73.

27. For replanting of smaller corms of blue dicks, see Bean and Saubel 1972:48. For replanting small bulbs of lilies, see Heffner 1984:52. For regeneration through harvesting or dividing underground stems, see Sauer 1952.

28. For Chumash harvesting after seeding, see Timbrook 1993:56–57.

29. Anderson 1993b.

30. Anderson 1993b.

31. Quote is from Rice 1920b:115. In the *Jepson Manual*, Skinner (1993: 1200) notes that the Washington lily is found in "[c]onifer forest, esp gaps, burned clearcuts."

32. Murphey 1959:16.

33. Coville 1897, quoted in Spier 1930.

34. Edwards 1996.

35. Harrington 1932:66.

36. For Wintu burning of Indian potatoes, see Shepard 1989:411.

37. For a study on flowering in relation to light, see Stone 1951. For conditions necessary for resting bulb bud formation, see Mathew 1997:3.

38. For references to the fire followers, see Keeley et al. 1981; Sampson 1944; Sweeney 1956.

39. For burning to promote geophytes in the Kalahari Desert, see Silberbauer 1981:270. For San burning in South Africa, see Schapera 1930:140. For Australian Aborigine burning, see Gott 1983. For Indian burning of camas fields in the Pacific Northwest, see Marshall 1999; Turner and Kuhnlein 1983:211; Turner 1999.

40. For information on rare and endangered geophytes, see Rundel 1996; California Native Plant Society 2001; Skinner and Pavlik 1994.

41. On biology of rare geophytes, see Fiedler 1996:36. On human causes of rarity of lilaceous plants, see Fiedler and Ahouse 1992.

42. Thinning as a strategy to maintain bulb populations of certain species is conjectural and requires further research.

43. Anderson 1992a.

44. Bowcutt pers. comm. 2004; Bowcutt 1994–96:93. See Schenck and Gifford 1952 for Karuk harvest of redwood lily. See Hansen 1899 for former abundance of different lily species in the open pine woods and how some human disturbance favors them.

45. For a sample experiment that simulates indigenous tilling practices, see Anderson and Rowney 1999.

CHAPTER 11. CONTEMPORARY HARVESTING AND MANAGEMENT PRACTICES

1. Larry Myers (Pomo), Executive Secretary, California Native American Heritage Commission, pers. comm. 2003.

2. Hinton and Montijo 1994:21.

3. On Indian cultural museums, see Magallanes and Thompson 2003; Farmer 2003; Jeffrey-Hinton 2001; Apodaca 1990. On language revitalization programs, see Supahan and Supahan 1995; Whittemore 1997; Dominguez 1998; Hinton 1994b, 1999. On intertribal gatherings, see Blue Cloud 1998.

4. For information on contemporary gathering activities of California Indian tribes, see Ortiz and Staff 1991; Billy 1994; Dick et al. 1988; Lee 1998; Yamane 2001; back issues of *News from Native California* published by Heyday Books, and *Roots and Shoots*, newsletter of the California Indian Basketweavers Association. For a detailed account of the steps involved in acorn processing, see Ortiz 1991; Alvarez and Peri 1987.

5. On modern tools used in pruning and digging, see Allen 1972:18; Peri, Patterson, and Goodrich 1982:29, 35.

6. Hunter 1988; Ruppert 2003; see also Pacific Southwest Region Accomplishment Reports 1992–2004.

7. Mendelsohn 1983:59.

8. On prescribed burning of bear-grass, see Nicola 1995; Stauffer 1993; Hunter 1988; Lee 2000; Dondero 1992. For contemporary gathering of bear-grass and discussion of quality, see Heffner 1984:16; Ortiz 1998.

9. Latta 1977:538.

10. Peri and Patterson 1993:185.

11. Peri and Patterson 1993:188. On cutting back aboveground leaves, see Mathewson 1998:171.

12. For length of rhizomes in cultivated sand root beds, see Peri and Patterson 1979:55. For numbers of rhizomes in a coil and numbers of rhizomes harvested, see Peri, Patterson, and Goodrich 1982:89.

13. Anderson 1992b; Lorrie Planas, pers. comm. 2003; For Caltrans burning of deergrass at the request of tribal elders of Tule River Reservation, see Stewart 2000.

14. Maidu Cultural and Development Group 2002.

15. For a history of the Timbisha Shoshone and their relationship with the National Park Service and the BIA, see Crum 1998.

16. On mesquite seed meal, see Driver 1937:68–69. On roasting green pods, see Fowler 2003.

17. Fowler 1996.

18. On Timbisha Shoshone management of pinyon and its reflection on the landscape, see Fowler 1996.

19. On the launching of the cooperative management project, see Fowler 2003. For the objectives of the project, see Fowler et al. 2003.

20. Lyon and Cassidy 1998.

21. On herbicide-free collection areas, see Deal et al. 1994. Information on the fencing of willow and sedge sites in Sierra National Forest from Joanna Clines, Forest Botanist, pers. comm. 2003.

22. Anderson 1993b; Heffner 1984; Ortiz 1993a.

23. Margolin 2001; Alvarez 1994. For threats to Pincushion Mountain, see Greer 1992.

24. On ratification of treaties, see Heizer 1972.

25. Larry Myers (Pomo; pers. comm. 2004), executive secretary of the Native American Heritage Commission, says: "Approximately forty-six California Indian groups have applied for federal recognition. There are thirty-eight California Indian groups that have not applied for federal recognition. And there are 108 federally recognized Indian tribes within the state. California has the largest number of federally recognized tribes of any state in the United States, and it also has more groups applying for federal recognition than in any other part of the country." According to Carole Goldberg (1997:183–84), an attorney: "A Native Amer-

ican group does not need to be acknowledged or recognized by the federal government to be a tribe. Federal recognition is merely an affirmative act by the federal government to acknowledge its trust responsibility and its statutory and other obligations to provide services and programs to Indian groups. . . . But certain consequences flow from this recognition, the most important being the many services and programs in education, health, and welfare that the federal government provides to Indian people." On Indian sovereignty, see Deloria and Lytle 1984.

26. Peña 2002; Mathewson 1998:119–21; Ortiz 1993b:7–10; California Indian Basketweavers Association 1997.

27. Goode 1992.

28. Anderson 1993a; Heffner 1984, 1985.

29. Usher 1987.

30. Haderlie and Abbott 1980.

31. California Resources Agency and Department of Fish and Game 1992:60.

32. Thelander and Crabtree 1994:265.

33. Ortiz 1993b:205.

34. Noss, LaRoe, and Scott 1995.

35. Fowler 2000:117–18.

36. Martinez 1992.

37. Heffner 1984.

38. Heffner 1984.

39. Noss 1987.

40. Heffner 1984; Yamane 1997; Anderson 1991a.

41. Ross and Espina 2001; Margolin and Smith 2001; Navarro 2000; United Indian Health Service 1995.

42. On native nutrition, see Price [1939] 1998; Schmid 1994.

43. Powers [1877] 1976:417. Lucy Young in Egli 1992:51. Hooper 1920:357. In parts of precontact California infant mortality rates were high. There is archaeological evidence that during drought periods some tribes suffered nutritional stresses and human life spans were shortened. See Lambert 2004; Schulz 1981.

44. See Brand et al. 1990; Nabhan 1991.

45. Cowen 1990.

46. Perry 1988; Swezey and Heizer 1993:299–328; Forbes 1996; *News from Native California* Staff 2001:33; Ortiz 1999.

47. See Weiss 1984; Altieri, Anderson, and Merrick 1987.

48. For use of the Kaweah brodiaea, see Anderson 1992a. For use of purple amole, see Mills 1985b. For use of Torrey pine nuts, see Shipek 1991.

49. For use of showy Indian clover, see Collier and Thalman 1991. For burning clovers, see Peri and Patterson 1979; Anderson 1993b.

50. Shipek 1991.

51. Chesnut [1902] 1974.

52. Loud 1918:230.

53. For Kawaiisu use of *Perideridia pringlei*, see Zigmond 1981. For burning areas with yampah, see Anderson 1993b.

54. Anderson and Moratto 1996.

55. Linden 1991:46.

CHAPTER 12. RESTORING LANDSCAPES
WITH NATIVE KNOWLEDGE

1. For the definition of naturalness, see Harker et al. 1999:2. For the definition of ecological restoration, see Harris, Birch, and Palmer 1996:17.

2. For views on active management, see Agee 2002; Leopold et al. 1963:4.

3. For the view that impacts started with food production, see Harlan and de Wet 1973.

4. Fiedel 1987. See Anderson 1996b and McCann 1999a, 1999b, for indigenous shaping of the pre-Columbian landscape by Native Americans in different parts of North America.

5. For example, understanding the historical role of Indian-set fires in shaping Monterey pine forests is essential to determining the goal of restoration of these forests. Indians shortened the fire return frequencies to every 10 years or less. Natural fire return frequencies are estimated to have been every 30 to 135 years. See Storer et al. 2001.

6. Society for Ecological Restoration International Science and Policy Working Group 2002.

7. For use of restoration in enhancing biological diversity, see Jackson 1992. For Jordan's views, see Jordan 2003. For ideas on how to foster our ecological heritage through native plant gardening and restoration, see Lowry 1999.

8. Cronon 1995.

9. Higgs 2003:285.

10. See Kimmerer 2000; Anderson and Barbour 2003.

11. For methods used in the reconstruction of historic ecosystems, see Egan and Howell 2001.

12. Fleischner 1992:244.

13. For ways to reconstruct species lists of an area to be restored, see Packard 1997:48; Stevens 1995:95–99.

14. Kline 1997:36; Egan and Howell 2001; Lepofsky et al. 2003; Anderson 2001.

15. For a guide to Harrington's field notes on microfilm for tribes of California, see Mills 1985a; Mills 1986. The J. P. Harrington Database Project, first funded by the National Science Foundation in 2001, consists principally in transcribing and coding the linguistic and ethnographic notes on American Indian languages collected by Harrington during the first half of the twentieth century. The men and women he interviewed were often among the last speakers of their languages. During the 1980s, Harrington's original handwritten field notes, currently housed in the National Anthropological Archives at the Smithsonian Institution, were microfilmed, resulting in 477 reels from which the Harrington Database is being created. This project focuses first on the California notes and includes some languages of Oregon, Nevada, and parts of Washington. The notes can be printed from the database, which will make them more

accessible to Native American scholars for use in cultural and language revitalization, as well as for documentation of tribal histories and genealogies. The URL for the project is http://cougar/nas/NALC/JPH.html.

16. Ford 1981:2179.

17. Torrey and Gray 1856.

18. See Davidson-Hunt and Berkes 2003; Long, Tecle, and Burnette 2003. For manuals that provide restorationists with techniques for conducting ethnobotanical fieldwork, see Alexiades 1996; Martin 1995.

19. Anderson 1993b.

20. On using species inventories of remnant vegetation patches for restoration, see Stevens 1995:100. On roadsides as sites supporting early-successional-stage natives, see Apfelbaum et al. 1997:100. On the importance of restoring areas with plants that have a genetic makeup similar to those that once existed there, see Whisenant 1999:132–33.

21. Stevens 1995:95–96.

22. Blackburn and Anderson 1993a.

23. Anthropologists frequently use the term "complex" to refer to crop complexes. A crop complex implies a group of species with an apparent common geographic origin and a mutual environmental and cultural association in the area where it develops. See Ford 1985.

24. Murphree 1994.

25. See Nabhan 2002.

26. Purdy 1976:153.

27. White 1978; White and Pickett 1985; Rappaport and Reiger 1995.

28. See Arno 1985; Veblen and Lorenz 1986; Wilson and King 1995. See Williams 2003 for a bibliography on Native American land use of fire and its effect on the land.

29. White and Pickett 1985:7; Kaufmann et al. 1994, in Averill et al. 1994.

30. See Kimmerer and Lake 2001. Sugihara, van Wagtendonk, and Fites-Kaufman state in *Fire in California Ecosystems* (in press) that "O'Neill et al. (1986) describe fire as a perturbation that ensures landscape diversity and preserves seed sources for recovery from any major disturbance. They state that viewing ecosystems on the arbitrary scale of the forest stand results in seeing fire as a catastrophic disturbance. If, however, fire is viewed at the scale appropriate to the frequency of occurrence, it can be seen as an essential ecosystem process that retains the spatial diversity of the landscape and permits recovery from disturbance."

31. See Brody 1998 for a good illustration of how an anthropologist was able to work with Native Americans to map their areas of collecting, hunting, and camping, and from that information and other oral histories, produce a chart of the year's activities in time and space. Brody's work is also an illustration of the cyclical nature of Native American hunting and gathering, which in many ways reflects the cycles of a dynamic landscape.

32. For sample experiments at the population scale, see Anderson 1993a; Stevens 1999.

33. For reliance of modern pharmaceuticals on native knowledge, see Farnsworth 1988; for a review of indigenous domestication of world crops, see Wilkes 1995; for use of native knowledge for modern genetics, see Reis and Schultes 1995:13. For sharing of native knowledge of agriculture with newcomers, see Jordan and Kaups 1989:232. For some of the gifts native peoples have given Westerners with regard to biological wealth, see Weatherford 1991.

34. Greaves 1996:26.

35. For reacquisition of lands within reservation boundaries, see Suagee and Stearns 1994:103. For ideas for co-management collaboration with private landowners or public land managers, see Stevens 1997; Egan 2003.

36. For examples of forest management on tribal lands that incorporate indigenous cultural values, see *Journal of Forestry* (November 1997), which is devoted to that topic (Staebler and Atwater 1997).

37. For a philosophical discussion of the values of restoration, see Ouderkirk 1992:29. See Mills 1995 for a discussion of reintegration of humans in nature through cooperation. See Baldwin, De Luce, and Pletsch 1994:4 on the need for a new paradigm for how humans relate to nature.

38. Whittaker and Woodwell 1972.

CODA: INDIGENOUS WISDOM IN THE MODERN WORLD

1. For the definition of *tend*, see *American Heritage Dictionary of the English Language* 1992:1849. For Martin quote, see Martin 1999:24.

2. For the etymology of *tend*, see *American Heritage Dictionary of the English Language* 1992:2129.

3. Johnson 1999:434.

4. Berry 1977:130.

5. For early managers' and ecologists' views on the destructive nature of fires, see Sterling 1905; Olmstead 1911; California State Board of Forestry 1892. On contemporary managers' and ecologists' views on the benefits of fire to many of California's ecosystems and the benefits of prescribed burning, see Biswell 1989; Sugihara et al. in press.

6. For the beginnings of domestication in the northeastern United States, see Smith 1992, 1995. For beginnings of domestication in the southwestern United States, see Adams 1987; Adams and Adams 1998; Bohrer 1991; Nabhan and Felger 1978.

7. Grumbine 1992:238.

8. Oelschlaeger 1991:24.

9. For the importance of intergenerational use of and connection with places, see Hailstone 1992–93; Gardner-Loster 1998–99; LaPena and LaPena 1996–97.

Bibliography

Abel-Vidor, S., D. Brovarney, and S. Billy. 1996. *Remember Your Relations: The Elsie Allen Baskets, Family and Friends*. Ukiah, Calif.: Grace Hudson Museum.

Abrahamson, W. G., and D. C. Hartnett. 1990. Pine flatwoods and dry prairies. In *Ecosystems of Florida*, ed. R. L. Myers and J. J. Ewel, 103–49. Orlando: University of Central Florida Press.

Adam, D. P., and G. J. West. 1983. Temperature and precipitation estimates through the last glacial cycle from Clear Lake, California, pollen data. *Science* 219:168–70.

Adams, K. R. 1987. Little barley (*Hordeum pusillum* Nutt.) as a possible New World domesticate. In *Specialized Studies in the Economy, Environment and Culture of La Ciudad*, Pt. III, ed. J. E. Kisselburg, G. E. Rice and B. L. Shears. Arizona State University Anthropological Field Studies 20.

Adams, K. R., and R. K. Adams. 1998. How does our Agave grow? Reproductive biology of a suspected ancient Arizona cultivar, *Agave murpheyi* Gibson. *Desert Plants* 14(2):11–20.

Adovasio, J. M. 1974. Prehistoric North American basketry. In *Collected Papers on Aboriginal Basketry*, ed. D. R. Tuohy and D. L. Rendall, 98–148. Anthropological Papers No. 16. Carson City: Nevada State Museum.

Advisory Council on California Indian Policy. 1997. *ACCIP Termination Report: The Termination Policy and Its Lingering Effects on California Indians*. Unpublished report.

Agee, J. K. 2002. The fallacy of passive management: Managing for firesafe forest reserves. *Conservation Biology in Practice* 3(1):18–25.

Aginsky, B. W. 1943. Culture element distributions. XXIV: Central Sierra. *University of California Anthropological Records* 8(4).

Alexiades, M. N., ed. 1996. *Selected Guidelines for Ethnobotanical Research: A Field Manual*. Bronx: New York Botanical Garden.

Allaby, M., ed. 1985. The *Oxford Dictionary of Natural History*. Oxford: Oxford University Press.

Allen, E. 1972. *Pomo Basketmaking: A Supreme Art for the Weaver.* Happy Camp, Calif.: Naturegraph.

Allen, R. W. 1971. *An Examination of the Botanical References in the Accounts Relating to Drake's Encampment at Nova Albion in 1579.* Research Report for Drake Navigators Guild, Point Reyes, Calif.

Altieri, M. A., M. K. Anderson, and L. C. Merrick. 1987. Peasant agriculture and the conservation of crop and wild plant resources. *Journal of the Society of Conservation Biology* 1:49–58.

Alvarez, M. 1994. Mount Shasta: A question of power. *News from Native California* 8(3):4–7.

Alvarez, S. H., and D. W. Peri. 1987. Acorns: The staff of life. *News from Native California* 1(4):10–14.

Alverson, W. S., W. Kuhlmann, and D. M. Waller. 1994. *Wild Forests: Conservation Biology and Public Policy.* Washington, D.C.: Island Press.

American Friends Service Committee. 1960. *Indians of California: Past and Present.* Pamphlet. San Francisco, Calif.

Anderson, E. N. 1996. *Ecologies of the Heart: Emotion, Belief, and the Environment.* Oxford: Oxford University Press.

Anderson, M. K. 1988. Southern Sierra Miwok plant resource use and management of the Yosemite region. Master's thesis, University of California, Berkeley.

———. 1991a. California Indian horticulture: Management and use of redbud by the Southern Sierra Miwok. *Journal of Ethnobiology* 11(1):145–57.

———. 1991b. "We are still here": Tribal recognition for the Southern Sierra Miwok. *Yosemite* 53(4)1–5. Yosemite Association, Yosemite National Park, Calif.

———. 1992a. At home in the wilderness. *News from Native California* 6(2):19–21.

———. 1992b. Restoring deer grass. *News from Native California* 6(2):40.

———. 1993a. The experimental approach to assessment of the potential ecological effects of horticultural practices by indigenous peoples on California wildlands. Ph.D. dissertation, University of California, Berkeley.

———. 1993b. *Indian Fire-based Management in the Sequoia–Mixed Conifer Forests of the Central and Southern Sierra Nevada.* Final Report to the Yosemite Research Center, Yosemite National Park. U.S. Department of the Interior, National Park Service, Western Region. Cooperative Agreement Order No. 8027–002.

———. 1993c. The mountains smell like fire. *Fremontia: Journal of the California Native Plant Society* 21(4):15–20.

———. 1993d. Native Californians as ancient and contemporary cultivators. In *Before the Wilderness: Environmental Management by Native Californians,* ed. T. C. Blackburn and M. K. Anderson, 151–74. Menlo Park, Calif.: Ballena Press.

———. 1996a. The ethnobotany of deergrass, *Muhlenbergia rigens* (Poaceae): Its uses and fire management by California Indian tribes. *Economic Botany* 50(4):409–22.

———. 1996b. Tending the wilderness. *Restoration and Management Notes* 14(2):154–66.

———. 1997. From tillage to table: The indigenous cultivation of geophytes for food in California. *Journal of Ethnobiology* 17(2):149–69.

———. 1999. The fire, pruning, and coppice management of temperate ecosystems for basketry material by California Indian tribes. *Human Ecology* 27(1):79–113.

———. 2001. The contribution of ethnobiology to the reconstruction and restoration of historic ecosystems. In *The Historical Ecology Handbook: A Restorationist's Guide to Reference Ecosystems,* ed. D. Egan and E. A. Howell, 55–72. Washington, D.C.: Island Press.

Anderson, M. K., and M. Barbour. 2003. Simulated indigenous management: A new model for ecological restoration in national parks. *Ecological Restoration* 21(4):269–77.

Anderson, M. K., M. G. Barbour, and V. Whitworth. 1997. A world of balance and plenty: Land, plants, animals, and humans in a pre-European California. In *Contested Eden: California before the Gold Rush,* ed. R. A. Gutierrez and R. J. Orsi, 12–47. Berkeley: University of California Press.

Anderson, M. K., and M. J. Moratto. 1996. Native American Land-Use Practices and Ecological Impacts. In *Sierra Nevada Ecosystem Project: Final Report to Congress.* Vol. 2: *Assessments and Scientific Basis for Management Options.* Davis: University of California, Centers for Water and Wildland Resources.

Anderson, M. K., and G. P. Nabhan. 1991. Gardeners in Eden. *Wilderness Magazine* 55(194):27–30.

Anderson, M. K., and D. L. Rowney. 1998. California geophytes: Their ecology, ethnobotany, and conservation. *Fremontia: Journal of the California Native Plant Society* 26(1):12–18.

———. 1999. The edible plant *Dichelostemma capitatum:* Its vegetative reproduction response to different indigenous harvesting regimes in California. *Restoration Ecology* 7(3):231–40.

Anderson, R. S., and S. L. Carpenter. 1991. Vegetation change in Yosemite Valley, Yosemite National Park, California, during the protohistoric period. *Madroño* 38(1):1–13.

Apfelbaum, S. I., B. J. Bader, F. Faessler, and D. Mahler. 1997. Obtaining and processing seeds. In *The Tallgrass Restoration Handbook for Prairies, Savannas, and Woodlands,* ed. S. Packard and C. F. Mutel, 99–134. Washington, D.C.: Island Press.

Apodaca, P. 1990. California tribes look into creating museums. *News from Native California* 4(4):18–19.

Archuleta, M. L., B. J. Child, and K. Tsianina Lomawaima, eds. 2000. *Away from Home: American Indian Boarding School Experiences.* Phoenix, Ariz.: Heard Museum.

ARCIA. Commissioner of Indian Affairs. 1849–1903. Annual Reports to the Secretary of the Interior. U.S. Government Printing Office, Washington, D.C. Reprint. New York: AMS Press, 1976–77.

Arno, S. F. 1985. Ecological effects and management implications of Indian fires. In *Proceedings of the Symposium and Workshop on Wilderness Fire, November 15–18, 1983, Missoula, Montana,* tech. coords. James E. Lotan et al., 81–86. General Technical Report INT-182. Ogden, Utah: U.S. Department of Agriculture, Forest Service, Intermountain Forest and Range Experiment Station.

Arno, S. F., and S. Allison-Bunnell. 2002. *Flames in Our Forest: Disaster or Renewal?* Washington, D.C.: Island Press.

Arnold, J. F. 1950. Changes in ponderosa pine–bunchgrass ranges in northern Arizona resulting from pine regeneration and grazing. *Journal of Forestry* 48:118–26.

Asisara, L. 2001. The killing of Fr. Andres Quintana at Mission Santa Cruz. In *Chronicles of Early California, 1535–1846: Lands of Promise and Despair,* ed. R. M. Beebe and R. M. Senkewicz, 284–92. Santa Clara, Calif.: Santa Clara University; Berkeley, Calif.: Heyday Books.

Audubon, J. J. 1989. *The Quadrupeds of North America.* Secaucus, N.J.: Wellfleet Press.

Averill, R. D., L. Larson, J. Saveland, P. Wargo, J. Williams, and M. Bellinger. 1994. *Disturbance Processes and Ecosystem Management: Executive Summary.* U.S. Department of Agriculture, Forest Service.

Bade, W. F. 1905. Forest in Tuolumne. *Sunset Magazine* 14(April):597–603.

Baker, A. J., trans. 1930. Fray Benito de la Serra's account of the Hezeta expedition to the Northwest Coast in 1775. *California Historical Society Quarterly* 9(2):201–42.

Baker, M. A. 1981. The ethnobotany of the Yurok, Tolowa, and Karok Indians of northwest California. Master's thesis, Humboldt State University.

Bakker, E. 1994. California: The great mosaic. In *Life on the Edge: A Guide to California's Endangered Natural Resources: Wildlife,* ed. C. G. Thelander and M. Crabtree, 6. Santa Cruz, Calif.: BioSystems Books; Berkeley, Calif.: Heyday Books.

Baldwin, A. D., J. De Luce, and C. Pletsch, eds. 1994. *Beyond Preservation: Restoring and Inventing Landscapes.* Minneapolis: University of Minnesota Press.

Bancroft, H. H. 1966. *History of California.* Vol. 3: 1825–1840. In *The Works of Hubert Howe Bancroft.* 20 vols. Santa Barbara, Calif.: Wallace Hebberd. [Orig. pub. San Francisco: History Company, 1886.]

Barbour, M. G., J. H. Burk, and W. D. Pitt. 1980. *Terrestrial Plant Ecology.* Menlo Park, Calif.: Benjamin/Cummings.

Barbour, M. G., S. Lydon, M. Borchert, M. Popper, V. Whitworth, and J. Evarts. 2001. *Coast Redwood: A Natural and Cultural History.* Los Olivos, Calif.: Chacuma Press.

Barbour, M. G., B. Pavlik, F. Drysdale, and S. Lindstrom. 1993. *California's Changing Landscapes: Diversity and Conservation of California Vegetation.* Sacramento: California Native Plant Society.

Barbour, M. G., A. Solomeshch, C. Witham, R. Holland, R. Macdonald, S. Cilliers, J. A. Molina, J. Buck, and J. Hillman. 2003. Vernal pool vegetation of California: Variation within pools. *Madroño* 50(3):129–46.

Barker, J. 1955. *San Joaquin Vignettes*. Ed. W. H. Boyd and G. J. Roders. Bakersfield, Calif.: Kern County Historical Society.

Barnes, W. C. 1908. The U.S. Forest Service. *Outwest* 29(2):88–109.

Barney, M., and N. Frischknecht. 1974. Vegetation changes following fire in the pinyon-juniper type of west-central Utah. *Journal of Range Management* 27(2):91–96.

Barras, J. 1984. *Their Places Shall Know Them No More*. Tehachapi, Calif.: Judy and Bud Barras.

Barrett, R., and P. Weiss. 1994. The nuts and bolts of wildlife management: Conversation with Reginald Barrett, interview conducted by Peter Weiss. In *Life on the Edge: A Guide to California's Endangered Natural Resources: Wildlife*, ed. C. G. Thelander and M. Crabtree, 45–58. Santa Cruz, Calif.: BioSystems Books; Berkeley, Calif.: Heyday Books.

Barrett, S. A. 1908. Pomo Indian basketry. *University of California Publications in American Archaeology and Ethnology* 7(3):134–308.

———. 1910. The material culture of the Klamath Lake and Modoc Indians of northeastern California and southern Oregon. *University of California Publications in American Archaeology and Ethnology* 5(4):239–92.

———. 1952. Material aspects of Pomo culture. *Bulletin of the Public Museum of the City of Milwaukee* 20(1):1–260.

Barrett, S. A., and E. W. Gifford. 1933. Miwok material culture: Indian life of the Yosemite region. *Bulletin of the Milwaukee Public Museum* 2(4):118–375. Reprint. Yosemite National Park, Calif.: Yosemite Association.

Barrows, D. P. 1967. *Ethno-botany of the Coahuilla Indians*. Banning, Calif.: Malki Museum Press.

Basgall, M. E., and P. D. Bouey. 1991. *The Prehistory of North-Central Sonoma County: Archaeology of the Warm Springs Dam-Lake Sonoma Locality*. Report on file at the Northwest Information Center, Department of Anthropology, Sonoma State University.

Basgall, M. E., and K. R. McGuire. 1988. *The Archaeology of CA-INY-30 Prehistoric Culture Change in the Southern Owens Valley, California*. Davis, Calif.: Far Western Anthropological Research Group, for California State Department of Transportation District 9, Bishop, Calif. Unpublished report.

Bates, C. D. 1982a. *Coiled Basketry of the Sierra Miwok: A Study of Regional Variation*. San Diego Museum Papers No. 15. San Diego. Calif.: San Diego Museum of Man.

———. 1982b. Feathered regalia of central California: Wealth and power. *Occasional Papers of the Redding Museum* No. 2. Redding, Calif.: Redding Museum and Art Center.

———. 1984. Traditional Miwok basketry. *American Indian Basketry and Other Native Arts* 4:(13)3–18.

Bates, C. D., and M. J. Lee. 1990. *Tradition and Innovation: A Basket History of the Indians of the Yosemite–Mono Lake Area*. Yosemite National Park, Calif.: Yosemite Association.

Baumhoff, M. A. 1963. Ecological determinants of aboriginal California populations. *University of California Publications in American Archaeology and Ethnology* 49(2):155–236.

Baxley, W. H. 1865. *What I Saw on the West Coast of South and North America and at the Hawaiian Islands.* New York: D. Appleton & Company.

Beale, E. F. 1957. The report of Edward F. Beale, superintendent of Indian Affairs in California, respecting the condition of Indian affairs in that state. [Report submitted originally to Commissioner of Indian Affairs, Washington, D.C., February 25, 1853.] In *Central Route to the Pacific,* ed. G. H. Heap, 315–36. Glendale: Arthur H. Clark.

Bean, L. J., and H. W. Lawton. 1993. Some explanations for the rise of cultural complexity in native California with comments on proto-agriculture and agriculture. In *Before the Wilderness: Environmental Management by Native Californians,* ed. T. C. Blackburn and K. Anderson, 27–54. Menlo Park, Calif.: Ballena Press.

Bean, L. J., and K. S. Saubel. 1961. Cahuilla ethnobotanical notes: The aboriginal uses of oak. In *Archaeological Survey Annual Reports 1960–61,* 237–45. University of California, Los Angeles.

———. 1972. *Temalpakh: Cahuilla Indian Knowledge and Usage of Plants.* Banning, Calif.: Malki Museum Press.

Beard, Y. S. 1979. *The Wappo: A Report.* Banning, Calif.: Malki Museum Press.

Bee, R. L. 1983. Quechan. In *Handbook of North American Indians.* Vol. 10: *Southwest,* ed. A. Ortiz, 86–98. Washington, D.C.: Smithsonian Institution.

Beemer, Eleanor. 1980. *My Luiseño Neighbors: Excerpts from a Journal Kept in Pauma Valley Northern San Diego County, 1934 to 1974.* Ramona, Calif.: Acoma Books.

Beetle, Alan A. 1947. Distribution of the native grasses of California. *Hilgardia* 17(9): 309–57.

Belcher, Sir E. 1843. *Narrative of a Voyage Round the World, Performed in Her Majesty's Ship Sulphur during the Years 1836–1832.* 2 vols. London.

Bell, H. 1927. *Reminiscences of a Ranger: Or Early Times in Southern California.* Santa Barbara, Calif.: Wallace Hebberd.

Benz, C., H. Feldman, K. Hanni, and R. J. Jameson. 1994. Southern sea otter. In *Life on the Edge: A Guide to California's Endangered Natural Resources: Wildlife,* ed. C. G. Thelander and M. Crabtree, 500–504. Santa Cruz, Calif.: BioSystems Books; Berkeley, Calif.: Heyday Books.

Berkes, F. 1999. *Sacred Ecology: Traditional Ecological Knowledge and Resource Management.* Philadelphia: Taylor and Francis.

Berlin, B. 1992. *Ethnobiological Classification: Principles of Categorization of Plants and Animals in Traditional Societies.* Princeton: Princeton University Press.

Berry, W. 1977. *The Unsettling of America: Culture and Agriculture.* San Francisco, Calif.: Sierra Club Books.

Bethel, R., P. V. Kroskrity, C. Loether, and G. A. Reinhardt. 1984. *A Practical Dictionary of Western Mono.* Los Angeles: American Indian Studies Center, University of California, Los Angeles.

Bettinger, R. L. 1976. The development of pinyon exploitation in central eastern California. *Journal of California Anthropology* 3(1):81–95.

Bibby, B. 1996. *The Fine Art of California Indian Basketry.* Sacramento, Calif.: Crocker Art Museum.

Bicknell, S. H. 1989. Strategy for reconstructing presettlement vegetation. *Supplement to Bulletin of the Ecological Society of America, Program and Abstracts,* 70(2):62.

———. 1990. *Montana de Oro State Park: Presettlement Vegetation Mapping and Ecological Status of Eucalyptus Final Report.* California State Department of Parks and Recreation Interagency Agreement No. 86–06–336.

———. 1992. Late prehistoric vegetation patterns at six sites in coastal California. *Supplement to Bulletin of the Ecological Society of America,* Program and Abstracts, 73(2):112.

Bicknell, S. H., and A. Austin. 1991. *Lake Earl Project Presettlement Vegetation Final Report.* California State Department of Parks and Recreation Interagency Agreement No. 4–100–8401.

Bicknell, S. H., A. T. Austin, D. J. Bigg, and R. Parker Godar. 1993. *Fort Ross State Historic Park Prehistoric Vegetation Final Report.* California State Department of Parks and Recreation Interagency Agreement No. 4–100–0252.

Bicknell, S. H., D. J. Bigg, and A. T. Austin. 1993. *Salt Point State Park Prehistoric Vegetation Final Report.* California State Department of Parks and Recreation Interagency Agreement No. 88–11–013.

Bicknell, S. H., D. J. Bigg, R. P. Godar, and A. T. Austin. 1993. *Sinkyone Wilderness State Park Prehistoric Vegetation Final Report.* California State Department of Parks and Recreation Interagency Agreement No. 4–100–0252.

Bicknell, S. H., R. P. Godar, D. J. Bigg, and A. T. Austin. 1993. *Mount Tamalpais State Park Prehistoric Vegetation Final Report.* California State Department of Parks and Recreation Interagency Agreement No. 88–11–013.

Bicknell, S. H., C. A. Hansen, and E. M. Mackey. 1987. *Jug Handle State Reserve First Coastal Terrace Presettlement Vegetation Mapping Project Final Report.* California State Department of Parks and Recreation Interagency Agreement No. 4–100–6355.

———. 1988. *Patrick's Point State Park Presettlement Vegetation Mapping and Soils Classification Final Report.* California State Department of Parks and Recreation Interagency Agreement No. 84–04–130.

Bicknell, S. H. and E. M. Mackey. 1998. Mysterious nativity of California's sea fig. *Fremontia: A Journal of the California Native Plant Society* 26(1):3–11.

Bidwell, J. 1828. *Echoes of the Past about California.* Chicago: Donelly and Sons Co.

———. 1866. Annual address. *Transactions of the California State Agronomy Society* 1864–65:202–13.

Billy, S. 1994. So the spirit can move freely. In *All Roads Are Good: Native Voices on Life and Culture,* ed. T. Winch and C. Wilson, 196–207. Washington, D.C.: Smithsonian Institution Press.

Biswell, H. H. 1972. Fire ecology in ponderosa pine–grassland. In *Proceedings of the 12th Annual Tall Timbers Fire Ecology Conference,* 69–96. Tallahassee, Fla.: Tall Timbers Research Station.

————. 1989. *Prescribed Burning in California Wildlands Vegetation Management.* Berkeley, Calif.: University of California Press.

Blackburn, T., ed. 1975. *December's Child: A Book of Chumash Oral Narratives.* Berkeley, Calif.: University of California Press.

Blackburn, T. C., and K. Anderson. 1993a. Introduction: Managing the domesticated environment. In *Before the Wilderness: Environmental Management by Native Californians,* ed. T. C. Blackburn and K. Anderson, 15–25. Menlo Park, Calif.: Ballena Press.

Blackburn, T. C., and K. Anderson, eds. 1993b. *Before the Wilderness: Environmental Management by Native Californians.* Menlo Park, Calif.: Ballena Press.

Bledsoe, A. J. 1956. *Indian Wars of the Northwest: A California Sketch.* Oakland, Calif.: Biobooks.

Bleich, V., L. Chow, S. Torres, and J. Wehausen. 1994. California bighorn sheep. In *Life on the Edge: A Guide to California's Endangered Natural Resources: Wildlife,* ed. C. G. Thelander and M. Crabtree, 98–99. Santa Cruz, Calif.: BioSystems Books; Berkeley, Calif.: Heyday Books.

Blue Cloud, M. 1998. The strawberry festival: A California tradition. *News from Native California* 12(1):21–22.

Bocek, B. R. 1984. Ethnobotany of Costanoan Indians, California, based on collections by John P. Harrington. *Economic Botany* 38(2):240–55.

Bodie Weekly Standard. 1877. Paiutes on the warpath. November 7.

Bohrer, V. L. 1983. New life from ashes: The tale of the burnt bush *(Rhus trilobata). Desert Plants* 5(3):122–24.

————. 1991. Recently recognized cultivated and encouraged plants among the Hohokam. *Kiva* 56(3):227–35.

Bolton, H. E. 1911. Expedition to San Francisco Bay in 1770: Diary of Pedro Fages. *Publications of the Academy of Pacific Coast History* 2(3):143–59.

————, ed. 1916. *Original Narratives of Early American History: Spanish Exploration in the Southwest, 1542–1706.* New York: Charles Scribner's Sons.

————. 1930. Moraga's account of the founding of San Francisco. In *Anza's California Expeditions,* Vol. 3, 409–20. Berkeley, Calif.: University of California Press.

————. [1927] 1971. *Fray Juan Crespi: Missionary Explorer on the Pacific Coast, 1769–1774.* New York: AMS Press.

Bonnicksen, T. M. 2000. *America's Ancient Forests: From the Ice Age to the Age of Discovery.* New York: John Wiley.

Borthwick, J. D. 1948. *3 Years in California.* Oakland, Calif.: Biobooks.

Boscana, G. [1933] 1978. *Chinigchinich: Historical Account of the Beliefs, Usages, Customs and Extravagancies of the Indians of this Mission of San Juan Capistrano called the Acagchemem Tribe.* Annotated by J. P. Harrington. Translated by A. Robinson. Morongo Indian Reservation, Banning, Calif.: Malki Museum Press.

Bosqui, E. 1904. *Memoirs.* San Francisco: privately printed.

Bossard, C. C., J. M. Randall, and M. C. Hoshovosky, eds. 2000. *Invasive Plants of California's Wildlands.* Berkeley: University of California Press.

Botkin, D. B. 1990. *Discordant Harmonies: A New Ecology for the Twenty-first Century.* Oxford: Oxford University Press.

Bowcutt, F. 1994–1996. A floristic study of Sinkyone Wilderness State Park, Mendocino County, California. *The Wasmann Journal of Biology* 511(1–2): 64–143.

Bowman, R. I. 1990. Evolution and biodiversity in California. In *California's Wild Heritage: Threatened and Endangered Animals in the Golden State,* ed. Peter Steinhart, 3–7. San Francisco: Sierra Club Books.

Brace, C. L. 1869. *The New West: Or California in 1867–1868.* New York: G. P. Putnam and Son.

Bradley, H. C. 1911. The passing of our mountain meadows. *Sierra Club Bulletin* 8(1):39–42.

Braham, H., H. Feldman, K. Hanni, and J. Heyning. 1994. Sperm whale. In *Life on the Edge: A Guide to California's Endangered Natural Resources: Wildlife,* ed. C. G. Thelander and M. Crabtree, 494–98. Santa Cruz, Calif.: BioSystems Books; Berkeley, Calif.: Heyday Books.

Brand, J. C., B. J. Snow, G. P. Nabhan, and A. S. Truswell. 1990. Plasma glucose and insulin responses to traditional Pima Indian meals. *American Journal of Clinical Nutrition* 51:416–20.

Brandes, Ray. 1970. *The Costanso Narrative of the Portola Expedition.* Newhall, Calif.: Hogarth Press.

Brennan, L. A. and S. M. Hermann. 1994. Prescribed fire and forest pests: Solutions for today and tomorrow. *Journal of Forestry* 92(11):34–37.

Brewer, W. H. 1966. *Up and Down California in 1860–1864.* Berkeley: University of California Press.

Bright, W. 1978. Karok. In *Handbook of North American Indians.* Vol. 8: *California,* ed. R. F. Heizer, 180–89. Washington, D.C.: Smithsonian Institution.

Brodo, I. M., S. D. Sharnoff, and S. Sharnoff. 2001. *Lichens of North America.* New Haven, Conn.: Yale University Press.

Brody, H. 1998. *Maps and Dreams.* Prospect Heights, IL: Waveland Press.

Brown, G. E. 1977. *The Pruning of Trees, Shrubs, and Conifers.* London: Faber and Faber.

Brown, J. H. 1949. *Early Days of San Francisco, California.* Oakland, Calif.: Biobooks.

Brown, L. R. 1979. Insects feeding on California oak trees. In *Proceedings of the Symposium on the Ecology, Management, and Utilization of California Oaks.* Claremont, Calif., June 26–28. General Technical Report PSW-44. Berkeley, Calif.: USDA Forest Service, Pacific Southwest Forest and Range Experiment Station.

Brown, R. 1876. *The Races of Mankind: Being a Popular Description of the Characteristics, Manners and Customs of the Principal Varieties of the Human Family.* London: Cassell, Petter & Galpin.

Browne, J. R. 1877. The Indian reservations of California. In *The Indian Miscellany; Containing Papers on the History, Antiquities, Arts, Languages, Religions, Traditions and Superstitions of the American Aborigines; with Descriptions of their Domestic Life, Manners, Customs, Traits, Amusements and Exploits; Travels and Adventures in the Indian Country; Incidents of Border Warfare; Missionary Relations, etc.*, ed. W. W. Beach, 303–22. Albany, N.Y.: J. Munsell.

Brubaker F. 1926. Plants used by Yosemite Indians. *Yosemite Nature Notes* 5(10):73–79.

Bryant, E. 1985. *What I Saw in California.* Lincoln: University of Nebraska Press.

Buchmann, S. L., and G. P. Nabhan. 1996. *The Forgotten Pollinators.* Washington D.C.: Island Press/Shearwater Books.

Buckley, G. P., ed. 1992. *Ecology and Management of Coppice Woodlands.* London: Chapman and Hall.

Buell, J. H. 1940. Effect of season of cutting on sprouting of dogwood. *Journal of Forestry* 38(8):649–50.

Bullock, S. H. 1980. Dispersal of a desert palm by opportunistic frugivores. *Principes* 24(1):29–32.

Bunnell, L. H. 1880. *Discovery of the Yosemite, and the Indian War of 1851.* Chicago: F. Revell.

Burcham, L. T. 1940. Orchards of the red man. Unpublished essay.

———. 1982. *California Range Land.* Center for Archaeological Research at Davis Publication No. 7. University of California, Davis.

Burgess, J. W. 1860. Testimony taken before the Joint Special Committee on the Mendocino Indian War. *Majority and Minority Reports of the Joint Committee on the Mendocino War.* Bancroft Library, University of California, Berkeley.

Bye, R. A. 1972. Ethnobotany of the Southern Paiute Indians in the 1870's: With a note on the early ethnobotanical contributions of Dr. Edward Palmer. In *Great Basin Cultural Ecology: A Symposium*, ed. D. D. Fowler, 87–104. Publications in Social Science No. 8. Reno, Nev.: Desert Research Institute.

Cain, E. M. 1961. *The Story of Early Mono County.* San Francisco: Fearon.

Cajete, G. 2000. *Native Science: Natural Laws of Interdependence.* Santa Fe, New Mex.: Clear Light.

Calambokidis, J., D. Demaster, H. Feldman, and K. Hanni. 1994. Humpback whale. In *Life on the Edge: A Guide to California's Endangered Natural Resources: Wildlife*, ed. C. G. Thelander and M. Crabtree, 487–91. Santa Cruz, Calif.: BioSystems Books; Berkeley, Calif.: Heyday Books.

Caldwell, G. 1995. *St. Francis Turned on His Head: A Summary Assessment of Mission Impact on the Indian Population of Alta California, 1769–1834.* Report prepared for the Education Task Force, Advisory Council on California Indian Policy.

California Indian Basketweavers Association. 1997. Report to the Board of Directors. Pesticide Strategy Meeting, September 6, 1996, Sacramento, Calif.

California Native Plant Society. 2001. *Inventory of Rare and Endangered*

Plants of California, 6th ed. Ed. D. P. Tibor. Special Publications No. 1. Sacramento: California Native Plant Society.

California Resources Agency and Department of Fish and Game. 1992. *Annual Report on the Status of California State Listed Threatened and Endangered Animals and Plants*. Sacramento.

California State Board of Forestry. 1892. *Fourth Biennial Report of the California State Board of Forestry for the Years 1891–92 to Governor H. H. Markham*. Sacramento, Calif.: A.J. Johnston, Supt. State Printing.

Callaghan, C. A. 1984. *Plains Miwok Dictionary*. University of California Publications in Linguistics 105. Berkeley: University of California Press.

Callahan, B., ed. 1979. *A Jaime de Angulo Reader*. Berkeley, Calif.: Turtle Island.

Callicott, J. B., and M. P. Nelson, eds. 1998. *The Great New Wilderness Debate*. Athens: University of Georgia Press.

Calvin, R. 1951. *Lieutenant Emory Reports*. Albuquerque: University of New Mexico Press. [Reprint of Lieutenant W. H. Emory's Notes of a Military Reconnoissance.]

Canham, C. D., and P. L. Marks. 1985. The response of woody plants to disturbance: patterns of establishment and growth. In *The Ecology of Natural Disturbance and Patch Dynamics*, eds. S. T. A. Pickett and P. S. White, 197–216. Orlando, Fla: Academic Press.

Caprio, A. C., and T. W. Swetnam. 1995. Historic fire regimes along an elevational gradient on the west slope of the Sierra Nevada, California. In *Proceedings of a Symposium on Fire in Wilderness and Park Management, March 30–April 1, 1993, Missoula, Montana.*, tech. coords. J. K. Brown, R. W. Mutch, C. W. Spoon, and R. H. Wakimoto, 173–79. General Technical Report INT-320. Ogden, Utah: U.S. Department of Agriculture, Forest Service, Intermountain Research Station.

Carlson, P. C., G. W. Tanner, J. M. Wood, and S. R. Humphrey. 1993. Fire in key deer habitat improves browse, prevents succession, and preserves endemic herbs. *Journal of Wildlife Management* 57(4):914–28.

Carpenter, H. M. 1893. Among the Diggers of thirty years ago. *Overland Monthly* 21(124):389–99.

Carson, J. H. 1950. *Recollections of the California Mines*. Oakland, Calif.: Biobooks.

Castillo, E. D. 1978. The impact of Euro-American exploration and settlement. In *Handbook of North American Indians*. Vol 8: *California*, ed. R. F. Heizer, 99–127. Washington, D.C.: Smithsonian Institution.

——. 1998. Short overview of California Indian history. http://ceres.ca.gov/nahc/califindian.html.

Cermak, R. W. 1996. Sugar pine in the history of the West Coast. In *Sugar Pine: Status, Values, and Roles in Ecosystems*, ed. B. B. Kinloch Jr., M. Marosy, and M. E. Huddleston, 10–21. Proceedings of a Symposium presented by the California Sugar Pine Management Committee, March 30–April 1, 1992. University of California, Davis, Division of Agricultural and Natural Resources Publication 3362.

Chalfant, W. A. 1933. *The Story of Inyo*. Bishop, Calif.: Published by the author.

Chamisso, A. V. 1999. A voyage around the world, 1816. In *A World Transformed: Firsthand Accounts of California Before the Gold Rush,* ed. J. Paddison, 135–54. Berkeley, Calif.: Heyday Books.

Chang, C. 1996. Ecosystem responses to fire and variations in fire regimes. In *Sierra Nevada Ecosystem Project: Final Report to Congress.* Vol. 2: *Assessments of Scientific Basis for Management Options,* 1071–99. Davis: University of California, Centers for Water and Wildland Resources.

Chapman, C. E. 1911. Expedition on the Sacramento and San Joaquin Rivers in 1817: Diary of Fray Narcisco Duran. Ed. C. E. Chapman. *Publications of the Academy of Pacific Coast History* 2(5):331–49. University of California, Berkeley.

———. 1939. *California: The Spanish Period.* New York: Macmillan.

Chesnut, V. K. [1902] 1974. Plants used by the Indians of Mendocino County, California. *Contributions from the United States National Herbarium,* Government Printing Office 1900–1902. Vol. 7:295–422. Reprint. Fort Bragg, Calif.: Mendocino County Historical Society.

Christensen, N. L. 1988. Succession and natural disturbance: paradigms, problems, and preservation of natural ecosystems. In *Ecosystem Management for Parks and Wilderness,* ed. J. K. Agee and D. R. Johnson, 62–86. Seattle: University of Washington Press.

Christensen, N. L., L. Cotton, T. Harvey, R. Martin, J. McBride, P. Rundel, and R. Wakimoto. 1987. Final Report. Review of fire management programs for sequoia–mixed conifer forests of Yosemite, Sequoia and Kings Canyon National Parks. Yosemite Research Library, Yosemite National Park.

Clar, C. R. 1959. *California Government and Forestry from Spanish Days until the Creation of the Department of Natural Resources in 1927.* Sacramento, Calif.: Division of Forestry Department of Natural Resources.

Clark, C. U., trans. 1959. Excerpts from the journals of Prince Paul of Wurtemberg, year 1850. *Southwestern Journal of Anthropology* 15(3):291–99.

Clark, G. 1894. Letter to the Board of Commissioners of the Yosemite Valley and Mariposa Big Tree Grove. August 30. Yosemite Research Library, Yosemite National Park, Calif.

———. [1904] 1987. *Indians of the Yosemite Valley and Vicinity: Their History, Customs and Traditions.* Walnut Creek, Calif.: Diablo Books.

Clark, W. V. T., ed. 1973. *The Journals of Alfred Doten, 1849–1903.* Vol. 1. Reno: University of Nevada Press.

Clarke Memorial Museum. 1985. *The Hover Collection of Karuk Baskets.* Eureka, Calif.: Clarke Memorial Museum.

Clary, D. A. 1986. *Timber and the Forest Service.* Lawrence, Kans.: University Press of Kansas.

Clyman, J. 1926a. James Clyman: His diaries and reminiscences [book 5], June 8, 1845, on the Oregon-California Trail. *California Historical Society Quarterly* 5(2):110–37.

———. 1926b. James Clyman: His diaries and reminiscences [book 7], December, 1845. *California Historical Society Quarterly* 5(3):255–88.

Cogswell, H. L., and G. Christman. 1977. *Water Birds of California*. Berkeley: University of California Press.

Cohen, M. P. 1984. *The Pathless Way: John Muir and American Wilderness*. Madison: University of Wisconsin Press.

Coles, J. M. 1989. The world's oldest road. *Scientific American* 261(5):100–106.

Collier, E. T., and S. B. Thalman, eds. 1991. *Interviews with Tom Smith and Maria Copa: Isabel Kelly's Ethnographic Notes on the Coast Miwok Indians of Marin and Southern Sonoma Counties, California*. Miwok Archeological Preserve of Marin Occasional Papers No. 6. San Rafael, Calif.

Collings, J. L. 1979. Profile of a Chemehuevi weaver. *American Indian Art* 4(4):60–67.

Collins, P. W. 1991. Interaction between island foxes *(Urocyon littoralis)* and Indians on islands off the coast of Southern California: I. Morphologic and archaeological evidence of human assisted dispersal. *Journal of Ethnobiology* 11(1):51–82.

Colton, Rev. W. 1860. *The Land of Gold; Or, Three Years in California*. New York: D. W. Evans & Co.

———. 1948. *The California Diary*. Oakland, Calif.: Biobooks.

———. 1949. *Three Years in California: Together with Excerpts from the Author's Deck and Port, Covering His Arrival in California, and a Selection of his Letters from Monterey*. Stanford, Calif: Stanford University Press.

Conklin, H. C. 1962. The lexicographical treatment of folk taxonomies. *International Journal of American Linguistics* 28:119–41.

Connell, J. H. 1978. Diversity in tropical rain forests and coral reefs. *Science* 199:1302–10.

Connolly, T. J., J. M. Erlandson, and S. E. Norris. 1995. Early Holocence basketry and cordage from Daisy Cave, San Miguel Island, California. *American Antiquity* 60(2):309–18.

Cook, S. F. 1943. The conflict between the California Indians and white civilization, I: The Indian versus the Spanish Mission. *Ibero-Americana* 21. Berkeley, Calif.

———. 1960. Colonial expeditions to the interior of California: Central Valley, 1800–1820. *University of California Anthropological Records* 16(6):239–92. Berkeley, Calif.

———. 1971. The aboriginal population of Upper California. In *The California Indians: A Source Book*, ed. R. F. Heizer and M. A. Whipple, 66–72. Berkeley: University of California Press.

———. 1976. *The Conflict between the California Indian and White Civilization*. Berkeley: University of California Press.

———. 1978. Historical demography. In *Handbook of North American Indians*. Vol. 8: *California*, ed. R. F. Heizer, 91–98. Washington, D.C.: Smithsonian Institution.

Cooper, W. S. 1922. *The Broad-sclerophyll Vegetation of California. An Ecological Study of the Chaparral and Its Related Communities*. Washington, D.C.: Carnegie Institution of Washington, Publication no. 319.

Cornett, J. W. 1985a. A new locality for desert fan palms in California. *Desert Plants* 7(3):164.

———. 1985b. Reading fan palms. *Natural History* 94(10):64–73.

———. 1986. Spineless petioles in *Washingtonia filifera* (Arecaceae). *Madroño* 33(1):76–78.

———. 1987a. A giant boring beetle. *Environment Southwest* 518:21–25.

———. 1987b. Indians and the desert fan palm. *Masterkey* 12–17.

———. 1989a. The desert fan palm—not a relict. Abstract. In *Mojave Desert Quaternary Research Center Third Annual Symposium Proceedings*, ed. C. A. Warren and J. S. Schneider. San Bernardino: San Bernardino County Museum Association.

———. 1989b. *Desert Palm Oasis*. Santa Barbara, Calif.: Palm Springs Desert Museum and Companion Press.

———. 1995. *Indian Uses of Desert Plants*. Palm Springs, Calif.: Palm Springs Desert Museum.

Costo, R., and J. H. Costo. 1987. *The Missions of California: A Legacy of Genocide*. San Francisco, Calif.: Indian Historian Press.

Coville, F. V. 1892. The Panamint Indians of California. *American Anthropologist* 5(4):351–57.

———. 1897. Notes on the plants used by the Klamath Indians of Oregon. *Contributions to the U.S. National Herbarium* 5(2):87–108.

Cowan, R. G. 1977. *Ranchos of California*. Los Angeles: Historical Society of Southern California.

Cowen, R. 1990. Seeds of protection. *Science News* 137:350–51.

Crampton, C. G., ed. 1957. *The Mariposa Indian War 1850–1851: Diaries of Robert Eccleston: The California Gold Rush, Yosemite, and the High Sierra*. Salt Lake City: University of Utah Press.

Cronise, T. F. 1868. *The Natural Wealth of California*. San Francisco: H. H. Bancroft.

Cronon, W. 1995. The trouble with wilderness; or, getting back to the wrong nature. In *Uncommon Ground: Toward Reinventing Nature*, ed. W. Cronon, 69–90. New York: W. W. Norton.

Crosby, A. W., Jr. 1972. *The Columbian Exchange: Biological and Cultural Consequences of 1492*. Westport, Conn.: Greenwood Press.

———. 1986. *Ecological Imperialism: The Biological Expansion of Europe, 900–1900*. New York: Cambridge University Press.

Crum, S. 1998. A tripartite state of affairs: the Timbisha Shoshone Tribe, the National Park Service, and the Bureau of Indian Affairs, 1933–1994. *American Indian Culture and Research Journal* 22(1):117–36.

———. 2001. Deeply attached to the land: The Owens Valley Paiutes and their rejection of Indian removal, 1863–1937. *News from Native California* 14(4):18–20.

Crumley, C. L. 1994. Historical ecology: A multidimensional ecological orientation. In *Historical Ecology: Cultural Knowledge and Changing Landscapes*, ed. C. L. Crumley. Santa Fe, New Mex.: School of American Research Press.

Culin, S. [1907] 1975. *Games of the North American Indians.* New York: Dover.

Cunningham-Summerfield, B. 1998. Yalulu. *News from Native California* 2(4):35–37.

Curtis, E. S. 1924a. The Maidu. In *The North American Indian.* Vol. 14, ed. F. W. Hodge, 99–126. New York: Johnson Reprint Corporation.

———. 1924b. The Miwok. In *The North American Indian.* Vol. 14, ed. F. W. Hodge, 129–47. New York: Johnson Reprint Corporation.

———. 1924c. The Mono. In *The North American Indian.* Vol. 15, ed. F. W. Hodge, 55–66. New York: Johnson Reprint Corporation.

———. 1924d. The Washoe. In *The North American Indian.* Vol. 14, ed. F. W. Hodge, 89–98. New York: Johnson Reprint Corporation.

Cutter, D. C. 1960. *Malaspina in California.* San Francisco: John Howell Books.

Dane, G. E., trans. 1935. The founding of the presidio and mission of our Father Saint Francis. *California Historical Society Quarterly* 14(2):102–10.

Dangberg, G. 1968. *Washo Tales.* Occasional Paper of the Nevada State Museum. Carson City.

Darby, H. C. 1956. The clearing of the woodland in Europe. In *Man's Role in Changing the Face of the Earth,* ed. W. L. Thomas, Jr., 183–216. Chicago: University of Chicago Press.

Dasmann, R. F. 1965. *The Destruction of California.* New York: Macmillan.

———. 1985. Preface. In *Wild California: Vanishing Lands, Vanishing Wildlife,* by A. S. Leopold and T. A. Blake, ix. Berkeley: University of California Press.

———. 1991. The importance of cultural and biological diversity. In *Biodiversity: Culture, Conservation, and Ecodevelopment,* ed. M. Oldfield and J. Alcorn, 7–15. Boulder, Colo.: Westview Press.

———. 1999. Environmental changes before and after the Gold Rush. In *A Golden State: Mining and Economic Development in Gold Rush California,* ed. J. J. Rawls and R. J. Orsi, 105–22. Berkeley: University of California Press.

Daugherty, A. E. 1979. *Marine Mammals of California.* University of California Sea Grant Advisory Program.

Davidson-Hunt, I., and F. Berkes. 2003. Learning as you journey: Anishinaabe perceptions of socio-ecological environments and adaptive learning. *Conservation Ecology* 8(1):5.

Davis, J. T. 1961. Trade routes and economic exchange among the Indians of California. *University of California Archeological Survey Reports.* Department of Anthropology, University of California, Berkeley.

Davis, L. 1988. On this earth: Hupa land domains, images and ecology on "deddeh ninnisan." Ph.D. dissertation, University of California, Berkeley.

Davis, W. H. 1929. *Seventy Years in California.* San Francisco: J. Howell.

Davis, W. H. 1971. Indian insurrections and treachery. In *Sketches of Early California: A Collection of Personal Adventures,* comp. D. DeNevi, 33–42. San Francisco: Chronicle Books.

Dawson, L. E. 1973. *The Indian Basketry.* [In the series of keepsakes consisting of twelve folders issued by the Book Club of California in 1973.] San Francisco: Greenwood Press.

Dawson, R., and G. Brechin. 1999. *Farewell, Promised Land: Waking from the California Dream.* Berkeley: University of California Press.

Dawson, W. L. 1923. *The Birds of California.* 4 vols. San Diego: South Moulton Co.

d'Azevedo, W. L. n.d. The Washoe. Unpublished manuscript.

d'Azevedo, W. L., ed. 1986. Washoe. In *Handbook of North American Indians.* Vol. 11: Great Basin, ed. W. L. d'Azevedo, 466–98. Washington, D.C.: Smithsonian Institution.

Deal, K., J. Rood, C. Harasek, and D. McLemore. 1994. Eldorado National Forest: Basket material gathering. Pacific Southwest Region. *Working Together: California Indians and the Forest Service Accomplishment Report 1994.* San Francisco, Calif.: U.S. Forest Service.

de Angulo, J. 1935. Pomo creation myth. *Journal of American Folk-Lore* 48(189):203–62.

De Hertogh, A., and M. Le Nard, eds. 1993. *The Physiology of Flower Bulbs.* Amsterdam: Elsevier.

Deloria, V., Jr. 1985. *American Indian Policy in the Twentieth Century.* Norman: University of Oklahoma Press.

———. 1994. *God Is Red: A Native View of Religion.* Golden, Colo.: Fulcrum.

Deloria, V., Jr., and C. M. Lytle. 1984. *The Nations Within: The Past and Future of American Indian Sovereignty.* New York: Pantheon Books.

de Massey, E. 1926. *A Frenchman in the Goldrush Part V.* Trans. M. E. Wilbur. *California Historical Society Quarterly* 5(2):149.

Dempster, L. T. 1993. *Apocynum* dogbane, Indian hemp. In *The Jepson Manual: Higher Plants of California,* ed. J. C. Hickman, 68. Berkeley: University of California Press.

Denevan, W. M. 1992. The pristine myth: The landscape of the Americas in 1492. *Annals of the Association of American Geographers* 82(3):369–85.

DeQuille, D. 1963. *Washoe Rambles.* Los Angeles: Westernlore Press.

Derby, G. H. 1932. The topographical reports of Lieutenant George H. Derby. *California Historical Society Quarterly* 11(3):247–73.

Deur, D., and N. J. Turner. In press. *Keeping It Living: Indigenous Plant Management on the Northwest Coast.* Seattle: University of Washington Press.

Diamond, J. 1992. *The Third Chimpanzee.* New York: Harper Collins.

Dick, L. E., L. Planas, J. Polanich, C. D. Bates, and M. J. Lee. 1988. *Strands of Time: Yokuts, Mono, and Miwok Basketmakers.* Fresno, Calif.: Fresno Metropolitan Museum.

Dillon, R. H. 1965. *J. Ross Browne: Confidential Agent in Old California.* Norman: University of Oklahoma Press.

———. 1967. *Captain John Sutter: Sacramento Valley's Sainted Sinner.* Santa Cruz, Calif.: Western Tanager.

Dixon, R. B. 1905. The Northern Maidu: The Huntington California Expedition. *Bulletin of the American Museum of Natural History* 17(3):119–346.

———. 1908. Notes on the Achomawi and Atsugewi Indians of northern California. *American Anthropologist* 10(2):208–20.

Dobyns, H. F. 1983. Disease transfer at contact. *Annual Reviews of Anthropology* 22:273–91.

Doebley, J. F. 1984. "Seeds" of wild grasses: A major food of Southwestern Indians. *Economic Botany* 38(1):52–64.

Dominguez, L. 1998. A vision of a living language. *News from Native California* 12(1):17.

Donahue, D. L. 1999. *The Western Range Revisited: Removing Livestock from Public Lands to Conserve Native Biodiversity.* Norman: University of Oklahoma Press.

Dondero, J. 1992. *A Burning Desire to Do the Right Thing.* Report from Orleans Ranger District, Six Rivers National Forest.

Doolittle, W. E. 2000. *Cultivated Landscapes of Native North America.* Oxford: Oxford University Press.

Doten, A. 1973. *The Journals of Alfred Doten, 1849–1903.* Ed. W. Van Tilburg Clark. Reno: University of Nevada Press.

Downs, J. F. 1966. *The Two Worlds of the Washo.* New York: Holt, Rinehart and Winston.

Driver, H. E. 1936. Wappo ethnography. *University of California Publications in American Archaeology and Ethnology* 36(3):179–220.

———. 1937. Culture element distributions VI: Southern Sierra Nevada. *University of California Anthropological Records* 1(2):53–154.

———. 1939. Culture element distributions X: Northwest California. *University of California Anthropological Records* 1(6):297–433.

Driver, H. E., and W. C. Massey. 1957. Comparative studies of North American Indians. *Transactions of the American Philosophical Society* 47(2)1–456.

Drucker, P. 1937. The Tolowa and their southwest Oregon kin. *University of California Publications in American Archaeology and Ethnology* 36(4):221–300.

———. 1939. Culture element distributions V: Southern California. *University of California Anthropological Records* 1:1–51.

Du Bois, C. A. 1932. Tolowa notes. *American Anthropologist* 34(2):248–62.

———. 1935. Wintu ethnography. *University of California Publications in American Archaeology and Ethnology* 36(1):1–148.

Duhaut-Cilly, A. 1929a. Duhaut-Cilly's account of California in the years 1827–28. *California Historical Society Quarterly* 8(3):214–50.

———. 1929b. Duhaut-Cilly's account of California in the years 1827–1828. *California Historical Society Quarterly* 8(4):306–36.

Duncan III, J. W. 1964. Maidu ethnobotany. Master's thesis, Sacramento State University.

Dutcher, B. H. 1893. Pinon gathering among the Panamint Indians. *American Anthropologist* 6:377–80.

Eastwood, A. n.d. Some plants of Tulare County, with the Indian names and uses. Unpublished manuscript. G.H.M. Acc. No. 20,108. Grace Hudson Museum and Sun House, Ukiah, Calif.

Edwards, S. W. 1996. A rancholabrean-age, late-Pleistocene bestiary for California botanists. *Four Seasons* 10(2):5–34.

Egan, D., ed. 2003. Special Issue: Native American Land Management Practices in National Parks. *Ecological Restoration* 21(4):237–58.

Egan, D., and E. A. Howell, eds. 2001. *The Historical Ecology Handbook: A Restorationist's Guide to Reference Ecosystems.* Washington, D.C.: Island Press.

Egli, I. R. 1992. *No Rooms of Their Own: Women Writers of Early California.* Berkeley, Calif.: Heyday Books in association with Rick Heide.

Ehrlich, P. R. 2000. *Human Genes, Cultures, and the Human Prospect.* Washington, D.C.: Island Press/Shearwater Books.

Ehrlich, P. R., and A. Ehrlich. 1981. *Extinction: The Causes and Consequences of the Disappearance of Species.* New York: Ballantine Books.

Eigenheer, R. A. 1976. Early perceptions of agricultural resources in the Central Valley of California. Ph.D. dissertation, University of California, Davis.

Elliot, G., and C. Littlefield. 1994. Greater sandhill crane. In *Life on the Edge: A Guide to California's Endangered Natural Resources: Wildlife,* ed. C. G. Thelander and M. Crabtree, 166–70. Santa Cruz, Calif.: BioSystems Books; Berkeley, Calif.: Heyday Books.

Elliott, H. W., and J. D. Wehausen. 1974. Vegetational succession on coastal rangeland of Point Reyes Peninsula. *Madroño* 22:231–38.

Embody, E. 1907. Inyo County Indians. *Los Angeles Times,* March 17.

Engelhardt, Fr. Z. 1929. *Mission San Antonio de Padua.* Santa Barbara, Calif.: O.F.M. Mission Santa Barbara.

Engstrand, I. H. 1997. Seekers of the "northern mystery": European exploration of California and the Pacific. In *Contested Eden: California before the Gold Rush,* ed. R. A. Gutiérrez and R. J. Orsi, 78–110. Berkeley: University of California Press.

Ernst, E. F. 1949. Vanishing meadows in Yosemite Valley. *Yosemite Nature Notes* 28(5):34–41.

Espinosa, J. L., and M. Garcia. 2001. 1840: Indian attacks near Guadalupe. In *Chronicles of Early California, 1535–1846: Lands of Promise and Despair,* ed. R. M. Beebe and R. M. Senkewicz, 420–22. Santa Clara, Calif.: Santa Clara University; Berkeley, Calif.: Heyday Books.

Essene, F. 1942. Culture element distributions XXI: Round Valley. *University of California Anthropological Records* 8(1):1–97.

Evans, J. W. 1990. *Powerful Rockey: The Blue Mountains and the Oregon Trail, 1811–1883.* Enterprise, Ore.: Pika Press.

Everett, R. L. 1987. Plant response to fire in the pinyon-juniper zone. In *Proceeding of the Pinyon-Juniper Conference, January 13–16, 1986, Reno, Nevada,* comp. R. L. Everett, 152–57. General Technical Report INT-215. Ogden, Utah: U.S. Department of Agriculture, Forest Service, Intermountain Research Station.

Evermann, B. W. 1916. The California valley elk. *California Fish and Game* 2(2):1–8.

Faber, P. M., ed. 1997. *California's Wild Gardens.* Sacramento, Calif.: California Native Plant Society.

Fagan, B. H. 2003. *Before California: An Archaeologist Looks at Our Earliest Inhabitants.* Lanham, Md.: Rowman and Littlefield.

Farmer, J. F. 1993. Preserving Diegueño basket weaving. In *Native American Basketry of Southern California,* ed. Christopher L. Moser, 141–47. Riverside, Calif.: Riverside Museum Press.

———. 2003. Agave roast at the Malki Museum. *News from Native California* 16(4):4–5.

Farnham, T. J. 1846. *Life and Adventures in California.* New York: W. H. Graham.

Farnsworth, N. R. 1988. Screening plants for new medicines. In *Biodiversity,* ed. E. O. Wilson, 83–97. Washington, D.C.: National Academy Press.

Farris, G. J. 1982. Aboriginal use of pine nuts in California: An ethnological, nutritional, and archaeological investigation into the use of the seeds of *Pinus lambertiana Doug.* and *Pinus sabiniana Doug.* by the Indians of northern California. Ph.D. dissertation, University of California, Davis.

———. 1993. Quality food: the quest for pine nuts in northern California. In *Before the Wilderness: Environmental Management by Native Californians,* ed. T. C. Blackburn and K. Anderson, 229–40. Menlo Park, Calif.: Ballena Press.

Faulk, O. B., and L. E. Faulk. 1988. The Modoc. In *Indians of North America,* ed. F. W. Porter III. New York: Chelsea House.

Fenenga, G. L., and E. M. Fisher. 1978. The Cahuilla use of Piyatem, larvae of the white-lined sphinx moth *(Hyles lineata),* as food. *Journal of California Anthropology* 5(1):84–90.

Fiedel, S. J. 1987. *Prehistory of the Americas.* Cambridge, Mass.: Cambridge University Press.

———. 1999. Older than we thought: Implications of corrected dates for paleo-Indians. *American Antiquity* 64(1):95–115.

Fiedler, P. L. 1996. *Rare Lilies of California.* Sacramento: California Native Plant Society.

Fiedler, P. L., and J. J. Ahouse. 1992. Hierarchies of cause: Toward an understanding of rarity in vascular plant species. In *Conservation Biology: The Theory and Practice of Nature Conservation, Preservation, and Management,* ed. P. L. Fiedler and S. K. Jain, 23–47. New York: Chapman and Hall.

Fillmore, L. 1995–96. Washoe pinenut camp, 1991: A tribute to Goldie. *News from Native California* 9(2):23.

Fitzgerald, R. T., and T. L. Jones. 1999. The milling stone horizon revisited: New perspectives from northern and central California. *Journal of California and Great Basin Anthropology* 21(1):67–93.

Fleck, R. F. 1985. *Henry Thoreau and John Muir among the Indians.* Hamden, Conn.: Archon Books.

Fleischner, T. L. 1992. Preservation is not enough: The need for courage in wilderness management. In *Wilderness Tapestry: An Eclectic Approach to Preservation,* ed. S. I. Zeveloff, L. M. Vause, and W. H. McVaugh, 236–53. Reno: University of Nevada Press.

Fletcher, A. C. 1888. *Indian Education and Civilization.* U.S. Bureau of Education Special Report prepared in answer to Senate Resolution of February 23, 1885. 48th Cong., 2d sess. Washington, D.C.: Government Printing Office.

Fletcher, T. C. 1987. *Paiute, Prospector, Pioneer: A History of the Bodie–Mono Lake Area in the Nineteenth Century.* Lee Vining, Calif.: Artemisia Press.

Forbes, A. [1839] 1937. *California: A History of Upper and Lower California.* San Francisco: John Henry Nash. [Orig. pub. London: Smith, Elder, and Co.]

Forbes, J. 1982. *Native Americans of California and Nevada.* Happy Camp, Calif.: Naturegraph.

———. 1996. Who owns the water? *News from Native California* 9(3):16.

———. 2001. Nature and culture: problematic concepts for Native Americans. In *Indigenous Traditions and Ecology,* ed. J. A. Grim, 103–24. Cambridge, Mass.: Harvard University Press.

Font, P. 1971. The Colorado Yumans in 1775. In *The California Indians: A Source Book,* 2d ed., ed. R. F. Heizer and M. A. Whipple, 247–61. Berkeley: University of California Press.

Ford, J., and D. Martinez, eds. 2000. Traditional ecological knowledge, ecosystem science, and environmental management. Invited Feature. *Ecological Applications* 10(5):1249–1341.

Ford, R. I. 1981. Ethnobotany in North America: An historical phytogeographic perspective. *Canadian Journal of Botany* 59:2178–88.

———. 1985. Patterns of food production in North America. In *Prehistoric Food Production in North America,* ed. R. I. Ford, 341–64. Anthropological Papers No. 75. Ann Arbor: Museum of Anthropology, University of Michigan.

Fortner, H. J. 1978. *The Limu Eater: A Cookbook of Hawaiian Seaweed.* Honolulu: University of Hawaii Sea Grant Program.

Foster, G. M. 1944. A summary of Yuki culture. *University of California Anthropological Records* 5(3):155–244.

Foster, J. W. 1999. Ajumawi fish traps: Harvesting and managing suckers in the springs of the Pit River drainage. In *Proceedings of the Society for California Archaeology.* Vol. 13, ed. J. Reed, G. Greenway, and K. McCormick, 266–72. Thirty-third Annual Meeting of the Society for California Archaeology, Sacramento, Calif., April 23–25, 1999. Fresno: Society for California Archaeology.

Fowler, C. 1986. Subsistence. In *Handbook of North American Indians.* Vol. 11: *Great Basin,* ed. W. L. d'Azevedo, 64–97. Washington, D.C.: Smithsonian Institution.

———. 1996. Historical perspectives on Timbisha Shoshone land management practices, Death Valley, California. In *Case Studies in Environmental Archaeology,* ed. E. J. Reitz, L. A. Newsom, and S. J. Scudder, 87–101. New York: Plenum Press.

———. 2000. "We live by them": Native knowledge of biodiversity in the Great Basin of western North America. In *Biodiversity and Native America,* eds. P. E. Minnis and W. J. Elisens, 99–132. Norman: University of Oklahoma Press.

———. 2003. Timbisha Shoshone mesquite and pinyon management: Applied ethnoecology in Death Valley National Park. Departmental lecture: Graduate group in geography, University of California, Davis.

Fowler, C., P. Esteves, G. Goad, B. Helmer, and K. Watterson. 2003. Caring for the trees: Restoring Timbisha Shoshone land management practices in Death Valley National Park. *Ecological Restoration* 21(4):302–6.

Fowler, C. S., and S. Liljeblad. 1986. Northern Paiute. In *Handbook of North American Indians*. Vol. 11: *Great Basin*, ed. W. L. d'Azevedo, 435–65. Washington, D.C.: Smithsonian Institution.

Fox, S. 1985. *The American Conservation Movement: John Muir and His Legacy*. Madison: University of Wisconsin Press.

Franco, H. [Lalo]. 1993. "That place needs a good fire." *News from Native California* 7(2):17–19.

Frémont, J. C. 1887. *Memoirs of My Life: A Retrospect of Fifty Years, Covering the Most Eventful Periods of Modern American History*. Vol. 1. Chicago: Belford, Clarke & Company.

Frenkel, R. E. 1970. *Ruderal Vegetation along Some California Roadsides*. Berkeley: University of California Press.

Fulkerson, M. L. 1995. *Weavers of Tradition and Beauty: Basketmakers of the Great Basin*. Reno: University of Nevada Press.

Fuster, V. 2001. 1775: Rebellion at San Diego. In *Chronicles of Early California, 1535–1846: Lands of Promise and Despair*, ed. R. M. Beebe and R. M. Senkewicz, 186–92. Santa Clara, Calif.: Santa Clara University; Berkeley, Calif.: Heyday Books.

Gardner-Loster, J. 1998–99. The continuing thread. *News from Native California* 12(2):8–15.

Garner, W. R. 1970. *Letters from California, 1846–1847*. Ed. D. M. Craig. Berkeley: University of California Press.

Garth, T. R. 1945. Emphasis on industriousness among the Atsugewi. *American Anthropologist* 47(4):554–66.

———. 1953. Atsugewi ethnography. *University of California Anthropological Records* 14(2):129–212.

———. 1978. Atsugewi. In *Handbook of North American Indians*. Vol. 8: *California*, ed. R. F. Heizer, 236–43. Washington, D.C.: Smithsonian Institution.

Gayton, A. H. 1946. Culture-environment integration: External references in Yokuts life. *Southwestern Journal of Anthropology* 2(3):252–68.

———. 1948a. Yokuts and Western Mono ethnography I: Tulare Lake, Southern Valley, and Central Foothill Yokuts. *Anthropological Records* 10(1): 1–138.

———. 1948b. Yokuts and Western Mono ethnography II: Northern Foothill Yokuts and Western Mono. *Anthropological Records* 10(2):139–302.

Gerlach, J. D. 1998. How the West was lost: Reconstructing the invasion dynamics of yellow star-thistle and other plant invaders of western rangelands and natural areas. In *Proceedings of the California Exotic Pest Plant Council Symposium, 3, 1997*, ed. M. Kelly, E. Wagner, and P. Warner, 67–72.

Gibbs, G., comp. 1853. Journal of the expedition of Colonel Redick McKee, United States Indian agent, through North-Western California. In *Historical and Statistical Information Respecting the History, Conditions and Prospects of the Indian Tribes of the United States*. Vol. 3: *Information Respecting the History, Condition, and Prospects of the Indian Tribes of the United States*, ed. H. R. Schoolcraft, 99–177. Philadelphia: Lippincott, Grambo.

Gibson, L. P. 1969. Monograph of the Genus *Curculio* in the New World (Coleoptera: Curculionidae). Part I: United States and Canada. *Miscellaneous Publications of the Entomological Society of America* 6(5):241–85.

Gifford, E. W. 1916. Miwok moieties. *University of California Publications in American Archaeology and Ethnology* 12(4):139–94.

———. 1932. The Northfork Mono. *University of California Publications in American Archaeology and Ethnology* 31(2):15–65.

———. 1939. Karok field notes. Pt. 1. Ethnological Document No. 174 in Department and Museum of Anthropology, University of California; Manuscript in University Archives, Bancroft Library, Berkeley.

———. [1939] 1965. The Coast Yuki. Repr. *Sacramento Anthropological Society Papers* 2. Sacramento: Sacramento State College. [Orig. pub. as *Anthropos* 34:292–375.]

———. 1971. Californian balanophagy. In *The California Indians: A Source Book*, 2d ed., ed. R. F. Heizer and M. A. Whipple, 301–5. Berkeley: University of California Press.

Gilmore, M. R. 1931. Dispersal by Indians: A factor in the extension of discontinuous distribution of certain species of native plants. *Papers of the Michigan Academy of Science, Arts, and Letters* 13:89–94.

———. 1977. *Uses of Plants by the Indians of the Missouri River Region*. Lincoln: University of Nebraska Press.

Goddard, I. 1996. Introduction. In *Handbook of North American Indians*. Vol. 17: *Languages*, ed. W. C. Sturtevant, 1–16. Washington, D.C.: Smithsonian Institution.

Goddard, P. E. 1903–4. Life and culture of the Hupa. *University of California Publications in American Archaeology and Ethnology* 1(1):1–88.

Goetzmann, W. H. 2000. *Exploration and Empire: The Explorer and the Scientist in the Winning of the American West*. Austin: Texas State Historical Association.

Goldberg, C. 1997. Acknowledging the repatriation claims of unacknowledged California tribes. *American Indian Culture and Research Journal* 21(3): 183–90.

Goldberg-Ambrose, C., and T. C. Steward. 1997. *Planting Tail Feathers: Tribal Survival and Public Law 280*. Los Angeles: American Indian Studies Center, University of California, Los Angeles.

Goldschmidt, W. R. 1951. Nomlaki ethnography. *University of California Publications in American Archaeology and Ethnology* 42(4):303–443.

———. 1978. Nomlaki. In *Handbook of North American Indians*. Vol. 8: *California*, ed. R. F. Heizer, 341–49. Washington, D.C.: Smithsonian Institution.

Gómez-Pompa, A., and A. Kaus. 1992. Taming the wilderness myth. *Bioscience* 42(4):271–79.

González, R. 2001. 1824: The Chumash revolt. In *Chronicles of Early California, 1535–1846: Lands of Promise and Despair,* ed. R. M. Beebe and R. M. Senkewicz, 323–28. Santa Clara, Calif.: Santa Clara University; Berkeley, Calif.: Heyday Books.

Goode, R. W. 1992. Cultural traditions endangered. Unpublished report to the U.S. Forest Service Sierra National Forest.

Goodrich, J., C. Lawson, and V. P. Lawson. 1980. *Kashaya Pomo Plants.* American Indian Monograph Series No. 2. Los Angeles: American Indian Studies Center, University of California, Los Angeles.

Goodyear, F. H. 2000. Foreword to *Away from Home: American Indian Boarding School Experiences 1879–2000,* ed. M. L. Archuleta, B. J. Child, and K. T. Lomawaima, 9–10. Phoenix, Ariz.: Heard Museum.

Gordon-Cumming, C. F. 1883. Wild tribes of the Sierra. *National Review* 2(8):412–21.

Gott, B. 1983. Murnong—*Microseris scapigera:* A study of a staple food of Victorian Aborigines. *Australian Aboriginal Studies* 2:2–18.

Grant, C. 1978. Eastern Coastal Chumash. In *Handbook of North American Indians.* Vol. 8: *California,* ed. R. F. Heizer, 509–19. Washington, D.C.: Smithsonian Institution.

Gray, T. B. 1993. *The Stanislaus Indian Wars.* Modesto, Calif.: McHenry Museum Press.

Greaves, T. 1996. Tribal rights. In *Valuing Local Knowledge: Indigenous People and Intellectual Property Rights,* ed. S. B. Brush and D. Stabinsky, 25–40. Washington, D.C.: Island Press.

Green, W. S. 1895. The Digger Indian. *Overland Monthly* 25:282–84.

Greene, C. S. 1892. Rabbit driving in the San Joaquin Valley. *Overland Monthly* 20:49–58.

Greenlee, J. M., and J. H. Langenheim. 1990. Historic fire regimes and their relation to vegetation patterns in the Monterey Bay area of California. *American Midland Naturalist* 124(2):239–53.

Greer, C. 1992. Pincushion is target. *Mountain Press* 18(48):1, 8.

Griggs, F. T. 2000. Vina Plains preserve: Eighteen years of adaptive management. *Fremontia* 27(4):48–51.

Grinnell, J., H. C. Bryant, and T. I. Storer. 1918. *The Game Birds of California.* Berkeley: University of California Press.

Grinnell, J., J. S. Dixon, and J. M. Linsdale. 1937. *Fur-bearing Mammals of California: Their Natural History, Systematic Status, and Relations to Man.* 2 vols. Berkeley: University of California Press.

Grumbine, R. E. 1992. *Ghost Bears: Exploring the Biodiversity Crisis.* Washington, D.C.: Island Press.

Gutherie, R. D. 1993. Listen to the birds? The use of avian remains in Channel Islands archaeology. In *Archaeology on the Northern Channel Islands of California: Studies of Subsistence, Economics, and Social Organization,* ed.

M. A. Glassow, 153–67. Archives of California Prehistory No. 34. Salinas, Calif.: Coyote Press.

Guthrie, D. A. 1993. New information on the prehistoric fauna of San Miguel Island. In *Third Channel Islands Symposium*, ed. F. G. Nocaberg, 405–16. Santa Barbara, Calif.: Santa Barbara Museum of Natural History.

Haas, L. 1995. *Conquests and Historical Identities in California, 1769–1936.* Berkeley: University of California Press.

Haderlie, E. C., and D. P. Abbott. 1980. Bivalvia: The clams and allies. In *Intertidal Invertebrates of California*, ed. R. H. Morris, D. P. Abbott, and E. C. Haderlie, 355–411. Stanford: Stanford University Press.

Hafleigh, H. 2000. Indian vaqueros of California. *News from Native California* 13(4):25–40.

Hailstone, V. 1992–93. The past is but the beginning of the beginning. *News from Native California* 7(1):4–5.

Hanes, T. L. 1988. California chaparral. In *Terrestrial Vegetation of California*, expanded ed., ed. M. G. Barbour and J. Majors, 355–411. Special Publication No. 9. Sacramento: California Native Plant Society.

Hansen, G. 1899. The lilies of the Sierra Nevada. *Erythea: A Journal of Botany, West American and General* 7:21–23.

Haq, N. 1993. Lupins (*Lupinus* species). In *Underutilized Crops: Pulses and Vegetables*, ed. J. T. Williams, 103–30. London: Chapman and Hall.

Hardison, J. R. 1976. Fire and disease. In *Proceedings of the Annual Tall Timbers Fire Ecology Conference*, October 16–17, Portland, Ore., 223–34. Tallahassee, Fla.: Tall Timbers Research Station.

Hargreaves, B. J. 1996. Tubers and bulbs as an adaptive strategy in the Kalahari. In *The Biodiversity of African Plants*, 88–91. Proceedings of the XIVth AETFAT Congress, August 22–27, Wageningen, the Netherlands. Dordrecht: Kluwer Academic.

Harker, D., G. Libby, K. Harker, S. Evans, and M. Evans. 1999. *Landscape Restoration Handbook.* 2d ed. Washington, D.C.: Lewis.

Harlan, J. R. 1992. *Crops and Man.* 2d ed. Madison, Wis.: American Society of Agronomy and Crop Science Society of America.

———. 1995. *The Living Fields: Our Agricultural Heritage.* Cambridge: Cambridge University Press.

Harlan, J. R., and J. M. de Wet. 1973. On the quality of evidence for origin and dispersal of cultivated plants. *Current Anthropology* 14(1–2):51–55.

Harrington, J. P. 1916–17. Microfilm Reel 100 Yawelmani Grammar [former B.A.E. mss. 2973, 3041, 3047, 3048, 3054], frames 374–426. *The Papers of John Peabody Harrington in the Smithsonian Institution, 1907–1957.* Vol. 2. Smithsonian Institution, Washington, D.C.

———. 1932. Tobacco among the Karuk Indians of California. *Bureau of American Ethnology Bulletin* 94. Washington, D.C.

———. 1986. Kitanemuk notes. In *The Papers of John P. Harrington in the Smithsonian Institution, 1907–1957.* Vol. 3: *A Guide to the Field Notes: Native American History, Language and Culture of Southern California/Basin,* ed. E. L. Mills and A. J. Brickfield. White Plains, N.Y.: Kraus International Publications.

Harris, G. H. 1891. Root foods of the Seneca Indians. *Proceedings of the Rochester Academy of Science.* Vol. 1. Rochester, N.Y.: Rochester Academy of Science.

Harris, J. A., P. Birch, and J. Palmer. 1996. *Land Restoration and Reclamation: Principles and Practice.* Essex: Addison Wesley Longman.

Harshberger, J. W. 1928. Letter to Dr. John W. Hudson from John W. Harshberger dated September 23, 1928, G.H.M. Acc. No. 6046. Collection of the Grace Hudson Museum and Sun House, Ukiah, Calif.

Havard, V. 1896. Drink plants of the North American Indians. *Bulletin of the Torrey Botanical Club* 23(2):33–46.

Heady, H. 1988. Valley grassland. In *Terrestrial Vegetation of California,* expanded ed., ed. M. G. Barbour and J. Major, 491–514. Special Publication No. 9. Sacramento: California Native Plant Society.

Heady, H. F., T. C. Foin, M. M. Hektner, D. W. Taylor, M. G. Barbour, and W. J. Barry. 1988. Coast prairie and northern coastal scrub. In *Terrestrial Vegetation of California,* expanded ed., ed. M. G. Barbour and J. Major, 733–60. Special Publication No. 9. Sacramento: California Native Plant Society.

Hedges, K., and C. Beresford. 1986. *Santa Ysabel Ethnobotany.* Ethnic Technology Notes No. 20. San Diego, Calif.: San Diego Museum of Man.

Heffner, K. 1984. Following the smoke: Contemporary plant procurement by the Indians of northwest California. Unpublished document. Six Rivers National Forest, Eureka.

———. 1985. " . . . and another thing . . . ": Results of Hupa sensing on national forest resource management. Unpublished report. Six Rivers National Forest, Eureka, Calif.

Heinsen, V. 1972. *Mission San Antonio de Padua Herbs.* n.p.

Heizer, R. F. 1947. *Francis Drake and the California Indians, 1579.* Berkeley: University of California Press.

———. 1966. *Languages, Territories, and Names of California Indian Tribes.* Berkeley: University of California Press.

———. 1972. *The Eighteen Unratified Treaties of 1851–1852 between the California Indians and the United States Government.* Archaeological Research Facility, Department of Anthropology, University of California, Berkeley.

———. 1978. Natural forces and native world view. In *Handbook of North American Indians.* Vol. 8: *California,* ed. R. F. Heizer, 649–53. Washington, D.C.: Smithsonian Institution.

———, ed. [1974] 1993. *The Destruction of the California Indians.* Lincoln: University of Nebraska Press.

Heizer, R. F., and A. J. Almquist. 1971. *The Other Californians.* Berkeley: University of California Press.

Heizer, R. F., and A. B. Elsasser. 1980. *The Natural World of the California Indians.* Berkeley: University of California Press.

Heizer, R. F., and T. Kroeber, eds. 1979. *Ishi the Last Yahi.* Berkeley: University of California Press.

Helms, J. A., and R. D. Ratliff. 1987. Germination and establishment of *Pinus Contorta* var. *murrayana* (pinaceae) in mountain meadows of Yosemite National Park, California. *Madroño* 34(2):77–90.

Hendry, G. W. 1931. The adobe brick as a historical source. *Agricultural History* 5:110–27.

Henley, T. J. 1854. Office of the superintendent Indian affairs report. Report of the Secretary of the Interior for 1854. Senate Executive Document 1, 33rd Congr., 2d sess., 508–13.

Henshaw, H. W. 1876. Notes on the mammals taken and observed in California in 1875 by H. W. Henshaw. *Annual Report of the Geographical Survey West of the 100th Meridian* by G. M. Wheeler. Appendix JJ of the Annual Report, Chief of Engineers for 1876:305–12.

Hester, T. R. 1978. Salinan. In *Handbook of North American Indians.* Vol. 8: *California,* ed. R. F. Heizer, 500–504. Washington, D.C.: Smithsonian Institution.

Heusser, L. 1983. Contemporary pollen distribution in coastal California and Oregon. *Palynology* 7:19–42.

Hickman, J. C., ed. 1993. *The Jepson Manual: Higher Plants of California.* Berkeley: University of California Press.

Higgs, E. 2003. *Nature by Design: People, Natural Process, and Ecological Restoration.* Cambridge, Mass: MIT Press.

Hildebrandt, W. R., and T. L. Jones. 2002. Depletion of prehistoric pinniped populations along the California and Oregon coasts: Were humans the cause? In *Wilderness and Political Ecology: Aboriginal Influences and the Original State of Nature,* ed. C. E. Kay and R. T. Simmons, 111–40. Salt Lake City: University of Utah Press.

Hilgard, E. W., T. C. Jones, and R. W. Furnas. 1882. *Report on the Climatic and Agricultural Features and the Agricultural Practice and Needs of the Arid Regions of the Pacific Slope, with Notes on Arizona and New Mexico.* Washington, D.C.: U.S. Government Printing Office.

Hill, M. 1999. *Gold: The California Story.* Berkeley: University of California Press.

Hillman, G. C. and M. S. Davies. 1999. Domestication rate in wild wheats and barley under primitive cultivation: Preliminary results and archaeological implications of field measurements of selection coefficient. In *Prehistory of Agriculture: New Experimental and Ethnographic Approaches,* ed. P. C. Anderson, 70–102. Monograph 40. Los Angeles: Institute of Archaeology.

Hinton, L., ed. 1994a. *Flutes of Fire: Essays on California Indian Languages.* Berkeley, Calif.: Heyday Books.

———. 1994b. Preserving the future: A progress report on the master-apprentice language learning program. *News from Native California* 8(3):14–20.

———. 1999. The advocates. *News from Native California* 12(3):8–12.

Hinton, L., and Y. Montijo. 1994. Living California Indian languages. In *Flutes of Fire: Essays on California Indian Languages,* ed. L. Hinton, 21–33. Berkeley, Calif.: Heyday Books.

Hirt, P. W. 1994. *A Conspiracy of Optimism: Management of the National Forests since World War Two.* Lincoln: University of Nebraska Press.

History of Stanislaus County, California, with Illustrations Descriptive of Its Scenery, Farms, Residences, Public Buildings, Factories, Hotels, Business Houses, Schools, Churches, and Mines. 1881. San Francisco: Elliott and Moore.

Hitchcock, A. S. 1951. *Manual of the Grasses of the United States.* 2nd ed., revised by A. Chase. U.S. Department of Agriculture Publication 200. Washington, D.C.: U.S. Government Printing Office.

Hittel, T. H. 1897. *History of California.* Vol. 1. San Francisco: N. J. Stone.

Holing, D. 1988. *California Wild Lands: A Guide to the Nature Conservancy Preserves.* San Francisco: Chronicle Books.

Holliday, J. S. 1981. *The World Rushed In.* New York: Simon and Schuster.

———. 1999. *Rush for Riches: Gold Fever and the Making of California.* Berkeley: University of California Press and Oakland Museum.

Hooper, L. 1920. The Cahuilla Indians. *University of California Publications in American Archaeology and Ethnology* 16(6):316–80.

Hoover, R. L. 1973. *Chumash Fishing Equipment.* Ethnic Technology Notes No. 9. Menlo Park, Calif.: Ballena Press.

Hough, W. 1893. Primitive American armor. In *Report of the National Museum,* 625–51. Washington, D.C.

Howe, C. B. 1979. *Ancient Modocs of California and Oregon.* Portland, Ore.: Binford and Mort.

Howorth, P. C. 1978. *The Abalone Book.* Happy Camp, Calif.: Naturegraph.

Hudson, J. W. 1899. Catalogue of the Hudson collection of Indian Products. [Catalog accompanying the Hudson Ethnographic Collection made for the Smithsonian Institution.] G.H.M. Acc. No. 1982-7-1. Manuscript in the Collection of Grace Hudson Museum, Ukiah, Calif.

———. 1901a. Field notebook. G.H.M. Acc. No. 20,004. Collection of Grace Hudson Museum and Sun House, Ukiah, Calif.

———. 1901b. Field notebook. G.H.M. Acc. No. 20,002. Collection of Grace Hudson Museum and Sun House, Ukiah, Calif.

———. 1903. Correspondence, John Hudson to Grace Hudson, March 26. G.H.M. Acc. No. 4455. Collection of Grace Hudson Museum and Sun House, Ukiah, Calif.

———. 1904. Field notebook. G.H.M. Acc. No. 20,017. Collection of Grace Hudson Museum, Ukiah, Calif.

———. n.d.a. Field notebook. G.H.M. Acc. No. 20,011. Collection of Grace Hudson Museum and Sun House, Ukiah, Calif.

———. n.d.b. Manuscript on Pomoan Languages. G.H.M. Acc. Nos. 20,207a, 20,211, 21,029 (Childbirth), and 21,170. Collection of Grace Hudson Museum and Sun House, Ukiah, Calif.

———. n.d.c. *Vocabulary for Basketry.* G.H.M. Acc. No. 21,170. Collection of Grace Hudson Museum, Ukiah, Calif.

Hudson, T., and E. Underhay. 1978. *Crystals in the Sky: An Intellectual Odyssey Involving Chumash Astronomy, Cosmology and Rock Art.* Socorro, New Mexico: Ballena Press.

Hughes, J. D. 1983. *American Indian Ecology.* El Paso: Texas Western Press.

Hunn, E. S. 1990. *Nch'i-Wána: "The Big River" Mid-Columbia Indians and Their Land.* Seattle: University of Washington Press.

Hunn, E. S., and D. H. French. 1981. Lomatium: A key resource for Columbia Plateau native subsistence. *Northwest Science* 55(2):87–94.

Hunter, J. E. 1988. Prescribed burning for cultural resources. *Fire Management Notes* 49(2):8–9.

Hurtado, A. L. 1982. "Hardly a farm house—a kitchen without them": Indian and white households on the California borderland frontier in 1860. *Western Historical Quarterly* 13(3):245–70.

———. 1988. *Indian Survival on the California Frontier.* New Haven: Yale University Press.

Hutchings, J. 1962a. California quail—male and female. In *Scenes of Wonder & Curiosity from Hutchings' California Magazine, 1856–1861,* ed. R. R. Olmstead, 257–58. Berkeley, Calif.: Howell-North.

———. 1962b. The Farallone Islands. In *Scenes of Wonder & Curiosity from Hutchings' California Magazine, 1856–1861,* ed. R. R. Olmstead, 39–47. Berkeley, Calif.: Howell-North.

———. 1962c. Peter Lassen. In *Scenes of Wonder & Curiosity from Hutchings' California Magazine, 1856–1861,* ed. R. R. Olmstead, 385–87. Berkeley, Calif.: Howell-North.

———. 1962d. Reminiscences of Mendocino. In *Scenes of Wonder & Curiosity from Hutchings' California Magazine, 1856–1861,* ed. R. R. Olmstead, 242–56. Berkeley, Calif.: Howell-North.

———. 1990. *In the Heart of the Sierras: Yo Semite Valley and the Big Tree Groves.* Ed. P. Browning. Lafayette, Calif.: Great West Books.

Hutton, W. R. 1942. *Glances at California, 1847–1853; Diaries and Letters.* San Marino, Calif.: The Huntington Library.

Indian Board of Co-operation. 1919–20. Justice 70 years late. Bureau of Indian Affairs Manuscript 47, Box 6.15 "Indian Legal Cases." Holt-Atherton Center for Western Studies, University of the Pacific, Stockton, Calif.

Inglis, J. T., ed. 1993. *Traditional Ecological Knowledge: Concepts and Cases.* Ottawa: International Program on Traditional Ecological Knowledge and International Development Centre, Canadian Museum of Nature.

Inter-Tribal Council of Nevada. 1976. *Wa She Shu: A Washo Tribal History.* Inter-Tribal Council of Nevada, Reno. Salt Lake City: University of Utah.

Jack, K. R. 1916. An Indian's view of burning, and a reply. *California Fish and Game Journal* 2(4):194–96.

Jackson, D., and M. L. Spence. 1970. *The Expeditions of John Charles Frémont.* Vol. 1: *Travels from 1838 to 1844.* Chicago: University of Illinois Press.

Jackson, D. L., and L. L. Jackson. 2002. *The Farm as Natural Habitat: Reconnecting Food Systems with Ecosystems.* Washington, D.C.: Island Press.

Jackson, H., and A. Kinney. 1883. Report on the condition and needs of the Mission Indians of California, made by special agents Helen Jackson and Abbot Kinney to the Commissioner of Indian Affairs. Executive Document 49, 48th Cong., 1st sess., Message from the President of the United States, Transmitting: A Draft of Bill "For the Relief of the Mission Indian in the State of California," 7–37.

Jackson, L. L. 1992. The role of ecological restoration in conservation biology. In *Conservation Biology: The Theory and Practice of Nature Conservation, Preservation, and Management,* ed. P. L. Fiedler and S. K. Jain, 433–51. New York: Chapman and Hall.

Jameson, E. W., and H. J. Peeters. 1988. *California Mammals.* Berkeley: University of California Press.

Jeffrey-Hinton, C. 2001. Barona: A little museum with big ideas. *News from Native California* 14(4):10–11.

Jelinek, L. J. 1999. "Property of every kind": Ranching and farming during the Gold Rush era. In *A Golden State: Mining and Economic Development in Gold Rush California,* ed. J. J. Rawls and R. J. Orsi, 233–49. Berkeley: University of California Press.

Jensen, D. B., M. S. Torn, and J. Harte. 1993. *In Our Hands: A Strategy for Conserving California's Biological Diversity.* Berkeley: University of California Press.

Jepson, W. L. 1910. *The Silva of California.* Memoirs of the University of California. Vol. 2. Berkeley: The University Press.

———. 1911. *Flora of Western Middle California.* San Francisco: Cunningham, Curtiss and Welch.

———. 1923. *Trees of California.* Berkeley, Calif.: Sather Gate Bookshop.

———. 1943. *A Flora of California.* Vol. 3, Part 2. Berkeley: Jepson Herbarium and Library, University of California.

Jewell, D. 1971. Letter to R. Riegelhuth, Sequoia and Kings Canyon National Parks. On file, Research Office, Sequoia and Kings Canyon National Parks, Three Rivers, Calif.

Johnson, N. C. 1999. Humans as agents of ecological change—Overview. In *Ecological Stewardship: A Common Reference for Ecosystem Management.* Vol. 2, ed. R. C. Szaro, N. C. Johnson, W. T. Sexton, and A. J. Malk, 433–37. Oxford: Elsevier Science.

Johnson, P. J. 1978. Patwin. In *Handbook of North American Indians.* Vol. 8: *California,* ed. R. F. Heizer, 350–60. Washington, D.C.: Smithsonian Institution.

Johnson, S., G. Haslam, and R. Dawson. 1993. *The Great Central Valley: California's Heartland.* Berkeley: University of California Press.

Johnston, H. 1983. *They Felled the Redwoods: A Saga of Flumes and Rails in the High Sierra.* Glendale, Calif.: Trans-Anglo Books.

Johnston, V. R. 1970. *The Naturalist's America: Sierra Nevada.* Boston: Houghton Mifflin.

Johnston-Dodds, K. 2002. *Early California Laws and Policies Related to California Indians.* Prepared at the request of Senator John L. Burton, president pro tempore. Sacramento: California Research Bureau.

Johnstone, E. McD. 1888. *By Semi-Tropic Seas: Santa Barbara and Surroundings.* Buffalo, N.Y.: Matthews, Northrup and Co.

Jones, V. H. 1942. The location and delimitation of archaeological sites by means of divergent vegetation. *Society of American Archaeology, the Notebook* 2:64–65.

Jones and Stokes Associates. 1987. Sliding toward extinction: The state of California's natural heritage. Unpublished report to the California Senate Committee on Natural Resources and Wildlife.

Jordan, T. G., and M. Kaups. 1989. *The American Backwoods Frontier: An Ethnic and Ecological Interpretation.* Baltimore, Md.: Johns Hopkins University Press.

Jordan III, W. R. 2003. *The Sunflower Forest: Ecological Restoration and the New Communion with Nature.* Berkeley: University of California Press.

Kasch, C. 1947. The Yokayo Rancheria. *Quarterly of the California Historical Society* 26(3):209–16.

Kauffman, J. B., and R. E. Martin. 1987. Effects of fire and fire suppression on mortality and mode of reproduction of California black oak (*Quercus kelloggii* Newb.). In *Proceedings of the Symposium on Multiple-Use Management of California's Hardwood Resources, November 12–14, 1986, San Luis Obispo, California,* tech. coord. T. R. Plumb and N. H. Pillsbury, 122–26. General Technical Report PSW-100. U.S. Department of Agriculture, Forest Service, Pacific Southwest Forest and Range Experiment Station.

Kaufmann, M. R., R. T. Graham, D. A. Boyce Jr., W. H. Moir, L. Perry, R. T. Reynolds, R. L. Bassett, P. Mehlhop, C. B. Edminster, W. M. Block, and P. S. Corn. 1994. *An Ecological Basis for Ecosystem Management.* General Technical Report RM-246. Fort Collins, Colo.: U.S. Department of Agriculture, Forest Service, Rocky Mountain Forest and Range Experiment Station.

Keeley, J. E. 1986. Resilience of mediterranean shrub communities to fires. In *Resilience in Mediterranean-type Ecosystems,* ed. B. Dell, A. J. M. Hopkins, and B. B. Lamont, 95–112. Dordrecht: W. Junk.

———. 2002. Native American impacts on fire regimes of the California coastal ranges. *Journal of Biogeography* 29:303–20.

Keeley, J. E., and S. C. Keeley. 1989. Allelopathy and the fire-induced herb cycle. In *The California Chaparral,* ed. S. C. Keeley, 65–72. Science Series No. 34. Los Angeles: Natural History Museum of Los Angeles County.

Keeley, S. C., J. E. Keeley, S. M. Hutchinson, and A. W. Johnson. 1981. Postfire succession of herbaceous flora in southern California chaparral. *Ecology* 61(6):1608–21.

Keeling, R. 1992. *Cry for Luck: Sacred Song and Speech Among the Yurok, Hupa, and Karok Indians of Northwestern California.* Berkeley: University of California Press.

Keen, F. P. 1958. Cone and seed insects of western forest trees. *Technical Bulletin* 1169. Washington, D.C.: U.S. Department of Agriculture.

Kelly, I. 1932. Ethnography of the Surprise Valley Paiute. *University of California Publications in American Archaeology and Ethnology* 31(3):67–210.

———. 1978. Coast Miwok. In *Handbook of North American Indians.* Vol. 8: *California,* ed. R. F. Heizer, 414–25. Washington, D.C.: Smithsonian Institution.

Kelsey, H. 1986. *Juan Rodriguez Cabrillo*. San Marino, Calif.: Huntington Library.

Kemble, E. C. 1971. Confirming the gold discovery. In *Sketches of Early California: A Collection of Personal Adventures*, comp. D. DeNevi, 119–24. San Francisco, Calif.: Chronicle Books.

Keter, T. S. 1987. Indian burning: Managing the environment before 1865 along the North Fork. Paper presented to the Society for California Archaeology, April 16, Fresno, Calif.

Kickingbird, Esq., and E. R. Rhoades, M.D. 2000. The relation of Indian nations to the U.S. government. In *American Indian Health: Innovations in Health Care, Promotion, and Policy*, ed. E. R. Rhoades, M.D., 61–73. Baltimore, Md.: Johns Hopkins University Press.

Kilgore, B. M. 1973. Impact of prescribed burning on a sequoia–mixed conifer forest. In *Proceedings of the Annual Tall Timbers Fire Ecology Conference No. 12, Lubbock, Texas, June 8–9, 1972*, 345–75. Tallahassee, Fla.: Tall Timbers Research Station.

Kilgore, B. M., and D. Taylor. 1979. Fire history of a sequoia–mixed conifer forest. *Ecology* 60(1):129–42.

Kimbrell, A. 2002. *Fatal Harvest: The Tragedy of Industrial Agriculture*. Washington, D.C.: Island Press.

Kimmerer, R. W. 2000. Native knowledge for native ecosystems. *Journal of Forestry* 98(8):4–9.

Kimmerer, R. W., and F. K. Lake. 2001. The role of indigenous burning in land management. *Journal of Forestry* 99(11):36–41.

Kirk, A. 1994. Of whales and men: historical overview. In *Life on the Edge: A Guide to California's Endangered Natural Resources: Wildlife*, ed. C. G. Thelander and M. Crabtree, 468–73. Santa Cruz, Calif.: BioSystems Books; Berkeley, Calif.: Heyday Books.

Kirkpatrick, C. A. 1962. Salmon fishery on the Sacramento River. In *Scenes of Wonder & Curiosity from Hutchings' California Magazine, 1856–1861*, ed. R. R. Olmsted, 51–56. Berkeley, Calif.: Howell-North.

Kline, V. M. 1997. Planning a restoration. In *The Tallgrass Restoration Handbook: For Prairies, Savannas, and Woodlands*, ed. S. Packard and C. F. Mutel, 31–46. Washington, D.C.: Island Press.

Klinger, R. C., M. J. Kutilek, and H. S. Shellhammer. 1989. Population responses of black-tailed deer to prescribed burning. *Journal of Wildlife Management* 53(4):863–71.

Knap, A. H. 1975. *Wild Harvest: An Oudoorsman's Guide to Edible Wild Plants in North America*. Toronto: Pagurian Press.

Knapp, A. K., and T. R. Seastedt. 1986. Detritus accumulation limits productivity of tallgrass prairie. *Bioscience* 36(10):662–68.

Kneeland, S. 1872. *The Wonders of the Yosemite Valley and of California*. New York: Lee, Shepard, and Dillingham.

Kniffen, F. B. 1939. Pomo geography. *University of California Publications in American Archaeology and Ethnology* 36(6):353–400.

Knudsen, H. D., and R. Y. Sayler. 1992. Milkweed: The worth of a weed. In *New Crops, New Uses, New Markets*, 118–23. 1992 Yearbook of Agriculture. USDA.

Knudtson, P. M. 1977. *The Wintun Indians of California and Their Neighbors.* Happy Camp, Calif.: Naturegraph.

Kofoid, C. A. 1915. Marine biology on the Pacific coast. In *Nature and Science on the Pacific Coast,*124–32. San Francisco: Paul Elder and Co.

Korb, C. 1995. AICLS news. *News from Native California: An Inside View of the California Indian World* 9(2):41–42.

Kozlowski, T. T. 1971. *Growth and Development of Trees.* New York: Academic Press.

Krech III, S. 1999. *The Ecological Indian.* New York: W. W. Norton.

Kroeber, A. L. [1925] 1976. *Handbook of the Indians of California.* Bureau of American Ethnology Bulletin 78. Washington, D.C. Reprint. New York: Dover.

———. 1939a. Cultural and natural areas of native North America. *University of California Publications in American Archaeology and Ethnology* 38:1–242.

———. 1939b. Unpublished field notes on the Yurok. University Archives, Bancroft Library. University of California, Berkeley.

———. 1962. The nature of land-holding groups in aboriginal California. *University of California Survey Reports* 56:19–58.

———. 1971. The world renewal cult of northwest California. In *The California Indians: A Source Book,* 2d ed., ed. R. F. Heizer and M. A. Whipple, 464–71. Berkeley: University of California Press.

Kroeber, A. L., and S. A. Barrett. 1960. Fishing among the Indians of Northwestern California. *University of California Anthropological Records* 21(1): 1–210.

Kroeber, T. 1961. *Ishi in Two Worlds.* Berkeley: University of California Press.

Kroeber, T., and R. F. Heizer. 1968. *Almost Ancestors: The First Californians.* San Francisco: Sierra Club Books.

Krupp, E. C. 1987. Saluting the solstice. *News from Native California* 1(5): 10–13.

Kuhnlein, H. V., and N. J. Turner. 1991. *Traditional Plant Foods of Canadian Indigenous Peoples: Nutrition, Botany and Use.* Philadelphia: Gordon and Breach.

Kvasnicka, R. M. 1988. United States Indian treaties and agreements. In *Handbook of North American Indians.* Vol. 4: *History of Indian-White Relations,* ed. W. E. Washburn, 195–201. Washington, D.C.: Smithsonian Institution.

LaDuke, W. 1994. Traditional ecological knowledge and environmental futures. In *Endangered Peoples: Indigenous Rights and the Environment,* 126–48. Colorado Journal of International Environmental Law and Policy. Niwot: University Press of Colorado.

Lambert, P. M. 2004. Health in prehistoric populations of the Santa Barbara Channel Islands. In *Prehistoric California: Archaeology and the Myth of Paradise,* ed. L. M. Raab and T. L. Jones, 99–106. Salt Lake City: University of Utah Press.

Lang, J. 1991. The basket and world renewal. *Parabola* 26(4):83–85.

———, ed. 1994. *Ararapikva: Creation Stories of the People.* Berkeley, Calif.: Heyday Books.

———. 1996. Indian tobacco in northwestern California. *News from Native California* 9(3):28–36.

Langsdorff, G. H. von. 1927. *Langsdorff's Narrative of the Rezanov Voyage to Nueva California in 1806.* San Francisco: Press of T. C. Russell.

Lanner, R. M. 1981. *The Piñon Pine: A Natural and Cultural History.* Reno: University of Nevada Press.

———. 1996. *Made for Each Other: A Symbiosis of Birds and Pines.* New York: Oxford University Press.

LaPena, F. R. 1978. Wintu. In *Handbook of North American Indians.* Vol. 8: *California,* ed. R. F. Heizer, 324–40. Washington, D.C.: Smithsonian Institution.

———. 1997–98. Dancing and singing the sacredness of earth. *News from Native California* 11(2):17–19.

LaPena, F. R., and C. D. Bates, comps. 1981. *Legends of the Yosemite Miwok.* Yosemite National Park: Yosemite Natural History Association.

LaPena, S., and V. LaPena. 1996–97. All in the family. *News from Native California* 10(2):21–25.

LaRoe, E. T., G. S. Farris, C. E. Puckett, P. D. Doran, and M. J. Mac, eds. 1995. *Our Living Resources: A Report to the Nation on the Distribution, Abundance, and Health of U.S. Plants, Animals, and Ecosystems.* Washington, D.C.: U.S. Department of the Interior, National Biological Service.

Lathrop, E., and B. Martin. 1982. Fire ecology of deergrass *(Muhlenbergia rigens)* in Cuyamaca Rancho State Park, California. *Crossosoma* 8(5):1–4, 9–10.

Latta, F. F. 1933a. Interview with Henry Akers, July 17. Ethnographic Papers of Frank Forrest Latta. Yosemite Research Center, Yosemite National Park.

———. 1933b. Interview with Henry Akers, July 18. Ethnographic Papers of Frank Forrest Latta. Yosemite Research Center, Yosemite National Park.

———. 1934. Interview with Mrs. and Mr. Dick Francisco and Dan Williams, March 30. Ethnographic Papers of Frank Forrest Latta. Yosemite Research Center, Yosemite National Park.

———. 1962. Interview with Aida Icho (Wahnomkot), September 29. Ethnographic Papers of Frank Forrest Latta. Yosemite Research Center, Yosemite National Park.

———. 1977. *Handbook of Yokuts Indians.* Bakersfield, Calif.: Kern County Museum.

Lawlor, R. 1991. *Voices of the First Day: Awakening in the Aboriginal Dreamtime.* Rochester, Vt.: Inner Traditions International.

Lawton, H. W., P. J. Wilke, M. Dedecker, and W. M. Mason. 1993. Agriculture among the Paiute of Owens Valley. In *Before the Wilderness: Environmental Management by Native Californians,* ed. T. C. Blackburn and K. Anderson, 329–78. Menlo Park, Calif.: Ballena Press.

Le Conte, J. 1994. *A Journal of Ramblings through the High Sierras of California.* Yosemite National Park: Yosemite Association.

Ledig, F. T. 1999. Genic diversity, genetic structure, and biogeography of *Pinus sabiniana* Dougl. *Diversity and Distributions* 5:77–90.

Lee, G. D. 1998. *Walking Where We Lived: Memoirs of a Mono Indian Family.* Norman: University of Oklahoma Press.

Lee, J. T. 2000. *Emphasis on the Regeneration of the Bear Grass (Xerophyllum tenax) within the Wolf Timber Sale.* Technical Fire Management Report TFM-14. U.S. Department of Agriculture, Forest Service Region 5, Plumas National Forest, Mount Hough Ranger District.

Lee, M. H. 1989. *Indian of the Oaks.* San Diego, Calif.: San Diego Museum of Man.

Lee, R. B., and I. DeVore, eds. 1968. *Man the Hunter.* Chicago: Aldine.

Leonard, Z. 1978. *Adventures of a Mountain Man: The Narrative of Zenas Leonard.* Ed. M. M. Quaife. Lincoln: University of Nebraska Press.

Leopold, A. 1949. *A Sand County Almanac.* New York: Oxford University Press.

Leopold, A. S. 1977. *The California Quail.* Berkeley: University of California Press.

Leopold, A. S., and T. A. Blake. 1985. *Wild California: Vanishing Lands, Vanishing Wildlife.* Berkeley: University of California Press.

Leopold, A. S., S. A. Cain, I. N. Gabrielson, C. M. Cottam, and T. L. Kimball. 1963. The Leopold Report: Wildlife management in the national parks. Unpublished report to the U.S. Department of the Interior.

Lepofsky, D., E. K. Heyerdahl, K. Lertzman, D. Schaepe, and B. Mierendorf. 2003. Historical meadow dynamics in southwest British Columbia: A multidisciplinary analysis. *Conservation Ecology* 7(3):5.

Levy, R. 1978. Eastern Miwok. In *Handbook of North American Indians.* Vol. 8: *California,* ed. R. F. Heizer, 398–413. Washington, D.C.: Smithsonian Institution.

Lewallen, E., and J. Lewallen. 1995. *Sea Vegetable Gourmet Cookbook and Wildcrafter's Guide.* Mendocino, Calif.: Mendocino Vegetable Company.

Lewis, H. T. [1973] 1993. Patterns of Indian burning in California: Ecology and ethnohistory. Reprinted in *Before the Wilderness: Environmental Management by Native Californians,* ed. T. C. Blackburn and M. K. Anderson, 55–116. Menlo Park, Calif.: Ballena Press. [Orig. pub. as *Anthropological Papers 1.*]

———. 1989. Ecological knowledge of fire: Aborigines vs. park rangers in northern Australia. *American Anthropologist* 91:940–61.

———. 1991a. A parable of fire: Hunter-gatherers in Canada and Australia. In *Traditional Ecological Knowledge: A Collection of Essays,* ed. R. E. Johannes, 9–16. Gland, Switzerland: World Conservation Union.

———. 1991b. Technological complexity, ecological diversity, and fire regimes in northern Australia: Hunter-gatherer, cowboy ranger. In *Profiles in Cultural Evolution,* ed. A. T. Rambo and K. Gillogly, 261–88. Anthropological Papers No. 85. Ann Arbor: Museum of Anthropology, University of Michigan.

———. 1994. Management fires vs. corrective fires in northern Australia: An analogue for environmental change. *Chemosphere* 29:949–63.

Lewis, V. R. 1991. The temporal and spatial distribution of filbert weevil infested acorns in an oak woodland in Marin County, California. In *Proceedings of the Symposium on Oak Woodlands and Hardwood Rangeland Management, October 31–November 2, 1990, Davis, Calif.*, tech. coord. R. B. Standiford, 156–60. General Technical Report PSW-126. Berkeley, Calif.: U.S. Department of Agriculture, Forest Service, Pacific Southwest Research Station.

Lightfoot, K. G., and W. S. Simmons. 1998. Culture contact in protohistoric California: Social contexts of native and European encounters. *Journal of California and Great Basin Anthropology* 20(2):138–70.

Linden, E. 1991. Lost tribes, lost knowledge. *Time* 138(12):45–56.

Lindstrom, S. 1992. Great Basin fisherfolk: Optimal diet breadth. Modeling the Truckee River aboriginal subsistence fishery. Ph.D. dissertation, University of California, Davis.

Livingston, L. R. L. [1974] 1993. Letter from La Rhett L. Livingston dated August 17, 1856, to Major W. W. Mackall. In *The Destruction of California Indians*, ed. R. F. Heizer, 89–91. Lincoln: University of Nebraska Press.

Loeb, E. M. 1926a. The creator concept among the Indians of north central California. *American Anthropologist* 28(3):467–93.

———. 1926b. Pomo folkways. *University of California Publications in American Archaeology and Ethnology* 19(2):149–405.

Lomax, A., and C. M. Arensberg. 1977. A worldwide evolutionary classification of cultures by subsistence systems. *Current Anthropology* 18(4):659–708.

Long, J., A. Tecle, and B. Burnette. 2003. Cultural foundations for ecological restoration on the White Mountain Apache Reservation. *Conservation Ecology* 8(1):4.

Lonneberg, A. 1980. *Self and Savagery on the California Frontier: A Study of the Digger Stereotype.* Salinas, Calif.: Coyote Press.

Loud, L. L. 1918. Ethnography and archaeology of the Wiyot territory. *University of California Publications in American Archaeology and Ethnology* 14(3):221–436.

Lovegrove, B. G., and J. U. M. Jarvis. 1986. Coevolution between mole-rats (Bathyergidae) and a geophyte *Micranthus* (Iridaceae). *Cimbebasis* 8(9):80–85.

Lowie, R. H. 1939. Ethnographic notes on the Washo. *University of California Publications in American Archaeology and Ethnology* 36(5)301–52.

Lowry, J. L. 1999. *Gardening with a Wild Heart: Restoring California's Native Landscapes at Home.* Berkeley: University of California Press.

Lubbock, J. 1882. *On the Origin of Civilisation and Primitive Condition of Man: Mental and Social Condition of Savages.* 4th ed. London: Longmans, Green.

Lucas, J. M. 1945. *Indian Harvest: Wild Food Plants of America.* Philadelphia: J. B. Lippincott.

Lufkin, A., ed. 1991. *California's Salmon and Steelhead: The Struggle to Restore an Imperiled Resource.* Berkeley: University of California Press.

Lummis, C. F. 1929. *The Spanish Pioneers and the California Missions.* Chicago: A. C. McClury and Co.

Luther Burbank Society. 1914. *Luther Burbank: His Methods and Discoveries and their Practical Application.* Vol. 7. New York: Luther Burbank Press.

Lutke, F. P. 1989. *The Russian American Colonies, 1798–1867.* Vol. 3, ed. B. Dmytryshyn, E. A. P. Crownhart-Vaughan, and T. Vaughan. Portland: Oregon Historical Society Press.

Lynch, J. 1954. *With Stevenson to California.* Oakland, Calif.: Biobooks.

Lyon, D., and J. Cassidy. 1998. Shasta-Trinity National Forest: Native plants program and heritage program work. Pacific Southwest Region. *Working Together: California Indians and the Forest Service Accomplishment Report 1998.* San Francisco, Calif.: U.S. Forest Service.

Macdonald, K. B. 1988. Coastal salt marsh. In *Terrestrial Vegetation of California,* expanded ed., ed. M. G. Barbour and J. Major, 263–94. Special Publication No. 8. Sacramento: California Native Plant Society.

Machado, J. 2001. 1837: Indian attacks near San Diego. In *Chronicles of Early California, 1535–1846: Lands of Promise and Despair,* ed. R. M. Beebe and R. M. Senkewicz, 413–16. Santa Clara, Calif.: Santa Clara University; Berkeley, Calif.: Heyday Books.

Magallanes, F., and S. Thompson. 2003. What's cooking at the Malki Museum. *News from Native California* 16(4):6–7.

Maidu Cultural and Development Group. 2002. *Maidu Stewardship Project Action Plan for Plumas and Lassen National Forests.* Unpublished document.

Maloney, A. B. 1944. Fur brigade to the Bonaventura (continued): John Work's California expedition of 1832–33 for the Hudson's Bay Company. *California Historical Society Quarterly* 23(1):19–40.

Mann, C. C. 2002. 1491. *Atlantic Monthly* 289(3):41–53.

Margolin, M. 1981. *The Way We Lived: California Indian Reminiscences, Stories and Songs.* Berkeley, Calif.: Heyday Books.

———, ed. 1989. *Monterey in 1786: The Journals of Jean-François de La Pérouse.* Berkeley, Calif.: Heyday Books.

———. 1993. Foreword to *Indian Summer: Traditional Life among the Choinumne Indians of California's San Joaquin Valley,* by T. J. Mayfield, 9–17. Berkeley, Calif.: Heyday Books.

———. 1997–98. Traditional California Indian conservation. *News from Native California* 2(2):22–23.

———. 2001. The Big Valley roundhouse: Preserving the spirit. *News from Native California* 15(2):30–32.

Margolin, M., and L. Smith. 2001. Conservation with a healer. *News from Native California* 14(4):22–25.

Marryat, F. 1855. *Mountains and Molehills.* London: Longman, Brown, Green, and Longman.

Marshall, A. G. 1999. Unusual gardens: The Nez Perce and wild horticulture on the eastern Columbia Plateau. In *Northwest Lands, Northwest Peoples: Readings in Environmental History,* ed. D. D. Goble and P. W. Hirt, 173–87. Seattle: University of Washington Press.

Martin, A. C., H. S. Zim, and A. L. Nelson. 1951. *American Wildlife and Plants: A Guide to Wildlife Food Habits.* New York: Dover.

Martin, C. L. 1999. *The Way of the Human Being.* New Haven, Conn.: Yale University Press.

Martin, G. J. 1995. *Ethnobotany: A Methods Manual.* London: Chapman and Hall.

Martin, P. S. and R. G. Klein, eds. 1984. *Quaternary Extinctions: A Prehistoric Revolution.* Tucson: University of Arizona Press.

Martin, R. E., and D. B. Sapsis. 1992. Fires as agents of biodiversity—Pyrodiversity promotes biodiversity. In *Proceedings of the Symposium on Biodiversity of Northwestern California, October 1991,* ed. R. R. Harris and D. C. Erman, 150–57. Berkeley: Division of Agriculture and Natural Resources, University of California.

Martinez, D. 1992. Native American forestry practices. In *California Forest Pest Council Proceedings of the 41st Annual Meeting: The Status and Future of Pesticide Use in California,* November 18–19, 1992.

———. 1993. Managing a precarious balance: Wilderness versus sustainable forestry. *Winds of Change* 8(3):23–28.

———. 2003. Protected areas, indigenous people, and the western idea of nature. *Ecological Restoration* 21(4):247–50.

Mason, H. L. 1955. Do we want sugar pine? *Sierra Club Bulletin* 40(8):40–44.

———. 1957. *A Flora of the Marshes of California.* Berkeley: University of California Press.

Mason, J. 1970. *Point Reyes—The Solemn Land.* Point Reyes Station, Calif.: De-Wolf Printing.

Mason, J. A. 1912. The ethnology of the Salinan Indians. *University of California Publications in American Archaeology and Ethnology* 10(4):97–240.

Mason, J. D. 1881. *History of Amador County, California, with Illustrations and Biographical Sketches of Its Prominent Men and Pioneers.* Oakland, Calif.: Thompson and West, Pacific Press Publishing House.

Mason, O. T. 1889. The Ray Collection from the Hupa Reservation. In *Annual Report of the Smithsonian Institution for 1886,* Pt. 1, 205–39. Washington, D.C.

Mathew, B. 1997. *Growing Bulbs: The Complete Practical Guide.* Portland, Ore.: Timber Press.

Mathewson, M. 1985. Threads of life: Cordage and other fibers of the California tribes. Senior thesis, University of California, Santa Cruz.

———. 1998. The living web: Contemporary expressions of Californian Indian basketry. Ph.D. dissertation, University of California, Berkeley.

Mayfield, T. J. 1993. *Indian Summer: Traditional Life among the Choinumne Indians of California's San Joaquin Valley.* Berkeley, Calif.: Heyday Books.

McBride, J., and H. F. Heady. 1968. Invasion of grassland by *Baccharis pilularis* DC. *Journal of Range Management* 21:106–8.

McCann, J. M. 1999a. Before 1492: The making of the pre-Columbian landscape part I: The environment. *Ecological Restoration* 17(1–):15–30.

————. 1999b. Before 1492: The making of the pre-Columbian landscape part II: The vegetation and implications for restoration for 2000 and beyond. *Ecological Restoration* 17(3):107–19.

McCarthy, H. 1993. Managing oaks and the acorn crop. In *Before the Wilderness: Environmental Management by Native Californians,* ed. T. C. Blackburn and K. Anderson, 213–28. Menlo Park, Calif.: Ballena Press.

McCauley, J. 1910. How a grizzly stopped berrying. *Grizzly Bear* 6(3):5.

McCawley, W. 1996. *The First Angelinos: The Gabrielino Indians of Los Angeles.* Banning, Calif.: Malki Museum Press and Ballena Press.

McClaran, M. P., and J. W. Bartolome. 1989. Fire-related recruitment in stagnant *Quercus douglasii* populations. *Canadian Journal of Forest Research* 19:580–85.

McClenaghan, L. R., Jr., and A. C. Beauchamp. 1986. Low genic differentiation among isolated populations of the California fan palm *(Washingtonia filifera). Evolution* 40(2):315–22.

McCorkle, T. 1978. Intergroup conflict. In *Handbook of North American Indians.* Vol. 8: *California,* ed. R. F. Heizer, 694–700. Washington, D.C.: Smithsonian Institution.

McCullough, D. R. 1969. *The Tule Elk: Its History, Behavior, and Ecology.* Berkeley: University of California Press.

McGuire, K. R., and W. R. Hildebrandt. 1994. The possibilities of women and men: Gender and the California milling stone horizon. *Journal of California and Great Basin Anthropology* 16(1):41–59.

McIntyre, R., ed. 1995. *War against the Wolf: America's Campaign to Exterminate the Wolf.* Stillwater, Minn.: Voyageur Press.

McKinney, G., and L. Conner Castro. 1999–2000. A wintertime project: Milkweed string. *News from Native California* 13(2):22–23.

McMillin, J. H. 1956. The aboriginal human ecology of the mountain meadows area in southwestern Lassen County, California. Master's thesis, Sacramento State University.

McMinn, H. E. 1939. *An Illustrated Manual of California Shrubs.* Berkeley: University of California Press.

Medvitz, A. G. 1999. Population growth and its impacts on agricultural land in California: 1850 to 1998. In *California Farmland and Urban Pressures: Statewide and Regional Perspectives,* ed. A. G. Medvitz, A. D. Sokolow, and C. Lemp, 11–32. Davis: University of California Agricultural Issues Center.

Mendelsohn, P. 1983. Northwest California basketry. *Southwest Art:* 57–62.

Mensing, S. 1988. Blue oak *(Quercus douglasii)* regeneration in the Tehachapi Mountains, Kern County, California. Master's thesis, University of California, Berkeley.

Menzies, A. 1924. Menzies' California journal. *California Historical Society Quarterly* 2(4):265–340.

Merchant, C. 1993. *Major Problems in American Environmental History.* Lexington, Mass.: D. C. Heath.

Merriam, C. H. 1902a. Unpublished field notes: Mariposa area. September 18–19: 205–17. Washington, D.C.: Library of Congress.

———. 1902b. Unpublished field notes: Chuckchancy area. September 21: 238–43. Washington, D.C.: Library of Congress.

———. 1905. The Indian population of California. *American Anthropologist* 7(4):594–606.

———. 1918. Two new manzanitas from the Sierra Nevada of California. *Proceedings of the Biological Society of Washington* 31:101–4.

———. [1921] 1994. The elk's last stand. In *Life on the Edge: A Guide to California's Endangered Natural Resources: Wildlife*, ed. C. G. Thelander and M. Crabtree, 106–9. Santa Cruz, Calif.: BioSystems Books; Berkeley, Calif.: Heyday Books.

———. 1955. *Studies of California Indians.* Berkeley: University of California Press.

———. 1967. Ethnographic notes on California Indian Tribes III. Ethnological notes on Central California Indian Tribes, comp. and ed. R. F. Heizer. *Reports of the University of California Archaeological Survey* 68, Part III.

Merrill, Ruth E. 1923. Plants used in basketry by the California Indians. *University of California Publications in American Archaeology and Ethnology* 20(13):215–42.

Miller, B. W. 1988. *Chumash: A Picture of Their World.* Los Osos, Calif.: Sand River Press.

———. 1991. *The Gabrielino.* Los Osos, Calif.: Sand River Press.

Miller, F. E. 1928. *The Medicinal Plants of Yosemite National Park.* Unpublished manuscript. Yosemite National Park Research Library.

Miller, J. 1987. *Life amongst the Modocs: Unwritten History.* San Jose, Calif.: Urion Press.

Mills, E. L., ed. 1985a. *The Papers of John Peabody Harrington in the Smithsonian Institution 1907–1957.* Vol. 2: *A Guide to the Field Notes: Native American History, Language, and Culture of Northern and Central California.* White Plains, N.Y.: Kraus International Publications.

———, ed. 1985b. Purple amole. Salinan. Reel 84, frame 339. *The Papers of John Peabody Harrington in the Smithsonian Institution 1907–1957.* Vol. 2: *A Guide to the Field Notes: Native American History, Language, and Culture of Northern and Central California.* White Plains, N.Y.: Kraus International Publications.

———, ed. 1986. *The Papers of John Peabody Harrington in the Smithsonian Institution 1907–1957.* Vol. 3: *A Guide to the Field Notes: Native American History, Language, and Culture of Southern California/Basin.* White Plains, N.Y.: Kraus International Publications.

Mills, S. 1995. *In Service of the Wild: Restoring and Reinhabiting Damaged Land.* Boston: Beacon Press.

Minnis, P., and S. Plog. 1976. A study of the site specific distribution of *Agave parryi* in east-central Arizona. *Kiva* 41:299–308.

Minnis, P. E., and W. J. Elisens, eds. 2000. *Biodiversity and Native America*. Norman: University of Oklahoma Press.

Mitchell, A. R. J. D. 1957. *Jim Savage and the Tulareño Indians*. Los Angeles: Westernlore Press.

Mitsch, W. J., and J. G. Gosselink. 1993. *Wetlands*. 2d ed. New York: Van Nostrand Reinhold.

Moerenhout, J. A. 1934a. 1843–1856: The French consulate in California. Ed. A. P. Nasatir. *California Historical Society Quarterly* 13(1):56–79.

———. 1934b. 1843–1856: The French consulate in California. Ed A. P. Nasatir and G. E. Dane. *California Historical Society Quarterly* 13(3):262–80.

Möllhausen, B. 1858. *Diary of a Journey from the Mississippi to the Coasts of the Pacific with a United States Government Expedition*, trans. Mrs. Percy Sinnet. 2 vols. London: Longman, Brown, Green, Longmans, and Roberts.

Moncada, F. R. 1774. Diary of Captain Fernando Rivera y Moncada, Monterey Sept. to Dec. Trans. G. Tays. Manuscript, California Historical Society, San Francisco.

Monson, G., and L. Sumner, eds. 1980. *The Desert Bighorn: Its Life History, Ecology, and Management*. Tucson: University of Arizona Press.

Mooney, J. 1890. Notes on the Cosumnes tribes of California. *American Anthropologist* 3:259–62.

Moratto, M. J. 1984. *California Archaeology*. Orlando, Fla.: Academic Press.

Morgan, L. H. 1877. *Ancient Society: Or Researches in the Lines of Human Progress from Savagery through Barbarism to Civilization*. Chicago: Charles H. Kerr.

Morrison, L. L. 1962. *Warner: The Man and the Ranch*. Los Angeles: privately published.

Mount, J. F. 1995. *California Rivers and Streams: The Conflict between Fluvial Process and Land Use*. Berkeley: University of California Press.

Muir, J. [1911] 1944. *John Muir: My First Summer in the Sierra*. Boston: Houghton Mifflin.

———. 1961. *The Mountains of California*. Garden City, N.J.: Doubleday.

———. 1962. *The Yosemite*. New York: Doubleday.

Munz, P. A., and D. D. Keck. 1973. *A California Flora with Supplement*. Berkeley: University of California Press.

Murphey, E. Van Allen. 1959. *Indian Uses of Native Plants*. Fort Bragg, Calif.: Mendocino County Historical Society.

———. n.d. *Indian Conservation*. Notes in the Edith Van Allen Murphey Papers. Held Poage Memorial Home and Resident Library, Ukiah, Calif.

Murphree, M. W. 1994. The role of institutions in community-based conservation. In *Natural Connections: Perspectives in Community-based Conservation*, ed. D. Western and R. M. Wright, 403–27. Washington, D.C.: Island Press.

Mutch, R. W. 1994. Fighting fire with prescribed fire: a return to ecosystem health. *Journal of Forestry* 92(11):31–33.

Myers, J. E. 1978. Cahto. In *Handbook of North American Indians*. Vol. 8: *California*, ed. R. F. Heizer, 244–48. Washington, D.C.: Smithsonian Institution.

Nabhan, G. P. 1985. *Gathering the Desert*. Tucson: University of Arizona Press.

———. 1991. Desert legumes as a nutritional intervention for diabetic indigenous dweller of arid lands. *Arid Lands Newsletter* 31:11–13.

———. 2001. Advancing sustainable harvest practices of basketry materials of Arizona basketweavers. Lecture, Sustainable Harvest Lecture Series, November 21, Northern Arizona State University, Flagstaff.

———. 2002. *Coming Home to Eat: The Pleasures and Politics of Local Foods*. New York: W. W. Norton & Company.

Nabhan, G. P., and R. S. Felger. 1978. Teparies in Southwestern North America: A biogeographical and ethnohistorical study *of Phaseolus acutifolius*. *Economic Botany* 32(1):2–19.

Nabhan, G. P., and A. Rea. 1987. Plant domestication and folk-biological change: The upper Piman/devil's claw example. *American Anthropologist* 89(1): 57–73.

Nabhan, G. P., A. Whiting, H. Dobyns, R. Hevly, and R. Euler. 1981. Devil's claw domestication: Evidence from southwestern Indian fields. *Journal of Ethnobiology* 1(1):135–64.

Nabokov, P., and R. Easton. 1989. *Native American Architecture*. New York: Oxford University Press.

Nash, R. F. [1967] 2001. *Wilderness and the American Mind*. 4th ed. New Haven, Conn.: Yale University Press.

Nathan, P. D., trans., and L. B. Simpson, ed. 1962. *The Letters of José Señán, O.F.M.: Mission San Buenaventura 1796–1823*. Ventura, Calif.: Ventura County Historical Society and John Howell Books.

National Geographic Society. 1924. *The Book of Wild Flowers*. Washington, D.C.: National Geographic Society.

Navarro, L. D. 2000. The committee for traditional Indian health: A program of the California rural Indian health board. *News from Native California* 14(2):22–24.

Naveh, E. 1975. The evolutionary significance of fire in the mediterranean region. *Vegetatio* 29:199–208.

Nazarea, V. D., ed. 1999. *Ethnoecology: Situated Knowledge/Located Lives*. Tucson: University of Arizona Press.

Neely, W. 1971. Miwok uses of Yosemite plants. Unpublished manuscript. Yosemite National Park Research Library.

Neihardt, J. G. 1972. *Black Elk Speaks*. Lincoln: University of Nebraska Press.

Nelson, B. 1988. *Our Home Forever: The Hupa Indians of Northern California*. Salt Lake City: Howe Brothers.

Nelson, J. S. 1997. Interpersonal violence in prehistoric northern California: A bioarchaeological approach. Master's thesis, California State University, Chico.

Nelson, R. 1983. *Make Prayers to the Raven*. Chicago: University of Chicago Press.

News from Native California Staff. 2001. Salmon catch. *News from Native California* 15(2):33.

Nicholson, G. 1923. Papers and correspondence covering trips to the Yurok, Pomo, and other tribes. Grace Nicholson Papers, Huntington Library, San Marino, Calif.

Nicola, S. 1995. Beargrass burning spreads. *California Indian Basketweavers Association Newsletter* 13:6.

Nidever, G. 1976. The life and adventures of a pioneer of California since 1834. In *Original Accounts of the Lone Woman of San Nicolas Island*, ed. R. F. Heizer and A. B. Elsasser, 7–15. Ramona, Calif.: Ballena Press. [Reprinted from Reports of the University of California Archaeological Survey No. 55, Berkeley.]

Noble, I. R., and R. O. Slatyer. 1980. The use of vital attributes to predict successional changes in plant communities subject to recurrent disturbances. *Vegetatio* 43:5–21.

Nomland, G. A. 1935. Sinkyone notes. *University of California Publications in American Archaeology and Ethnology* 36(2):149–78.

———. 1938. Bear River ethnography. *University of California Anthropological Records* 2(2):91–124.

Nordhoff, C. 1873. *California: A Book for Travellers and Settlers.* New York: Harper and Brothers.

Norton, J. 1979. *When Our Worlds Cried: Genocide in Northwestern California.* San Francisco: Indian Historical Press.

Noss, R. F. 1987. From plant communities to landscapes in conservation inventories: A look at the Nature Conservancy (USA). *Biological Conservation* 41:11–37.

———. 1991. Sustainability and wilderness. Comment. *Conservation Biology* 5(1):120–22.

Noss, R. F., and A. Y. Cooperrider. 1994. *Saving Nature's Legacy: Protecting and Restoring Biodiversity.* Washington, D.C.: Island Press.

Noss, R. F., E. T. LaRoe III, and J. M. Scott. 1995. *Endangered Ecosystems of the United States: A Preliminary Assessment of Loss and Degradation.* Biological Report 28. U.S. Department of Interior, National Biological Service, Washington, D.C.

Nunis, D. B. 1968. *The Hudson's Bay Company's First Fur Brigade to the Sacramento Valley: Alexander McLeod's 1829 Hunt.* Sacramento, Calif.: Sacramento Book Collectors Club.

O'Brien, S. 1989. *American Indian Tribal Governments.* Norman: University of Oklahoma Press.

Odum, E. P. 1969. The strategy of ecosystem development. *Science* 164:262–70.

Oelschlaeger, M. 1991. *The Idea of Wilderness: From Prehistory to the Age of Ecology.* New Haven, Conn.: Yale University Press.

Ogden, A. 1941. *The California Sea Otter Trade, 1784–1848.* University of California Publications in History 26. Berkeley: University of California Press.

Olmstead, D. L., and O. C. Stewart. 1978. Achumawi. In *Handbook of North American Indians.* Vol. 8: *California,* ed. R. F. Heizer, 225–35. Washington, D.C.: Smithsonian Institution.

Olmstead, F. E. 1911. Fire and the forest—the theory of "light burning." *Sierra Club Bulletin* 8(1):43–47.

Olmstead, R. R., ed. 1962. *Scenes of Wonder & Curiosity from Hutchings' California Magazine 1856–1861*. Berkeley, Calif.: Howell-North.

O'Neal, J. 1953. Two blades of grass where thousands grew before . . . *Western Livestock Journal* 31(16):61, 89–104.

O'Neale, Lila M. 1995. *Yurok-Karok Basket Weavers*. Berkeley: Phoebe Apperson Hearst Museum of Anthropology, University of California, Berkeley.

O'Neill, R. V., D. L. De Angelis, J. B. Waide, and T. F. H. Allen. 1986. *A Hierarchical Concept of Ecosystems*. New Jersey: Princeton University Press.

Ornduff, R. 1974. *An Introduction to California Plant Life*. Berkeley: University of California Press.

Orr, P. C. 1956. Pleistocene man in Fishbone Cave, Pershing County, Nevada. *Department of Archeology, Nevada State Museum Bulletin* 2:1–20. Carson City.

———. 1968. *Prehistory of Santa Rosa Island*. Santa Barbara, Calif.: Santa Barbara Museum of Natural History.

Ortiz, B. 1991. *It Will Live Forever: Traditional Yosemite Indian Acorn Preparation*. Berkeley, Calif.: Heyday Books.

———. 1993a. Contemporary California Indian basketweavers and the environment. In *Before the Wilderness: Native Californians as Environmental Managers*, ed. T. C. Blackburn and K. Anderson, 195–211. Menlo Park, Calif.: Ballena Press.

———. 1993b. Pesticides and basketry. *News from Native California* 7(3):7–9.

———. 1998. Following the smoke: Karuk indigenous basketweavers and the Forest Service. *News from Native California* 11(3):21–29.

———. 1999. Willis Conrad and the art of dipping. *News from Native California* 12(5):19.

———. 1999–2000. So nice to work with: Dogbane cordage. *News from Native California* 15(2):24–27.

Ortiz, B., and staff, eds. 1991. California Indian basketweavers gathering June 28–30, 1991: A special report. *News from Native California* 6(1):13–36.

Ortiz, S. 1998. Introduction. In *Speaking for the Generations: Native Writers on Writing*, xi–xix. Tucson: University of Arizona Press.

Otter, F. L. 1963. *The Men of Mammoth Forest*. Ann Arbor, Mich.: Edward Brothers.

Ouderkirk, W. 1992. Wilderness restoration: A preliminary philosophical analysis. In *Wilderness Tapestry: An Eclectic Approach to Preservation*, ed. S. I. Zeveloff, L. M. Vause, and W. H. McVaugh, 16–33. Reno: University of Nevada Press.

Pacific Southwest Region. Accomplishment Reports. 1992–2004. *Working Together: California Indians and the Forest Service*. San Francisco, Calif.: U.S. Forest Service.

Packard, S. 1997. Restoration options. In *The Tallgrass Restoration Handbook: For Prairies, Savannas, and Woodlands*, ed. S. Packard and C. F. Mutel, 47–62. Washington, D.C.: Island Press.

Paddison, J. 1999. Introduction. In *A World Transformed: Firsthand Accounts of California Before the Gold Rush*, ed. J. Paddison, ix–xxi. Berkeley, Calif.: Heyday Books.

Paige, K. N. 1992. Overcompensation in response to mammalian herbivory: From mutualistic to antagonistic interactions. *Ecology* 73(6):2076–85.

Palazzo, T. L. 1994a. Charles Melville Scammon. In *Life on the Edge: A Guide to California's Endangered Natural Resources: Wildlife*, ed. C. G. Thelander and M. Crabtree, 466–67. Santa Cruz, Calif.: BioSystems Books; Berkeley, Calif.: Heyday Books.

———. 1994b. Rediscovery. In *Life on the Edge: A Guide to California's Endangered Natural Resources: Wildlife*, ed. C. G. Thelander and M. Crabtree, 505–9. Santa Cruz, Calif.: BioSystems Books; Berkeley, Calif.: Heyday Books.

Panshin, A. J., and C. de Zeeuw. 1980. *Textbook of Wood Technology*. 4th ed. New York: McGraw-Hill.

Parish, S. B. 1907. A contribution toward a knowledge of the genus *Washingtonia*. *Botanical Gazette* 44:408–34.

Patencio, F. 1943. *Stories and Legends of the Palm Springs Indians*. Los Angeles: Times-Mirror Press.

———. 1971. *Desert Hours with Chief Patencio*. Palm Springs, Calif.: Palm Springs Desert Museum.

Patterson, V. 1998. Change and continuity: Transformations of Pomo life. *Expedition* 40(1):3–14.

Pavlik, B. M., P. C. Muick, S. G. Johnson, and M. Popper. 1991. *Oaks of California*. Los Olivos, Calif.: Cachuma Press.

Pearce, R. H. 1988. *Savages of America: A Study of the Indian and the Idea of Civilization*. Baltimore, Md.: Johns Hopkins University Press.

Peattie, D. C. 1953. *A Natural History of Western Trees*. Boston: Houghton Mifflin.

Peña, L. 2002. Chemical forestry threatens tradition and health on the Yurok Reservation. *News from Native California* 16(1):16–18.

Penzer, N. M., ed. 1926. *The World Encompassed and Analogous Documents*. London: Argonaut Press.

Peri, D. W. 1985. Pomoan plant resource management. *Ridge Review* (Mendocino) 4:4.

Peri, D. W., and S. M. Patterson. 1979. Ethnobotanical resources of the Warm Springs Dam–Lake Sonoma Project Area Sonoma County, California. Final Report (unpublished) for the U.S. Army Corps of Engineers Contract No. DACW07–78-C-0043. Elgar Hill, Environmental Analysis and Planning, and Sonoma State University.

———. 1993. "The basket is in the roots, that's where it begins." In *Before the Wilderness: Environmental Management by Native Californians*, ed. T. C. Blackburn and K. Anderson, 175–94. Menlo Park, Calif.: Ballena Press.

Peri, D. W., S. M. Patterson, and J. L. Goodrich. 1982. Ethnobotanical Mitigation Warm Springs Dam—Lake Sonoma California. Ed. E. Hill and R. N. Lerner.

Unpublished report of Elgar Hill, Environmental Analysis and Planning, Penngrove, Calif.

Perkins, J. E. 1863. Sheep husbandry in California. *Transactions of the California State Agronomy Society* 1863:134–45.

Perlin, J. 1989. *A Forest Journey: The Role of Wood in the Development of Civilization.* Cambridge, Mass.: Harvard University Press.

Perlot, J. N. 1985. *Gold Seeker: Adventures of a Belgian Argonaut during the Gold Rush Years.* Trans. H. H. Bretnor. New Haven, Conn.: Yale University Press.

Perry, J. 1988. Yurok fishing. *News from Native California* 2(3):13–15.

Peterson, B. M. 1993. *California Vanishing Habitats and Wildlife.* Wilsonville, Ore.: Beautiful America Publishing Co.

Phillips, G. H. 1993. *Indians and Intruders in Central California, 1769–1849.* Norman: University of Oklahoma Press.

Pickersgill, B. 1989. Cytological and genetical evidence on the domestication and diffusion of crops within the Americas. In *Foraging and Farming: The Evolution of Plant Exploitation,* ed. D. R. Harris and G. C. Hillman, 426–39. London: Unwin Hyman.

Pielou, E. C. 1991. *After the Ice Age: The Return of Life to Glaciated North America.* Chicago: University of Chicago Press.

Pimienta-Barrios, E. 1993. Vegetable cactus *(Opuntia).* In *Underutilized Crops: Pulses and Vegetables,* ed. J. T. Williams, 177–91. London: Chapman and Hall.

Pitcairn, M. J. 2000. *Verbascum thapsus* L. In *Invasive Plants of California Wildlands,* ed. C. C. Bossard, J. M. Randall, and M. C. Hoshovsky, 321–26. Berkeley: University of California Press.

Pitkin, H. 1985. *Wintu Dictionary.* University of California Publications in Linguistics 95. Berkeley: University of California Press.

Polk, D. B. 1991. *The Island of California: A History of the Myth.* Lincoln: University of Nebraska Press.

Ponting, C. 1991. *A Green History of the World.* New York: St. Martin's Press.

Pope, S. T. [1916] 1979. Ishi's death. In *Ishi the Last Yahi: A Documentary History,* ed. R. F. Heizer and T. Kroeber, 225–36. Berkeley: University of California Press.

———. [1918] 1979. Yahi archery. In *Ishi the Last Yahi: A Documentary History,* ed. R. F. Heizer and T. Kroeber, 172–200. Berkeley: University of California Press.

Popenoe, J. H., K. A. Bevis, B. R. Gordon, N. K. Sturhan, and D. L. Hauxwell. 1992. Soil-vegetation relationships in Franciscan terrain of northwestern California. *Journal of the Soil Science Society of America* 56:1951–59.

Posey, D. A. 1984. A preliminary report on diversified management of tropical forest by the Kayapó Indians of the Brazilian Amazon. *Advances in Economic Botany* 1:112–26.

———. 1985. Indigenous management of tropical forest ecosystems: The case of the Kayapó Indians of the Brazilian Amazon. *Agroforestry Systems* 3:139–58.

Potts, M. 1977. *The Northern Maidu.* Happy Camp, Calif.: Naturegraph.

Powell, J. W. 1877. Linguistics. In *Contributions to North American Ethnology, III,* 439–613. United States Geographical and Geological Survey of the Rocky Mountain Region. Washington, D.C.

Powers, B. 1981. *Indian Country of the Tubatulabal.* Tucson, Ariz.: Westernlore Press.

Powers, O. B. 1856. Map of the Calaveras Wagon Road Route transmitted to Senate by S. H. Marlette, Feb. 16. Annual Report of the Surveyor-General of the State of California 1855. Pp. 187–91.

Powers, S. [1877] 1976. *Tribes of California.* Berkeley: University of California Press. [Reprint of 1877 *Contributions to North American Ethnology, III.* Department of the Interior, U.S. Geographical and Geological Survey of the Rocky Mountain Region.]

Prescott-Allen, R., and C. Prescott-Allen. 1983. *Genes from the Wild: Using Wild Genetic Resources for Food and Raw Materials.* London: Earthscan.

———. 1990. How many plants feed the world? *Conservation Biology* 4(4): 365–74.

Preston, W. L. 1981. *Vanishing Landscapes: Land and Life in the Tulare Lake Basin.* Berkeley: University of California Press.

———. 1997. Serpent in the garden: Environmental change in Colonial California. In *Contested Eden: California Before the Gold Rush,* ed. R. Gutiérrez and R. J. Orsi, 260–98. Berkeley: University of California Press.

Price, J. A. 1980. The Washo Indians: History, life cycle, religion, technology, economy and modern life. Nevada State Museum Occasional Papers 4. Carson City: Nevada State Museum.

Price, W. A. [1939] 1998. *Nutrition and Physical Degeneration.* 6th ed. New Canaan, Conn.: Keats Publishing.

Priestley, H. I. 1937. *A Historical, Political, and Natural Description of California by Pedro Fages, Soldier of Spain.* Berkeley: University of California Press.

Prucha, F. P., ed. 1990. *Documents of United States Indian Policy.* 2d expanded ed.. Lincoln: University of Nebraska Press.

Purdy, C. 1976. *My Life and My Times.* N.p.: privately published by E. E. Humphrey and M. E. Humphrey.

Pyne, S. J. 1982. *Fire in America: A Cultural History of Wildland and Rural Fire.* Princeton: Princeton University Press.

Quick, C. R. 1962. Resurgence of a gooseberry population after fire in mature timber. *Journal of Forestry* 60(2):100–103.

Quinn, A. 1997. *Hell with the Fire Out: A History of the Modoc War.* Boston: Faber and Faber.

Raab, L. M. 1996. Debating prehistory in coastal southern California: Resource intensification versus political economy. *Journal of California and Great Basin Anthropology* 18(1):64–80.

Raab, L. M., and K. Bradford. 1994. Advances in southern Channel Islands archaeology: 1983 to 1993. *Journal of California and Great Basin Anthropology* 16(2):243–70.

Rackham, O. 1967. The history and effects of coppicing as a woodland practice. *Proceedings of Monks Wood Experiment Station Symposium* 3:82–93.

Ramenofsky, A. F. 1987. *Vectors of Death: The Archaeology of European Contact.* Albuquerque: University of New Mexico Press.

Randall, J. M. 2000. *Ficus carica* L. In *Invasive Plants of California Wildlands,* ed. C. C. Bossard, J. M. Randall, and M. C. Hoshovsky, 193–98. Berkeley: University of California Press.

Randall, J. M., and M. C. Hoshovsky. 2000. California's wildland invasive plants. In *Invasive Plants of California Wildlands,* ed. C. C. Bossard, J. M. Randall, and M. C. Hoshovsky, 11–27. Berkeley: University of California Press.

Rappaport, D. J., and H. A. Reiger. 1995. Disturbance and stress effects on ecological systems. In *Complex Ecology: The Part-Whole Relation in Ecosystems,* ed. B. C. Patten and S. E. Jorgensen, 397–414. Englewood Cliffs, N.J.: Prentice Hall.

Ratliff, R. D. 1985. *Meadows in the Sierra Nevada of California: State of Knowledge.* Berkeley, Calif.: U.S. Department of Agriculture, Forest Service, Pacific Southwest Forest and Range Experiment Station.

Raup, R. M. 1959. *The Indian Health Program, 1800–1955.* Washington, D.C.: U.S. Division of Indian Health, Department of Health, Education and Welfare.

Rawls, J. J. 1976. Gold diggers: Indian miners in the California gold rush. *California Historical Quarterly* 54(1):28–45.

———. 1984. *Indians of California: The Changing Image.* Norman: University of Oklahoma Press.

———. 1999. Introduction. In *A Golden State: Mining and Economic Development in Gold Rush California,* ed. J. J. Rawls and R. J. Orsi, 1–23. Berkeley: University of California Press.

Ray, V. F. 1963. *Primitive Pragmatists: The Modoc Indians of Northern California.* Seattle: University of Washington Press.

Redding, G. H. H. 1880. An evening with Wintoon Indians. *Californian: A Western Monthly Magazine* 2(12):563–66.

Reid, R. L., ed. 1983. *A Treasury of the Sierra Nevada.* Berkeley, Calif.: Wilderness Press.

Reis, S. V., and R. E. Schultes. 1995. Preface. In *Ethnobotany Evolution of a Discipline,* ed. R. E. Schultes and S. V. Reis, 11–14. Portland, Ore.: Doscorides Press.

Rentz, E. D. 2003. The effects of fire on plant anatomical structure in native Californian basketry materials. Master's thesis, San Francisco State University.

Reynolds, H. G., and A. W. Sampson. 1943. Chaparral crown sprouts as browse for deer. *Journal of Wildlife Management* 7(1):119–23.

Reynolds, R. D. 1959. Effect of natural fires and aboriginal burning upon the forests of the central Sierra Nevada. M.A. thesis, University of California, Berkeley.

Rice, B. M. 1920a. Lupines. In *Popular Studies of California Wild Flowers,* ed. B. M. Rice and R. Rice, 16–19. San Francisco: Upton Bros. and Delzelle Publishers.

———. 1920b. The Washington lily. In *Popular Studies of California Wild Flowers*, ed. B. M. Rice and R. Rice, 114–16. San Francisco: Upton Bros. and Delzelle Publishers.

Richerson, P. J, M. Borgerhoff Mulder, and B. J. Vila. 1996. *Principles of Human Ecology*. Needham Heights, Mass.: Simon and Schuster Custom Publishing.

Ricklefs, R. E., and G. L. Miller. 2000. *Ecology*. 4th ed. New York: W. H. Freeman.

Risser, P. G. 1988. Diversity in and among grasslands. In *Biodiversity*, ed. E. O. Wilson, 176–80. Washington, D.C.: National Academy Press.

Roberts, H. H. 1932. The first salmon ceremony of the Karuk Indians. *American Anthropologist* 34(3):426–40.

Roberts, R. K. 1932. Conservation as formerly practiced by the Indians in the Klamath River region. *California Fish and Game* 18(4):283–90.

Robinson, G. 1988. *The Forest and the Trees: A Guide to Excellent Forestry*. Washington, D.C.: Island Press.

Robinson, W. W. 1948. *Land in California: The Story of Mission Lands, Ranchos, Squatters, Mining Claims, Railroad Grants, Land Scrip, Homesteads*. Berkeley: University of California Press.

Roof, J. 1981. Ecology in action: The great *Dichelostemma* caper. *Changing Seasons* 1(3):11–13.

Roos-Collins, M. 1990. *The Flavors of Home: A Guide to Wild Edible Plants of the San Francisco Bay Area*. Berkeley, Calif.: Heyday Books.

Ross, J. 2000. The boarding school experience from a healer's perspective. *News from Native California* 14(1):22–25.

Ross, J., and D. Espina. 2001. E-interview: Deanna Espina. *News from Native California* 14(4):27–28.

Ross, M. R. 1997. *Fisheries Conservation and Management*. Upper Saddle River, N.J.: Prentice Hall.

Rothrock, J. T. 1876. Appendix H 5. Report upon the operations of a special natural-history party and main field-party no. 1, California section, field-season of 1875, being the results of observations upon the economic botany and agriculture of portions of southern California. In G. M. Wheeler, *U.S. Geographical Surveys West of the One Hundredth Meridian*. Washington, D.C.: Government Printing Office.

Rothstein, M. 1999. California agriculture over time. In *California Farmland and Urban Pressures: Statewide and Regional Perspectives*, ed. A. G. Medvitz, A. D. Sokolow, and C. Lemp, 33–49. Davis: Agricultural Issues Center, Division of Agriculture and Natural Resources, University of California, Davis.

Rundel, P. W. 1996. Monocotyledonous geophytes in the California flora. *Madroño* 43(3):355–68.

Rundel, P. W., D. J. Parsons, and D. T. Gordon. 1988. Montane and subalpine vegetation of the Sierra Nevada and Cascade Ranges. In *Terrestrial Vegetation of California*, expanded ed., ed. M. G. Barbour and J. Majors, 559–600. Special Publication No. 9. Sacramento: California Native Plant Society.

Ruppert, D. 2003. Building partnerships between American Indian tribes and the National Park Service. *Ecological Restoration* 21(4):261–63.

Rusby, H. H. 1906. Wild foods of the United States in May. In *Country Life in America*, 66–69. New York: Doubleday, Page & Co.

Russell, C. P. 1927. Indians grazed on herbage. *Yosemite Native Notes* 6(6):42.

Salmón, E. 2000. Kincentric ecology: Indigenous perceptions of the human–nature relationship. *Ecological Applications* 10(5):1327–32.

Sample, L. L. 1950. Trade and trails in aboriginal California. *Reports of the University of California Archaeological Survey* No. 8. Berkeley, Calif.

Sampson, A. W. 1944. Plant succession on burned chaparral lands in northern California. *University of California Agriculture Experiment Substation Bulletin* 685. Berkeley.

Sánchez, J., and J. Piña. 2001. 1829: The revolt of Estanislao. In *Chronicles of Early California, 1535–1846: Lands of Promise and Despair*, ed. R. M. Beebe and R. M. Senkewicz, 366–74. Santa Clara, Calif.: Santa Clara University; Berkeley, Calif.: Heyday Books.

Sandos, J. A. 1997. Between crucifix and lance: Indian-white relations in California, 1769–1848. In *Contested Eden: California before the Gold Rush*, ed. R. A. Gutiérrez and R. J. Orsi, 196–229. Berkeley: University of California Press.

San Francisco Bulletin. 1861. Apprenticing Indians. March 2.

Sapir, E., and L. Spier. 1943. Notes on the culture of the Yana. *University of California Anthropological Records* 3(3):239–98.

Sauer, C. O. 1952. *Agricultural Origins and Dispersals*. Bowman Memorial Lectures. American Geographical Society. New York: George Grady Press.

———. 1967. *Land and Life: A Selection from the Writings of Carl Ortwin Sauer*. Ed. John Leighly. Berkeley: University of California Press.

———. [1952] 1969. *Seeds, Spades, Hearths, and Herds: The Domestication of Animals and Foodstuffs*. 2d ed. Cambridge, Mass.: MIT Press.

Savelle, G. D. 1977. Comparative structure and function in a California annual and native bunchgrass community. P.h.D. dissertation, University of California, Berkeley.

Sawyer, J. O., D. A. Thornburgh, and J. R. Griffin. 1988. Mixed evergreen forest. In *Terrestrial Vegetation of California*, expanded ed., ed. M. G. Barbour and J. Majors, 359–82. Special Publication No. 9. Sacramento: California Native Plant Society.

Scanland, J. M. 1894. The decline of the Mission Indians. *Overland Monthly* 24(144):634–39.

Schapera, I. 1930. *The Khoisan Peoples of South Africa*. London: Routledge and Kegan Paul.

Schenck, S. M., and E. W. Gifford. 1952. Karok ethnobotany. *Anthropological Records* 13(6):377–92.

Schilling, F. A. 1939. A lily with daggers. *The Desert Magazine* 2(7):8–10.

Schlichtemeier, G. 1967. Marsh burning for waterfowl. *Proceedings of the Sixth Annual Tall Timbers Fire Ecology Conference, March 6–7, 1967*. Tallahassee, Fla.: Tall Timbers Research Station.

Schlick, M. D. 1994. *Columbia River Basketry: Gift of the Ancestors, Gift of the Earth*. Seattle: University of Washington Press.

Schmid, R. F. 1994. *Native Nutrition: Eating According to Ancestral Wisdom.* Rochester, Vt.: Healing Arts Press.

Schmidt, C. L. 1993. Ceanothus, California-lilac. In *The Jepson Manual: Higher Plants of California,* ed. J. C. Hickman, 932–43. Berkeley: University of California Press.

Schultes, R. E., and S. V. Reis, eds. 1995. *Ethnobotany: Evolution of a Discipline.* Portland, Ore.: Dioscorides Press.

Schulz, P. D. 1981. Osteoarchaeology and subsistence change in prehistoric central California. Ph.D. dissertation, University of California, Davis.

Schulz, P. E. [1954] 1988. *Indians of Lassen Volcanic National Park and Vicinity.* Rev. ed. Mineral, Calif.: Loomis Museum Association.

Schumacher, P. 1875. Ancient graves and shell-heaps of California. In *Annual Report of the Smithsonian Institution for 1874,* 335–50. Washington, D.C.

Secrest, W. B. 2003. *When the Great Spirit Died: The Destruction of the California Indians, 1850–1860.* Sanger, Calif.: Word Dancer Press.

Shepard, A. 1989. *Wintu Texts.* Berkeley: University of California Press.

Shepard, P. 1996. *A Paul Shepard Reader: The Only World We've Got.* San Francisco: Sierra Club Books.

Shipek, F. C. 1977. A strategy for change: The Luiseño of southern California. Ph.D. dissertation, University of Hawaii.

———. 1987. *Pushed into the Rocks: Southern California Indian Land Tenure, 1769–1986.* Lincoln: University of Nebraska Press.

———. 1989. An example of intensive plant husbandry: The Kumeyaay of southern California. In *Foraging and Farming: The Evolution of Plant Exploitation,* ed. D. R. Harris and G. C. Hillman, 159–70 London: Unwin-Hyman.

———. 1991. *Delfina Cuero.* Menlo Park, Calif.: Ballena Press.

———. 1993. Kumeyaay plant husbandry: Fire, water, and erosion control systems. In *Before the Wilderness: Environmental Management by Native Californians,* ed. T.C. Blackburn and K. Anderson, 379–88. Menlo Park, Calif.: Ballena Press.

Shipley, W. F. 1978. Native languages of California. In *Handbook of North American Indians.* Vol. 8: *California,* ed. R. F. Heizer, 80–90. Washington, D.C.: Smithsonian Institution.

Shirley, D. [Clappe, Louise] 1937. *The Shirley Letters: Being Letters Written in 1851–1852 from the California Mines.* Salt Lake City: Gibbs-Smith. [Orig. pub. in *Pioneer* magazine, 1854 and 1855.]

Silberbauer, G. B. 1981. *Hunter and Habitat in the Central Kalahari Desert.* Cambridge: Cambridge University Press.

Silver, S. 1978. Shastan peoples. In *Handbook of North American Indians.* Vol. 8: *California,* ed. R. F. Heizer, 211–24. Washington, D.C.: Smithsonian Institution.

Simmons, W. S. 1997. Indian peoples of California. In *Contested Eden: California before the Gold Rush,* ed. R. A. Gutiérrez and R. J. Orsi, 48–77. Berkeley: University of California Press.

Simpson, G. G. 1961. *Principles of Animal Taxonomy.* New York: Columbia University Press.

Simpson, L. B., ed. 1938. *California in 1792: The Expedition of José Longino Martínez.* San Marino, Calif.: Huntington Library.

———. 1961. *Journal of José Longinos Martínez: Notes and Observations of the Naturalist of the Botanical Expedition in Old and New California and the South Coast, 1791–1792.* San Francisco: John Howell Books.

———. 1962. *The Letters of José Señán, O.F.M.: Mission San Buenaventura 1796–1823.* Ventura, Calif.: John Howell Books.

Skinner, C. N., and C. R. Chang. 1996. Fire regimes, past and present. In *Sierra Nevada Ecosystem Project Final Report to Congress.* Vol. 2: *Assessments and Scientific Basis for Management Options,* 1041–69. Davis: University of California, Centers for Water and Wildland Resources.

Skinner, M. W. 1993. Lilium. In *The Jepson Manual: Higher Plants of California,* ed. J. C. Hickman, 1198–1201. Berkeley: University of California Press.

Skinner, M. W., and B. M. Pavlik, eds. 1994. *Inventory of Rare and Endangered Vascular Plants of California.* Special Publication No. 1. Sacramento: California Native Plant Society.

Small, A. 1994. *California Birds: Their Status and Distribution.* Vista, Calif.: Ibis.

Smith, B. D. 1992. *Rivers of Change: Essays on Early Agriculture in Eastern North America.* Washington, D.C.: Smithsonian Institution.

———. 1995. *The Emergence of Agriculture.* New York: Scientific American Library.

Smith, C. R. 1978. Tubatulabal. In *Handbook of North American Indians.* Vol. 8: *California,* ed. R. F. Heizer, 437–45. Washington, D.C.: Smithsonian Institution.

Smith, K. 1990. You'll never go hungry: Food traditions of one Dry Creek Pomo/Bodega Miwok family. *News from Native California* 4(2):4–5.

Smith-Ferri, S. 1998. The development of the commerical market for Pomo Indian baskets. *Expedition* 40(1):15–22.

Snyder, G. 1990. *The Practice of the Wild.* San Francisco: North Point Press.

Society for Ecological Restoration International, Science & Policy Working Group. 2002. The SER primer on ecological restoration. http://www.ser.org/content/ecological_restoration_primer.asp.

Sousa, W. P. 1985. Disturbance and patch dynamics on rocky intertidal shores. In *The Ecology of Natural Disturbance and Patch Dynamics,* ed. S. T. A. Pickett and P. S. White, 101–24. San Diego, Calif.: Academic Press.

Sparkman, P. S. 1908. The culture of the Luiseño Indians. *University of California Publications in American Archaeology and Ethnology* 8(4):187–234.

Spaulding, E. S. 1949. *The Quails.* New York: Macmillan.

Spier, K. 1930. Klamath ethnography. *University of California Publications in American Archaeology and Ethnology* 30.

Spier, R. F. G. 1978. Foothill Yokuts. In *Handbook of North American Indians.* Vol. 8: *California,* ed. R. F. Heizer, 471–85. Washington, D.C.: Smithsonian Institution.

Spott, R. 1926. Speech of Robert Spott. *Transactions of the Commonwealth Club of California* 21(3):133. San Francisco.

Spurr, S. H., and B. V. Barnes. 1980. *Forest Ecology*. 3d ed. New York: John Wiley.

Stadtman, V. A., ed. 1967. *The Centennial Record of the University of California*. Berkeley: University of California.

Staebler, R. N., and S. Atwater, eds. 1997. Forestry on tribal lands. *Journal of Forestry* 95(11):1–48.

Standing Bear, Chief L. [1933] 1998. Indian wisdom. In *The Great New Wilderness Debate*, ed. J. B. Callicott and M. P. Nelson, 201–6. Athens: University of Georgia Press.

Stanger, F. M. 1967. *Sawmills in the Redwoods: Logging on the San Francisco Peninsula, 1849–1967*. San Mateo: San Mateo County Historical Association.

Star Newspaper. 1889. Flints and fire sticks: Savage ways of making fire: Drills and strike-a-lights. August 31.

Starr, K. 1973. *Americans and the California Dream, 1850–1915*. New York: Oxford University Press.

Stauffer, R. 1993. Bear grass burned in the Northwest. *California Indian Basketweavers Association* 5:1.

Steinhart, P. 1990. *California's Wild Heritage: Threatened and Endangered Animals in the Golden State*. Sacramento: California Department of Fish and Game.

Stephens, S. L. 1997. *Fire History of a Mixed Oak-pine Forest in the Foothills of the Sierra Nevada, El Dorado County, California*. General Technical Report PSW-GTR-160. Berkeley, Calif.: U.S. Department of Agriculture, Forest Service, Pacific Southwest Research Station.

Sterling, E. A. 1905. Attitude of lumbermen toward forest fires. In *Yearbook of the United States Department of Agriculture 1904*, 133–40. Washington, D.C.: Government Printing Office.

Stevens, M. L. 1999. The ethnoecology and autecology of white root *(Carex barbarae Dewey)*: Implications for restoration. Ph.D. dissertation, University of California, Davis.

Stevens, S., ed. 1997. *Conservation through Cultural Survival: Indigenous Peoples and Protected Areas*. Washington, D.C.: Island Press.

Stevens, W. K. 1995. *Miracle under the Oaks: The Revival of Nature in America*. New York: Pocket Books.

Steward, J. H. 1933. Ethnography of the Owens Valley Paiute. *University of California Publications in American Archaeology and Ethnology* 33(3): 233–350.

———. 1934. Two Paiute autobiographies. *University of California Publications in American Archaeology and Ethnology* 33(5):423–38.

———. 1935. *Indian Tribes of Sequoia National Park Region*. Berkeley, Calif.: U.S. Department of Interior, National Park Service.

———. 1938a. *Basin-Plateau Aboriginal Sociopolitical Groups*. Smithsonian

Institution Bureau of American Ethnology Bulletin 120. Washington, D.C.: U.S. Government Printing Office.

———. 1938b. Panatubiji', an Owens Valley Paiute. Smithsonian Institution Bureau of American Ethnology Bulletin 119. *Anthropological Papers* No. 6:183–95.

Stewart, B. 1997. *Common Butterflies of California*. Point Reyes Station, Calif.: West Coast Lady Press.

Stewart, J. 2000. The Highway 190 Springville "deer grass" patch mystery. *Tule River Times*, February 24. Porterville, Calif.

Stewart, O. C. 1935. Pomo unpublished field notes. In the possession of D. Theodoratus.

———. 1941. Culture element distributions, XIV: Northern Paiute. *University of California Anthropological Records* 41(3):361–446.

———. 2002. *Forgotten Fires: Native Americans and the Transient Wilderness*. Ed. H. T. Lewis and M. K. Anderson. Norman: University of Oklahoma Press.

St. John, T. V., and P. W. Rundel. 1976. The role of fire as a mineralizing agent in a Sierran coniferous forest. *Oecologia* (25):35–45.

Stone, E. C. 1951. Notes and comment: the stimulative effect of fire on the flowering of the golden brodiaea (*Brodiaea ixioides* Wats. var. *lugens* Jeps.). *Ecology* 32(3):534–37.

Storer, A. J., D. L. Wood, T. R. Gordon, and W. J. Libby. 2001. Restoring native Monterey pine forests. *Journal of Forestry* 99(5):14–18.

Storer, T. I., and L. P. Tevis. 1955. *California Grizzly*. Berkeley: University of California Press.

Strike, S. S. 1994. *Ethnobotany of the California Indians: Aboriginal Uses of California's Indigenous Plants*. Vol. 2. Champaign, Ill.: Koeltz Scientific Books.

Stromberg, M. R., P. Kephart, and V. Yadon. 2001. Composition, invasibility, and diversity in coastal California grasslands. *Madroño* 48(4):236–52.

Strong, W. D. [1972] 1987. *Aboriginal Society in Southern California*. Banning, Calif.: Malki Museum. [Reprinted from *University of California Publications in American Archaeology and Ethnology*, vol. 26, 1929.]

Suagee, D. B., and C. T. Stearns. 1994. Indigenous self-government, environmental protection, and the consent of the governed: a tribal environmental review process. In *Endangered Peoples: Indigenous Rights and the Environment*, 59–104, ed. Colorado Journal of International Environmental Law and Policy. Niwot: University Press of Colorado.

Sudworth, G. B. 1900. Stanislaus and Lake Tahoe Forest Reserves, California, and adjacent territory. *U.S. Geological Survey, 21st Annual Report*, Pt. V.

Sugihara, N. G., and M. G. Barbour. In press. Fire and California vegetation. In *Fire in California Ecosystems*, ed. N. G. Sugihara, J. W. van Wagtendonk, J. Fites-Kaufman, K. Shaffer, and A. E. Thode. Berkeley: University of California Press.

Sugihara, N. G., and J. R. McBride. 1996. Dynamics of sugar pine and associated species following non-stand-replacing fires in white fir-dominated mixed-conifer forests. In *Sugar Pine: Status, Values, and Roles in Ecosystems*, ed. B. B. Kinloch Jr., M. Marosy, and M. E. Huddleston, 39–44. Proceedings of a Symposium presented by the California Sugar Pine Management Committee, March 30–April 1, 1992. University of California Division of Agriculture and Natural Resources Publication No. 3362, Davis.

Sugihara, N. G., J. W. van Wagtendonk, and J. Fites-Kaufman. In press. Fire as an ecological process. In *Fire in California Ecosystems*, ed. N. G. Sugihara, J. W. van Wagtendonk, J. Fites-Kaufman, K. Shaffer, and A. E. Thode. Berkeley: University of California Press.

Sugihara, N. G., J. W. van Wagtendonk, J. Fites-Kaufman, K. Shaffer, and A. E. Thode, eds. In press. *Fire in California Ecosystems*. Berkeley: University of California Press.

Supahan, T., and S. Supahan. 1995. Teaching well, learning quickly. *News from Native California: An Inside View of the California Indian World* 9(2):35–37.

Sutter, J. A. 1939. *New Helvetia Diary: A Record of Events Kept by John A. Sutter and his Clerks at New Helvetia, California, from September 9, 1845, to May 25, 1848*. San Francisco: Grabhorn Press in association with the Society of California Pioneers.

Sutton, M. Q. 1988. *Insects as Food: Aboriginal Entomophagy in the Great Basin*. Anthropological Papers No. 33, ed. T. C. Blackburn. Palo Alto, Calif.: Ballena Press.

Sweeney, J. 1956. Responses of vegetation to fire: A study of the herbaceous vegetation following chaparral fires. *University of California Publications in Botany* 28:143–250.

Sweet, E. M., Jr. 1918. *Hoopa Valley Report*. U.S. Bureau of Indian Affairs Collection. MS 47, Box 2, File Folder 2.14. Holt-Atherton Department of Special Collections. Stockton: University of the Pacific.

Swetnam, T. W. 1993. Fire history and climate change in giant sequoia groves. *Science* 262:885–89.

Swezey, S. 1975. The energetics of subsistence-assurance ritual in native California. *Contributions of the University of California Archaeological Research Facility* 23:1–46.

Swezey, S. L., and R. F. Heizer. 1993. Ritual management of salmonid fish resources in California. In *Before the Wilderness: Environmental Management by Native Californians*, ed. T. C. Blackburn and M. K. Anderson, 299–327. Menlo Park, Calif.: Ballena Press.

Swiecki, T. J., E. A. Bernhardt, and R. A. Arnold. 1990. Impacts of diseases and arthropods on California's rangeland oaks. Contract 8CA74545 to California Department of Forestry and Fire Protection.

———. 1991. Insect and disease impacts on blue oak acorns and seedlings. In *Proceedings of the Symposium on Oak Woodlands and Hardwood Rangeland Management, October 31–November 2, 1990, Davis, Calif.*, tech. coord. R. B. Standiford, 149–55. General Technical Report PSW-126. Berkeley, Calif.: U.S. Department of Agriculture, Forest Service, Pacific Southwest Research Station.

Szczawinski, A. F., and N. J. Turner. 1980. *Edible Garden Weeds of Canada*. Edible Wild Plants of Canada No. 1. Ottawa: National Museum of Natural Sciences, National Museums Canada.

Tappeiner II, J. C., P. M. McDonald, and D. F. Roy. 1990. *Lithocarpus densiflorus* (Hook. & Arn.) Rehd. Tanoak. In *Silvics of North America*. Vol. 2: *Hardwoods*, tech. coord. R. M. Burns and B. H. Honkala, 417–25. Washington, D.C.: U.S. Department of Agriculture, Forest Service.

Tarakanoff, V. P. 1953. *Statement of My Captivity among the Californians*. Early California Travels 14. Trans. Ivan Petroff; notes A. Woodward. Los Angeles: Glen Dawson.

Taylor, A. H. 1990. Tree invasion in meadows of Lassen Volcanic National Park, California. *The Professional Geographer* 42(4):457–70.

Taylor, B. [1850] 1968. *Eldorado*. Vol. 1. New York: Putnam. Special Edition Palo Alto, Calif.: Lewis Osborne.

———. 1951. *New Pictures from California*. Oakland, Calif.: Biobooks.

Taylor, H. J. 1932. The death of the last survivor. *University of California Chronicle* 34:51–55.

Teggart, F. J. 1923. The Gold Rush: Extracts from the diary of C. S. Lyman. *California Historical Society Quarterly* 2(3):181–202.

Templin Richards, R. 1997. What the natives know: Wild mushrooms and forest health. *Journal of Forestry* 95(9):4–10.

Thelander, C. G., and M. Crabtree, eds. 1994. *Life on the Edge: A Guide to California's Endangered Natural Resources: Wildlife*. Santa Cruz, Calif.: BioSystems Books; Berkeley, Calif.: Heyday Books.

Thompson, L. [1916] 1990. *To the American Indian: Reminiscences of a Yurok Woman*. Berkeley, Calif.: Heyday Books.

Thornton, R. 1987. *American Indian Holocaust and Survival: A Population History since 1492*. Norman: University of Oklahoma Press.

Thrall, B., and A. Gayton. 1938. Unpublished manuscript on Chukchansi Yokuts culture. Ethnological Documents of the Department and Museum of Anthropology. University of California Archives, Bancroft Library, Berkeley.

Tietje, W. 1990. Acorns: Planning for oak-woodland wildlife. *Fremontia* 18(3): 80–81.

Timbrook, J. 1982. Use of wild cherry pits as food by the California Indians. *Journal of Ethnobiology* 2(2):162–76.

———. 1986. Chia and the Chumash: A reconsideration of sage seeds in southern California. *Journal of California and Great Basin Anthropology* 8 (1):50–64.

———. 1990. Ethnobotany of the Chumash Indians, California, based on collections by John P. Harrington. *Economic Botany* 44(2):236–53.

———. 1993. *Island Chumash Ethnobotany*. Reprinted from *Archaeology on the Northern Channel Islands of California: Studies of Subsistence, Economics, and Social Organization*, ed. Michael A. Glassow, 47–62. Salinas, Calif.: Coyote Press.

Timbrook, J, J. R. Johnson, and D. D. Earle. 1993. Vegetation burning by the Chumash. In *Before the Wilderness: Environmental Management by Native Californians*, ed. T. C. Blackburn and K. Anderson, 117–49. Menlo Park, Calif.: Ballena Press.

Torrey, J., and A. Gray. 1856. Report of the botany of the expedition. *Reports of Explorations and Surveys, to Ascertain the most Practicable and Economical Route for a Railroad from the Mississippi River to the Pacific Ocean*. Lieutenant A. W. Whipple. Vol. 3. Washington, D.C.

Toypurina. 2001. 1785: Rebellion at San Gabriel. In *Chronicles of Early California, 1535–1846: Lands of Promise and Despair*, ed. R. M. Beebe and R. M. Senkewicz, 247–49. Santa Clara, Calif.: Santa Clara University; Berkeley, Calif.: Heyday Books.

Trigger, B. G. 1980. Archaeology and the image of the American Indian. *American Antiquity* 45(4):662–76.

Tucker, J. M. 1993. *Quercus* (oak). In *The Jepson Manual: Higher Plants of California*, ed. J. C. Hickman, 658–63. Berkeley: University of California Press.

Turner, N. J. 1999. "Time to burn": Traditional use of fire to enhance resource production by aboriginal peoples in British Columbia. In *Indians, Fire and the Land in the Pacific Northwest*, ed. R. Boyd, 185–218. Corvallis: Oregon State University Press.

———. In press. *The Earth's Blanket: Cultural Teachings for Sustainable Living*. Vancouver: Douglas and McIntyre.

Turner, N. J., and B. S. Efrat. 1981. The ethnobotany of the Hesquiat Indians of Vancouver Island. *British Columbia Provincial Museum Cultural Recovery Paper* No. 2. Victoria: Queen's Printer.

Turner, N. J., and H. V. Kuhnlein. 1983. Camas (*Camassia* spp.) and riceroot (*Fritillaria* spp.): Two liliaceous "root" foods of the Northwest Coast Indians. *Ecology of Food and Nutrition* 13:199–219.

Turner, N. J., and A. F. Szczawinski. 1979. *Edible Wild Fruits and Nuts of Canada*. Edible Wild Plants of Canada No. 3. Toronto: Fitzhenry and Whiteside.

Turner, N. J., L. C. Thompson, M. T. Thompson, and A. Z. York. 1990. *Thompson Ethnobotany: Knowledge and Usage of Plants by the Thompson Indians of British Columbia*. Memoir No. 3. Victoria, Canada: Royal British Columbia Museum.

Underhill, R. 1941. *The Northern Paiute Indians of California and Nevada*. Department of the Interior, Bureau of Indian Affairs.

United Indian Health Service. 1995. California Natives and diabetes—You are at risk. *News from Native California* 9(2):48.

U.S. Department of Commerce. 1975. *Historical Statistics of the United States: Colonial Times to 1970, Part 1*. Washington, D.C.: Bureau of the Census.

U.S. Department of Interior. 1903. *Report of the Acting Superintendent of the Yosemite National Park in California to the Secretary of the Interior*. Washington, D.C.: Government Printing Office.

U.S. Forest Service. 1936. *The Western Range*. Senate Document 199. 74th Cong., 2d sess.

Usher, P. J. 1987. Indigenous management systems and the conservation of wildlife in the Canadian north. Special Issue: Sustainable Development in Northern Communites. *Alternatives* 14(1):3–9.

Vale, T. R., ed. 2002. *Fire, Native Peoples, and the Natural Landscape.* Washington, D.C.: Island Press.

Vancouver, G. 1999. A voyage of discovery to the North Pacific Ocean, 1792. In *A World Transformed: Firsthand Accounts of California before the Gold Rush,* ed. J. Paddison, 61–94. Berkeley, Calif.: Heyday Books.

Van Dyke, T. S. 1886. *Southern California: Its Valleys, Hills, and Streams: Its Animals, Birds, and Fishes: Its Gardens, Farms, and Climate.* New York: Fords, Howard and Hulbert.

Van Etten, C. 1994. *Meeks Bay Memories.* Reno, Nev.: Silver Syndicate Press.

Veblen, T. T., and D. C. Lorenz. 1986. Anthropogenic disturbance and recovery patterns in montane forests, Colorado front range. *Physical Geography* 7(1):1–24.

Voegelin, E. W. 1938. Tubatulabal ethnography. *University of California Anthropological Records* 2(1):1–84.

———. 1942. Culture element distributions XX: Northeast California. *Anthropological Records* 7(2):47–251.

Vogl, R. J. 1967. Wood rat densities in southern California manzanita chaparral. *Southern Naturalist* 12:176–79.

Wagner, H. R. 1923. The voyage of Pedro de Unamuno to California in 1587. *California Historical Society Quarterly* 2(2):140–60.

———. 1924. The voyage to California of Sebastián Rodríguez Cermeño in 1595. *California Historical Society Quarterly* 3(1):3–24.

———. 1928a. Spanish voyages to the Northwest Coast in the sixteenth century. Chapter IV: The voyage of Juan Rodríguez Cabrillo. *California Historical Society Quarterly* 7(1):20–77.

———. 1928b. Spanish voyages to the Northwest Coast in the sixteenth century. Chapter XI: Father Antonio de la Ascension's account of the voyage of Sebastián Vizcaíno. *California Historical Society Quarterly* 7(4):295–394.

———. 1929. *Spanish Voyages to the Northwest Coast of American in the Sixteenth Century.* San Francisco: California Historical Society.

Walker, A. M. 1970. *The Rough and the Righteous of the Kern River Diggins.* Exeter, Calif.: Vintage Resources.

Walker, P. L., and J. R. Johnson. 1994. The decline of the Chumash Indian population. In *Wake of Contact: Biological Responses to Conquest,* ed. C. S. Larson and G. R. Miller, 109–20. New York: Wiley-Liss.

Wallace, E. 1978. Sexual status and role differences. In *Handbook of North American Indians.* Vol. 8: *California,* ed. R. Heizer, 683–89. Washington, D.C.: Smithsonian Institution.

Warburton, A. D., and J. F. Endert. 1966. *Indian Lore of the North California Coast.* Santa Clara, Calif.: Pacific Pueblo Press.

Warner, J. J. n.d. Reminiscences of early life in California. Unidentified newspaper article in Missions of Alta California, Benjamin Hayes Collection, C-C 21, pt. 1, Bancroft Library, Berkeley, Calif.

Waterman, T. T. 1925. The village sites in Tolowa and neighboring areas in northwestern California. *American Anthropologist* 27(4):528–43.

Weatherford, J. 1991. *Native Roots: How the Indians Enriched America.* New York: Crown.

Webb, E. B. 1952. *Indian Life at the Old Missions.* Lincoln: University of Nebraska Press.

Weeden, N. F. 1996. *A Sierra Nevada Flora.* 4th ed. Berkeley, Calif.: Wilderness Press.

Weidensaul, S. 1999. *Living on the Wind: Across the Hemisphere with Migratory Birds.* New York: North Point Press.

Weiss, E. B. 1984. The planetary trust: Conservation and intergenerational equity. *Ecology Law Quarterly* 11:495–581.

Welch, W. R. 1931. Game reminiscences of yesteryears. *California Fish and Game* 17:255–63.

Wells, P. V. 1969. The relation between mode of reproduction and extent of speciation in woody genera of the California chaparral. *Evolution* 23: 264–67.

———. 1993. *Arctostaphylos,* manzanita. In *The Jepson Manual: Higher Plants of California,* ed. J. C. Hickman, 545–59. Berkeley: University of California Press.

West, G. J. 1990. Holocene fossil pollen records of Douglas fir in northwestern California: Reconstruction of past climate. In *Proceedings of the Sixth Annual Pacific Climate (PACLIM) Workshop,* March 5–8, 1989, ed. J. L. Betancourt and A. M. MacKay, 119–22. California Department of Water Resources, Interagency Ecological Studies Program Technical Report 23.

———. 1993. The late Pleistocene-Holocene pollen record and prehistory of California's north coast ranges. In *There Grows a Green Tree: Papers in Honor of David A. Fredrickson,* ed. G. White, P. Mikkelsen, W. Hildebrandt, and M. E. Basgall, 219–36. Publication No. 11. Davis, Calif.: Center for Archaeological Research at Davis, Department of Anthropology, University of California, Davis.

Wheat, M. M. 1967. *Survival Arts of the Primitive Paiutes.* Reno: University of Nevada Press.

Whisenant, S. G. 1999. *Repairing Damaged Wildlands: A Process-Orientated, Landscape-Scale Approach.* Cambridge: Cambridge University Press.

Whistler, K. W. 1976. Patwin folk-taxonomic structures. Master's thesis, University of California, Berkeley.

White, P. S. 1978. Pattern, process, and natural disturbance in vegetation. *Botanical Review* 45: 229–99.

White, P. S., and S. T. A. Pickett. 1985. Natural disturbance and patch dynamics: An introduction. In *The Ecology of Natural Disturbance and Patch Dynamics,* 3–13. San Diego, Calif.: Academic Press.

White, R. 1975. Indian land use and environmental change: Island County, Washington, a case study. *Arizona and the West* 17:327–38.

Whitson, T. D., ed. , L. C. Burrill, S. A. Dewey, D. W. Cudney, B. E. Nelson, R. D.

Lee, and R. Parker. 1996. *Weeds of the West.* Newark, Calif.: Western Society of Weed Science.

Whittaker, R. H., and G. M. Woodwell. 1972. Evolution of natural communities. In *Ecosystem Structure and Function,* ed. J. A. Wiens, 137–59. Corvallis: Oregon State University Press.

Whittemore, K. 1997. To converse with creation: Saving California Indian languages. *Native Americas: Akwe:kon's Journal of Indigenous Issues* 14(3): 46–53.

Wierzbicki, F. P. [1849] 1970. *California as It Is and as It May Be: Or a Guide to the Gold Region.* New York: Burt Franklin.

Wieslander, A. E., and C. H. Gleason. 1954. *Major Brushland Areas of the Coast Ranges and Sierra-Cascade Foothills of California.* Miscellaneous Paper PSW-15. Berkeley, Calif.: U.S. Department of Agriculture, Forest Service, Pacific Southwest Forest and Range Experiment Station.

Wilke, P. J. 1993. Bow staves harvested from juniper trees by Indians of Nevada. In *Before the Wilderness: Environmental Management by Native Californians,* ed. T. C. Blackburn and K. Anderson, 241–78. Menlo Park, Calif.: Ballena Press.

Wilken, D. H. 1993. California's changing climates and flora. In *The Jepson Manual: Higher Plants of California,* ed. J. C. Hickman, 55–58. Berkeley: University of California Press.

Williams, G. W. 2003. *References on the American Indian Use of Fire in Ecosystems.* Unpublished bibliography. Washington, D.C.: U.S. Department of Agriculture, Forest Service.

Wilson, B. F. 1970. *The Growing Tree.* Amherst: University of Massachusetts Press.

Wilson, J. B., and W. M. King. 1995. Human-mediated vegetation switches as processes in landscape ecology. *Landscape Ecology* 10:191–96.

Wilson, N. L. 1972. Notes on traditional Foothill Nisenan food technology. *University of California Center for Archaeological Research Publications* 3:32–38.

Winder, W. 1974. Letter from First Lieutenant William Winder to Captain H. S. Burton dated April 29, 1856. In *The Destruction of California Indians,* ed. R. F. Heizer, 87–89. Lincoln: University of Nebraska Press.

Wistar, I. J. 1937. Autobiography of Isaac Jones Wistar, 1827–1905. Philadelphia: Wistar Institute of Anatomy and Biology.

Wohlgemuth, E. 1996. Resource intensification in prehistoric central California: Evidence from archaeobotanical data. *Journal of California and Great Basin Anthropology* 18(1):81–103.

———. 2004a. The course of plant food intensification in native central California. Ph.D. dissertation, University of California, Davis.

———. 2004b. Nine thousand years of plant use in native central California: Implications of the archaeobotanical record for archaeologists, native peoples, and restoration practitioners. Paper presented at the 27th annual Ethnobiology Conference, University of California, Davis.

Wolfe, L. M., ed. [1938] 1979. *John of the Mountains: The Unpublished Journals of John Muir.* Madison: University of Wisconsin Press.

Wood, S. W. 1975. Holocene stratigraphy and chronology of mountain meadows, Sierra Nevada, California. Ph.D. dissertation, California Institute of Technology, Pasadena.

Wooman, G., and D. C. Johnson. [1974] 1993. Letter from Wooman and Johnson to Col. T. J. Henley, March 25, 1855. In *The Destruction of California Indians*, ed. R. F. Heizer, 27–29. Lincoln: University of Nebraska Press.

Work, G. 1995. When feral pigs are good for the ecology and other heresies. *Growing Native* 6(5):17–20.

Worster, D. W. 1994. *An Unsettled Country: Changing Landscapes of the American West*. Albuquerque: University of New Mexico Press.

Wright, H. A., L. F. Neunschwander, and C. M. Britton. 1979. *The Role and Use of Fire in Sagebrush-Grass and Pinyon-Juniper Plant Communities; State-of-the-Art Review*. General Technical Report INT-58. Ogden, Utah: U.S. Department of Agriculture, Forest Service, Intermountain Forest and Range Experiment Station.

Yamane, L. 1991. Baskets in museum collections: A California Indian perspective. *News from Native California* 5(4):7.

———. 1997. *Weaving a California Tradition: A Native American Basketmaker*. Minneapolis, Minn.: Lerner Publications.

———. 2001. My world is out there. In *The Sweet Breathing of Plants*, ed. L. Hogan and B. Peterson, 59–65. New York: North Point Press.

Yarnell, R. A. 1965. Implications of distinctive flora on pueblo ruins. *American Anthropologist* 67(3):662–74.

Young, J. A., and R. A. Evans. 1981. Demography and fire history of a western juniper stand. *Journal of Range Management* 34:501–6.

Young, L. 1992. Out of the past. In *No Rooms of Their Own: Women Writers of Early California*, ed. I. R. Egli, 47–58. Berkeley, Calif.: Heyday Books.

Young, S. P., and E. A. Goldman. 1946. *The Puma: Mysterious American Cat*. Washington, D.C.: American Wildlife Institute.

Yount, G. C. 1923. Chronicles of George C. Yount. *California Historical Society Quarterly* 2(1):3–68.

Zigmond, M. L. 1981. *Kawaiisu Ethnobotany*. Salt Lake City: University of Utah Press.

Zigmond, M. L., C. G. Booth, and P. Munro. 1991. *Kawaiisu: A Grammar and Dictionary, with Texts*. University of California Publications in Linguistics 119. Berkeley: University of California Press.

Zimmermann, M. H., and C. L. Brown. 1971. *Trees: Structure and Function*. New York: Springer-Verlag.

Zinke, P. J. 1988. The redwood forest and associated north coast forests. In *Terrestrial Vegetation of California*, expanded ed., edited by M. G. Barbour and J. Major, 679–98. Special Publication No. 9. Sacramento: California Native Plant Society.

Index

Page numbers in *italics* indicate an illustration. Page numbers in *italics* followed by *t* indicate a table (e.g. *52t*); numbers followed by *m* indicate a map (e.g. *36m*).

Compositor:	Integrated Composition Systems
Text:	10/13 Aldus
Display:	Tasse Wide
Index:	Victoria Baker
Illustrations:	Claudia Graham
Maps:	Bill Nelson
Printer and binder:	Thomson-Shore, Inc.